Darwin's Challenge Answered

By
J. Jay Rigney, O.D.

Darwin's theory as Darwin stated, "absolutely breaks down"

ISBN: 978-1-63302-069-6 (Paperback)
978-1-63302-073-3 (Hardcover)

Table of Contents

About the Author

Dr. Jan Jay Rigney graduated from Northeastern State University in Tahlequah, Oklahoma in 1984. At the time of the writing of this book, he will have been in private practice Optometry for thirty-one years. While practicing Optometry, he has had the opportunity to daily, for over 35 years, observe the design, function, and intricate interworkings of the human eye.

Dr. Rigney does not have any specialization in his educational background in evolution or taxonomy (the classification of the animal and plant kingdoms) other than what basic taxonomy education he learned in general biology and botany classes in his undergraduate studies, so he does not consider himself to be an *expert* regarding evolution. However, he has had extensive training and education in human anatomy, and human physiology, which included some basic education in genetics. Dr. Rigney has had Doctoral level training in human anatomy, human physiology, ocular anatomy and ocular physiology; and has worked clinically, exclusively with the eye daily, for over thirty-five years. Yet, he does not claim to be an expert on the eye either. Fortunately, as Dr. Rigney demonstrates in the writing of this book, it is not necessary to be an expert on the eye for **_anyone_** to conclude that the eye answers Darwin's challenge directly, specifically, and decisively.

Introduction

When Darwin postulated his theory of evolution, he did so **_entirely_** based on the scientific process of observation and conclusion. Darwin made observations and then "imagined"[1], "supposed"[1], and "believed"[1] *(these are direct quotes from Darwin)* when he reached the conclusions which formed his "theory" of evolution. Darwin's observations led him to conclude, "that probably all the organic beings which ever lived on this earth have descended from some one primordial form, into which life was first breathed."[2] (I am directly quoting Darwin). He reached this conclusion based **_entirely_** on observation.

Darwin's observations and subsequent conclusions also led him, "to the belief all animals and plants have descended from some one prototype"[2] (I am directly quoting Darwin.) Again, he arrived at this "belief" based entirely on observation!

However, Darwin was not absolutely convinced his theory was not in error. We know this for several reasons. Darwin **_himself_** questioned (rather extensively, I might add) some of his views[3], starting on page 145 and continuing through page 250 of his book, *The Origin of Species*[3] He writes four Chapters expressing questions of his theory. Please note; these again are direct quotes from Darwin - Darwin's written words. We do not have to guess what he thought, or question any inference. Darwin *writes*, "Firstly, why if species have descended from other species by insensibly fine gradations, do we not everywhere see innumerable transitional forms? Why is not all nature in confusion instead of the species being, as we see them, well defined?" "Secondly, is it possible that an animal having, for instance, the structure and habits of a bat, could have been formed by the modification of some animal with wholly different habits?" "Thirdly can instincts be acquired and modified through natural selection?" "Fourthly, how can we account for species, when crossed, being sterile and producing sterile offspring, whereas, when varieties are crossed, their fertility is unimpaired?"[3]

Another reason we know Darwin questioned his "belief"[1,4] is he uses the word "probably"[4] when he writes, "<u>probably</u> all the organic beings which ever lived on this earth have descended from some one primordial form, into which life was first breathed."[4] If he was <u>certain</u>, "all the organic beings which ever lived on this earth have descended from some one primordial form, into which life was first breathed"[4] he would have left the word "probably"[4] out of his *written* words. In fact, if he was *absolutely* positive, he would have written something such as, "<u>the evidence supports the conclusion</u> **that** all the organic beings which ever lived on this earth have descended from some one primordial form, into which life was first breathed."[4] Instead his written statement is "probably"[4].

Furthermore, note he uses the word "belief"[4] when he wrote, "to the belief that all animals and plants have descended from some one prototype"[4]. If he was ***convinced*** he would have used the word "conclude". Instead he wrote, "believe"[4]. If he didn't question his theory he would have written a statement to the affect of: "after all the evidence has been examined <u>the conclusion reached is</u>: all animals and plants have descended from some one prototype". However, because he isn't certain he ***chose*** the word "belief"[4]. Just because you "believe" something, doesn't mean it is factual or true. For example, not too long ago man *believed* the world was flat. We know now it is not.

So, we ***conclude*** he himself questioned his theory. (In writing this underlined statement I am using it as an example specifically to emphasize my point. I could have written, "so we ***believe*** he himself questioned his theory" choosing the word ***believe*** as Darwin did when he wrote. But because I am certain he himself questioned his theory I chose to write, "we conclude", not "we believe". (An author ***chooses*** words as he writes.)

While all these statements by Darwin reveal he questioned the validity of his own "theory", the most important reason we know he questioned his own "beliefs"[1,4] is he left open the door for his "theory" to be disproved by anticipating, and furthermore prompting a challenge to his "theory". This challenge, if met, would be so significant he himself said it would cause his "theory" to "absolutely break down"[5] (Again, Darwin's ***written*** words.)

This book is written to answer the challenge Darwin himself put forth. Darwin himself said, ***in writing,*** on page 159 of *The Origin of Species*; "If it could be demonstrated that any complex organ existed, which could not possibly have been formed by numerous, successive, slight modifications, my theory would absolutely break down."[5] Note; he himself calls it a *"theory"*, and thankfully he ***chose*** the word *"absolutely"* when he wrote, "my theory would absolutely break down[5]". He didn't say, "…my theory *may* break down". He didn't say, "…*might bring into question the validity of my theory"*. He didn't say, "…*further examination of my theory should be undertaken", or* anything of that nature. He said, ***"MY THEORY WOULD ASOLUTELY BREAK DOWN."*** [5] *These are his* ***WRITTEN WORDS,*** and he intentionally ***chose*** the word ***"ABSOLUTELY."*** [5] He could have chosen any other word, or he could have left this statement out altogether, but he purposefully *chose* the word _**absolutely**_. (Underlining by Rigney) And remember, Darwin himself _**wrote**_ it so we do not have to guess what he is saying or question as to what he may be inferring; he _**wrote**_ it!

Therefore, with the writing of this book, Darwin's challenge is answered. Utilizing (as Darwin did) the scientific process of observation and conclusion. When one carefully observes the intricate structure, and functions of the eye, the only logical scientific conclusion that can be reached is; the eye could _**NOT**_ possibly have evolved.

Examine the evidence for yourself - don't just believe what you have been told, read and learn. Then after learning you can make an informed decision for yourself.

As you read, keep this in mind. When I wrote this book, instead of presenting the information in anatomical order, I wanted those systems and subsystems which answer Darwin's challenge most effectively to be presented first. Therefore, if someone was to read only the first chapter I will have had a chance to present my strongest argument first and they will have had a chance to hear at least one of the twenty-five reasons why the eye directly, specifically, and decisively answers Darwin's challenge.

Please note as you read, we are looking at the strongest evidence which answers Darwin's challenge first; then proceeding downward. It will seem to flow a little better if you keep this in mind.

Also, as you read and observe the illustrations, let this concept be in the forefront of your mind and thought processes. You can observe your shoe and conclude it was created.[6] Even though you did not see someone create your shoe; you still _**KNOW**_ it was created. How is that? By observation! By simple observation, you _**KNOW**_ the pieces could not have all come together forming a shoe - just by chance. The observed evidence "demonstrates" your shoe was created - so you can come to the _**logical**_ conclusion; "my shoe demonstrates creation!" You can arrive at the TRUE conclusion simply by observation and conclusion. You can observe the parts, how they are shaped, how they are interconnected, how they are glued or stitched together, and how they are dependent upon each other for the shoe to *then* perform a specific function. You can logically conclude it was, *indeed,* created. The evidence observed ***proves*** the shoe was created. It is just as that when one observes the eye – except the eye is infinitely more powerful in demonstrating Creation than your shoe! The eye demonstrates what I term ***Interdependent Evidence of Creation***. Keep this concept at the forefront of your thinking as you observe the many intricate illustrations and the "interdependent evidence" observed when you examine the eye.

Observation and conclusion; it is a scientific method of *proving* a theory. Does the observed evidence ***prove*** Darwin's theory of Evolution? Or, does the observed evidence ***prove*** the theory *Interdependent Evidence of Creation*? You read, you decide for yourself. Don't just believe what you have heard or been told. Examine the evidence. What does the evidence tell you? A theory must be *proven* not just "believed."[7] Darwin states, "may we not believe"[7] when discussing his "theory", but where is the proof of his "theory"? The eye ***proves*** the theory *Interdependent Evidence of Creation, I.E.C.* therefore, the eye ***proves*** Creation. In fact, as you will see in Chapter 18, Darwin confirms my theory. He discusses Interdependent Evidence of Creation and the reality of the concept. Darwin himself confirms the theory *Interdependent Evidence of Creation* and he didn't even know it.

Purpose

The purpose of this book is to directly answer Darwin's challenge. Darwin himself said on page 159 of *The Origin of Species*, "If it could be demonstrated that any complex organ existed, which could not possibly have been formed by numerous, successive, slight modifications, my theory would absolutely break down."[1]

The eye is "a complex organ" and "demonstrates", as Darwin anticipates, it could *NOT* have "been formed by numerous, successive, slight modifications."[1]

The purpose of this book is to educate the reader on the structure and function of the eye. To make plain to the reader the intelligible evidence observed when one examines the possibility of evolution verses Creation of the eye.

It is wise, but also necessary, to become educated, if one is to discern between evolution and Creation. In order to come to a logical conclusion one must become educated; that is, you have to know and understand all the facts before you can then examine ***all*** the evidence. Only by examining ***all*** the evidence can one *prove* a theory.

The eye demonstrates more than 25 systems and subsystems all of which help the reader to discern between evolution and Creation of the eye. Every one of these 25 systems are totally, and wholly interdependent upon each and every other system. Furthermore, each and every one of the 25 systems must be fully developed and be fully functional; all 25 of them must function fully: all, at the same time in order for the eye to stay alive and see. Fully developed, simultaneous function, of totally and wholly interdependent fully developed, fully functional parts can *NOT* happen by evolution. The oberved evidence supports a new theory; *Interdependent Evidence of Creation,* or *I.E.C.*

After reading this book I think you will agree the eye itself ***proves*** it could not possibly have been formed by "numerous, successive, slight modifications"[1] as Darwin's *theory* "supposes"[2]. Therefore, ***Darwin's theory as he himself anticipated, "absolutely breaks down."*** [3]

The eye proves the theory; Interdependent Evidence of Creation, I.E.C. "What is *Interdependent Evidence of Creation*?" you might ask. After reading this book I think you will understand the theory, and agree the eye *proves* the theory. Just as the shoe itself proves it was created; the eye itself proves it was Created.

1. External Eye Muscles

All references and illustrations regarding External Eye Muscles, Chapter 1 of this book are based upon and taken from Henry Gray, F.R.S, *Anatomy, of the Human Body Twentieth Edition* Lea & Febiger (Philadelphia and New York © 1918)[1]

Each eye has six different eye muscles: one on top, the *superior rectus*; one on the bottom, the *inferior rectus*; one on the outside, the *lateral rectus*; one on the inside, the *medial rectus*; one which comes from the front of the eye socket on the bottom and inside by the nose but goes backward along the bottom, the *inferior oblique*; and one which comes from the back of the eye socket and runs along top, then goes *forward* to the front and passes through a loop in the frontal bone on the upper portion of the inside corner of the eye socket - then goes *backward* to the back of the eyeball where it attaches to the eyeball on the top rear portion of the eyeball, the *trochlear or superior oblique muscle*.

These six different eye muscles, (six for each eye – twelve in total) must connect to eight different bones of each eye socket (eight bones per eye, sixteen bones total) at the exact and correct spot. Each eye muscle must have the proper blood supply (arteries to carry blood to the eye muscles and veins to carry the blood back to the heart, lungs and intestines) in order to stay alive and to function. There is an artery and vein for each of the six eye muscles, twelve arteries and veins per eye or twenty-four total arteries and veins for muscles of both eyes.

The inferior rectus, medial rectus, superior rectus, and inferior oblique are all served by one nerve which has four branches—the *oculomotor nerve*. The lateral rectus muscle has its own nerve, the *abduscence nerve*. The superior oblique muscle has its own nerve, the *trochlear nerve*. These three main nerves to the eye muscles (six nerves for both eyes) come from different areas of the brain stem[2].

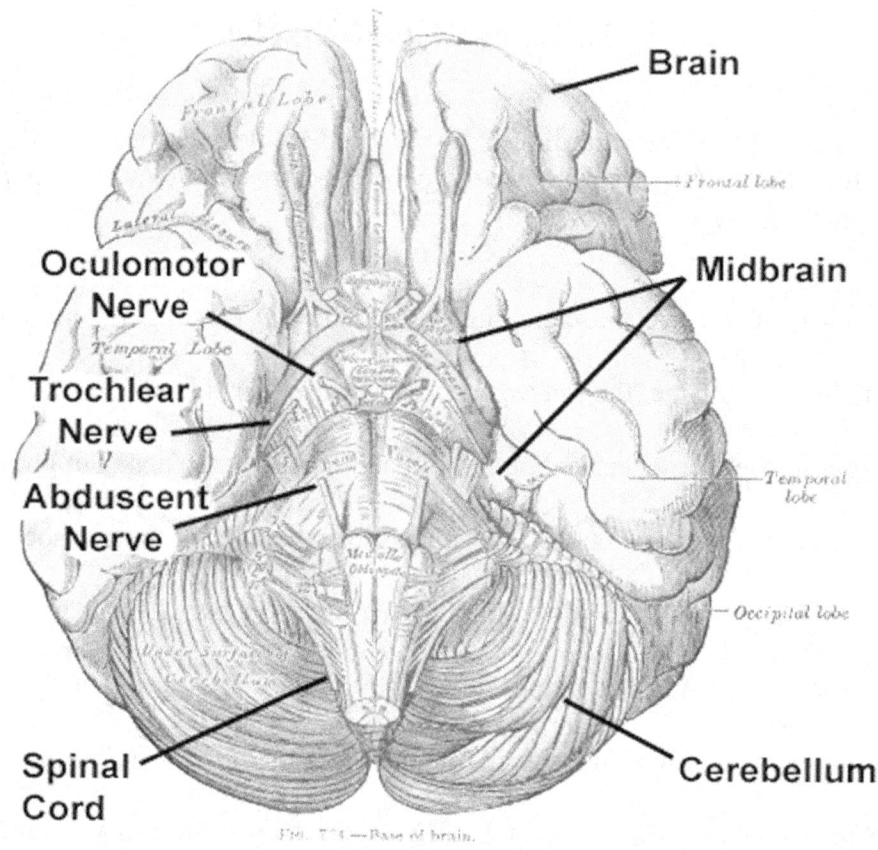

From Henry Gray, F.R.S, Henry Gray, F.R.S, *Anatomy of the Human Body* 20th Edition, Lea & Febiger (Philadelphia and New York © 1918) page 817[2] with modifications and additions by Rigney

There are three main nerves per eye. Because the oculomotor nerve branches into four smaller nerve branches, each eye muscle has its own nerve. As a result, there are six nerves for each eye, one for each eye muscle, twelve nerves in total.

The arteries, veins, and nerves all come from inside the skull, but the eyeball and all the eye muscles are outside the skull in the eye socket. The arteries, veins, and nerves must pass through the bones of the eye socket to supply the eye. There are holes (foramen) in the bones of the eye socket for the nerves, arteries, and veins to supply the eye muscles and other parts of the eye. There are seventeen holes (foramen) in the bones of the skull and eye socket <u>per eye</u> – plus the Foramen Magnum, which the spinal cord utilizes to exit the skull. The sympathetic pupil fibers for each eye utilize the Foramen Magnum. Therefore there are thirty-five foramen –holes in the bones of the skull, which ***must*** be present for all the arteries, veins, and nerves (and tear ducts) to pass through.[3,4,5] (We will discuss these thirty five holes that ***must*** be present in more detail in Chapter 12, Bones of the Eye, and Chapter 13, Foramen.)

The three nerves of the muscles which move the eye and the nerve which supplies the eyelid muscle all pass through the Foramen called the Superior Orbital Fissure.

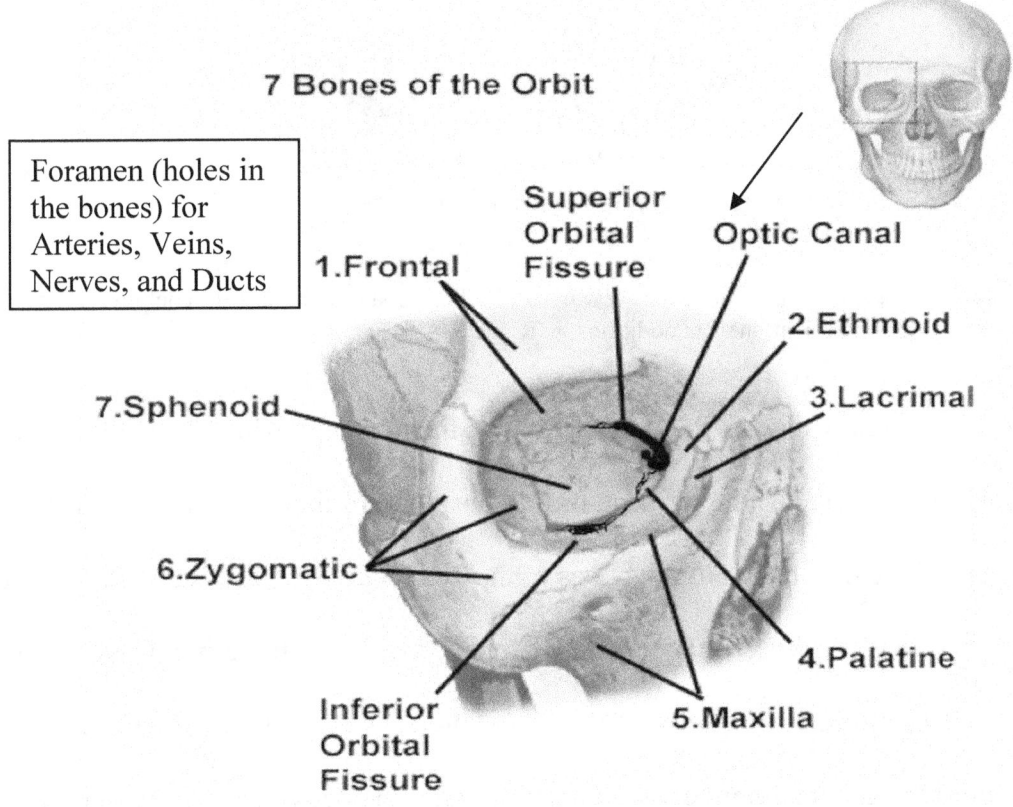

7 Bones of the Orbit

Foramen (holes in the bones) for Arteries, Veins, Nerves, and Ducts

Superior Orbital Fissure

Optic Canal

1.Frontal

2.Ethmoid

3.Lacrimal

7.Sphenoid

6.Zygomatic

4.Palatine

5.Maxilla

Inferior Orbital Fissure

From Henry Gray, F.R.S, *Anatomy of the Human Body* 20th Edition, Lea & Febiger (Philadelphia, & New York, © 1918) Page 186[3] with modifications and additions by Rigney

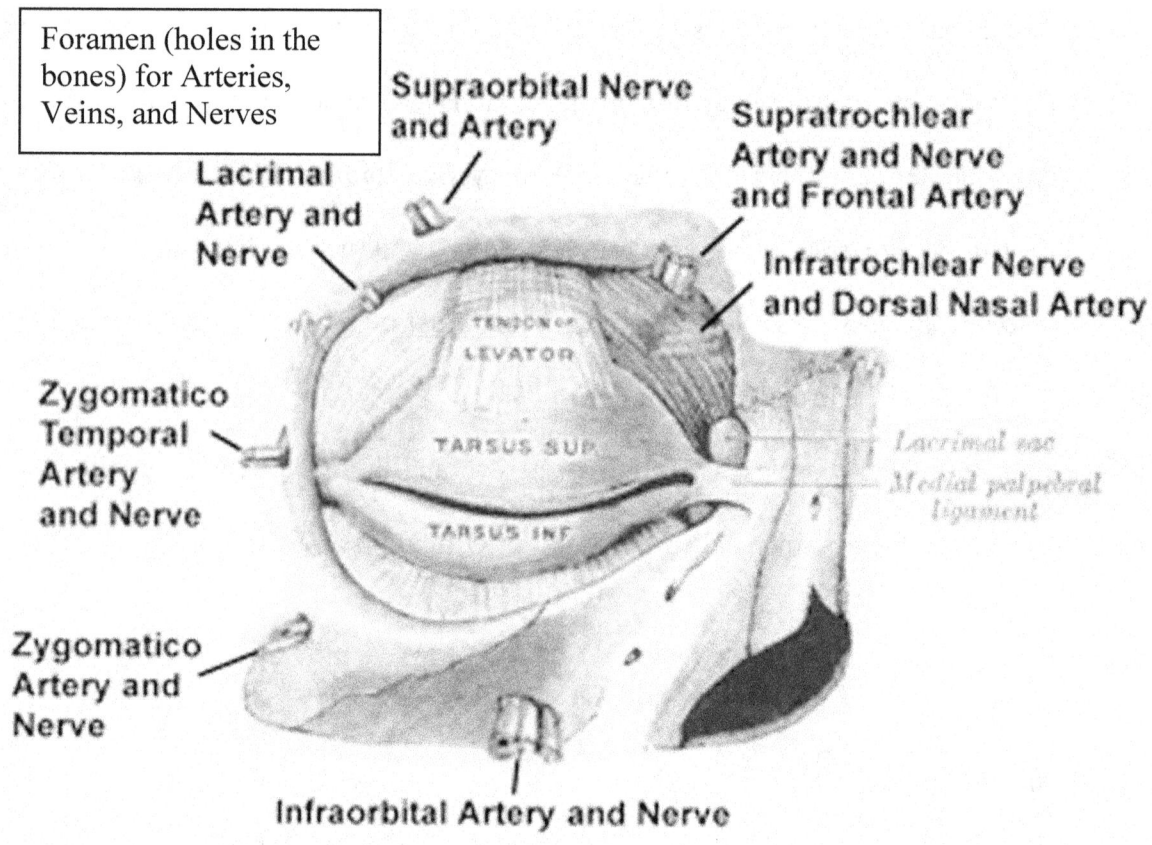

From Henry Gray, F.R.S, *Anatomy of the Human Body* 20th Edition, Lea & Febiger (Philadelphia & New York, © 1918) Page 1027[4] with modifications and additions by Rigney

The eye muscles and how they are connected, that is, the angles in which they are connected and where they attach on the eyeball (how far back or how far forward on the surface of the eyeball) and the place and angle where they attach to the bones of the eye socket- all truly show the hand of God.

If the twelve different eye muscles, by *chance,* happened to attach too long, or too short, or at an improper angle, (either on the eye itself or at the other end on the bones of the eye socket) the eyes will be out of alignment, causing double vision. Additionally, the eye could not move correctly in all directions.

The Eight Rectus Muscles (four per eye): Medial Rectus pulls the eye in. Lateral Rectus pulls the eye out. Superior Rectus pulls the eye up. Inferior Rectus pulls the eye down.

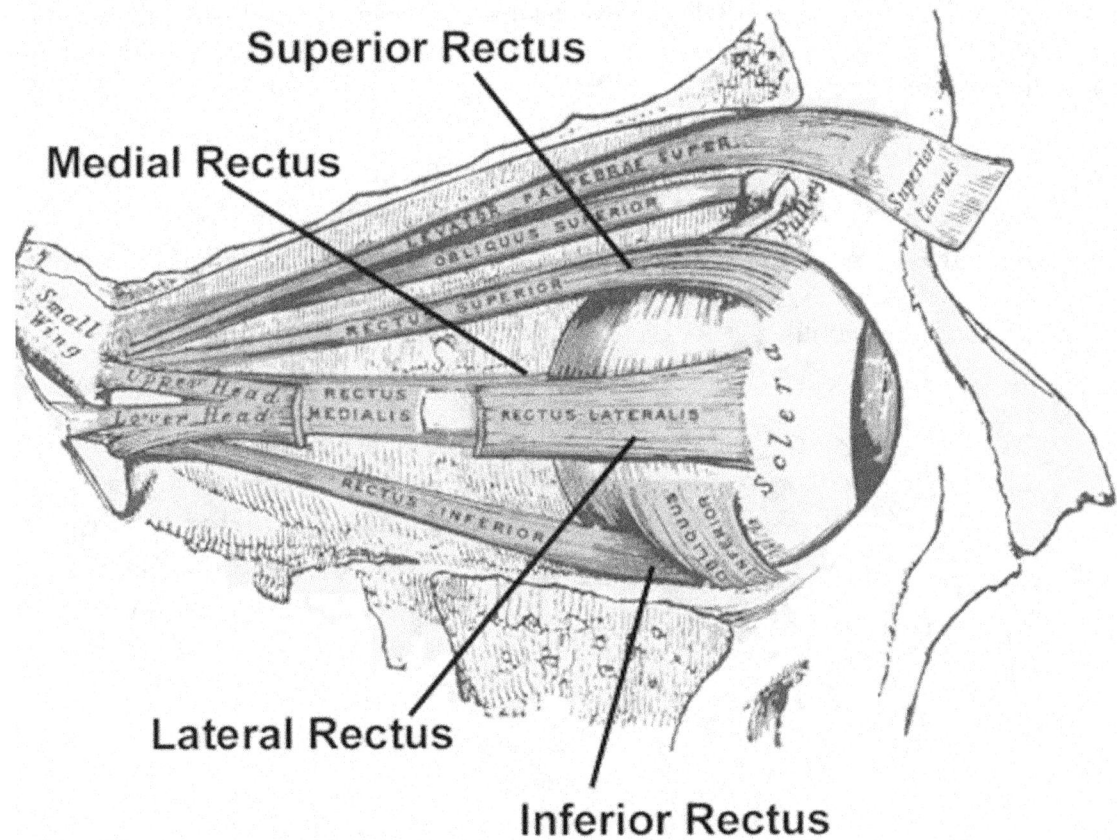

From Henry Gray, F.R.S, Henry Gray, F.R.S, *Anatomy of the Human Body* 20th Edition, Lea & Febiger (Philadelphia & New York © 1918) p.1022[5] with modifications and additions by Rigney

The Four Oblique Muscles (two per eye): Rotate the eyes and pulls the eye up or down when the eye is looking in. For example; the right Superior Oblique rotates the eye inward when the head is tilted to the right. The right Superior Oblique also pulls the right eye down when the right eye is looking down and in. The right Inferior Oblique rotates the right eye outward when the head is tilted to the left and pulls the right eye up when the right eye is looking in. (Right Superior Oblique and Right Inferior Oblique are shown.)

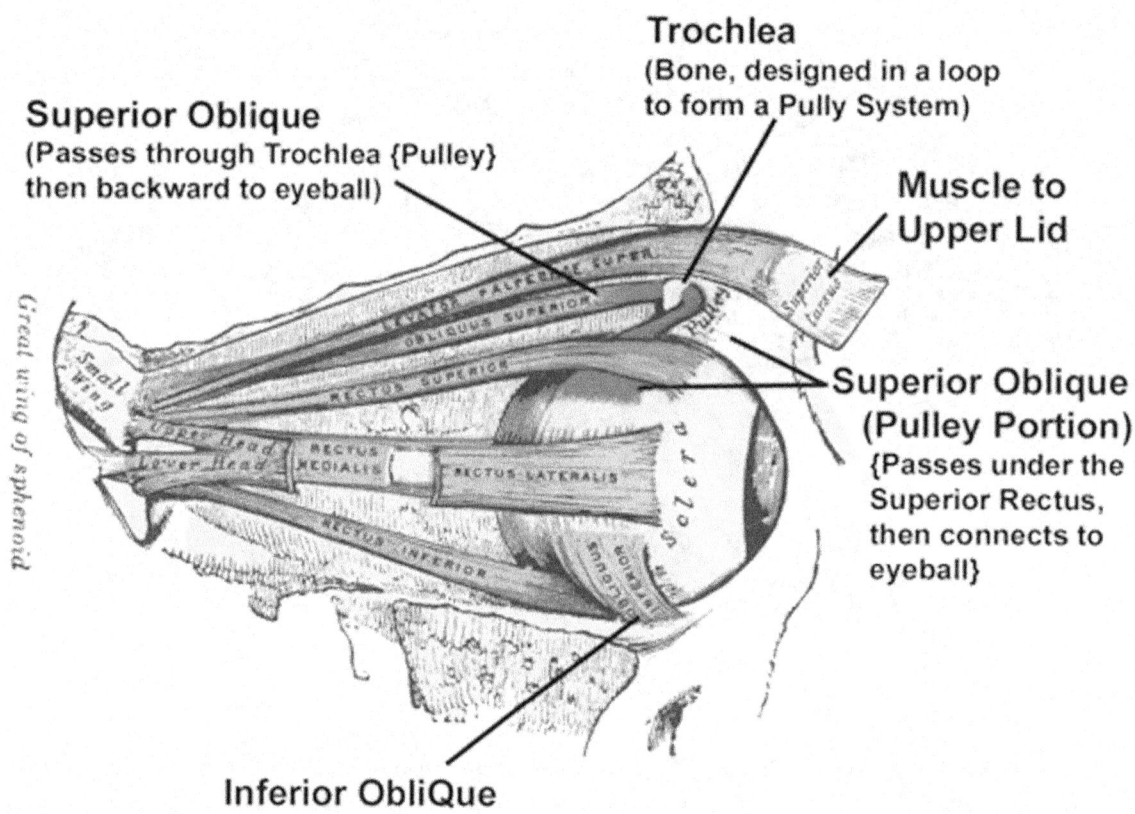

From Henry Gray, F.R.S, Henry Gray, F.R.S, *Anatomy of the Human Body* 20th Edition, Lea & Febiger (Philadelphia & New York, © 1918) p.1022[5] with modifications and additions by Rigney

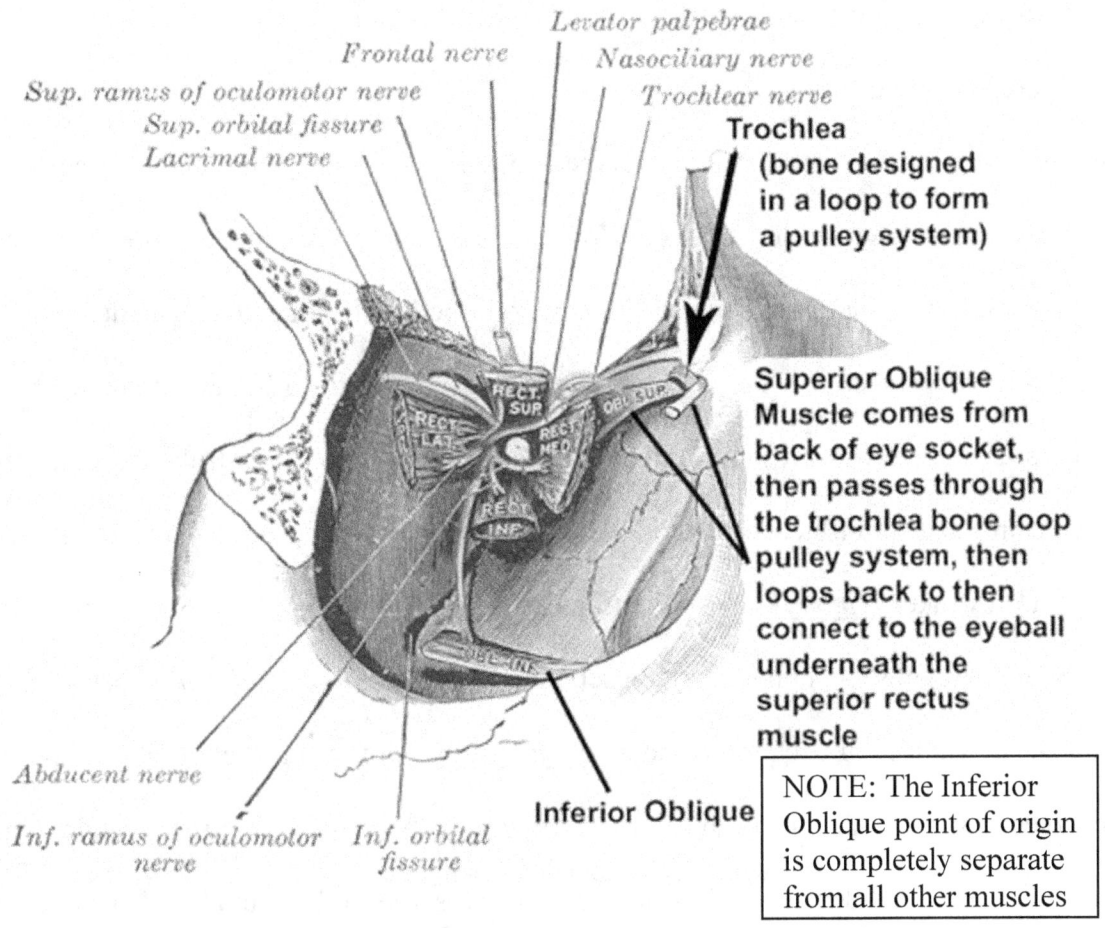

Frontal nerve
Levator palpebrae
Sup. ramus of oculomotor nerve
Nasociliary nerve
Sup. orbital fissure
Trochlear nerve
Lacrimal nerve

Trochlea (bone designed in a loop to form a pulley system)

Superior Oblique Muscle comes from back of eye socket, then passes through the trochlea bone loop pulley system, then loops back to then connect to the eyeball underneath the superior rectus muscle

Abducent nerve

Inf. ramus of oculomotor nerve   *Inf. orbital fissure*

Inferior Oblique

NOTE: The Inferior Oblique point of origin is completely separate from all other muscles

From Henry Gray, F.R.S, *Anatomy of the Human Body* 20th Edition, Lea & Febiger (Philadelphia & New York, © 1918) Page 1023[6] with modifications and additions by Rigney

(**NOTE:** The Trochlea bone-loop-pulley (arrow) *MUST* be present. The trochlea bone-loop pulley system is ***proof*** of design and creation. **NOTE:** the Inferior Oblique muscle point of origin is completely separate from all the other eye muscles. Evolution cannot explain the trochlea bone-loop pulley system or the separate origin of the Inferior Oblique Muscle; They are designed and engineered, ***NOT*** evolved.)

If the placement of each muscle were any different or if the angles of attachment were any different, or if the attachments were any further, or longer, or closer, or shorter; the eye could not move properly, it could not move in all directions, and you would see double. Additionally, each eye muscle *must* have an *opposing* eye muscle for the eye to line up straight. Both opposing eye muscles *must* be formed at the *very* same time. This is further explained on pages 23-27 of this chapter. Evolution cannot explain this process!

Now, pay *very* close attention to the trochlea-bone-loop-pulley-system and the superior oblique muscle. They are an engineering marvel! How anyone can believe in evolution after seeing the anatomy and design of the trochlear muscle is beyond belief! It is *clearly* evident it was not only designed but also *engineered.* One would have to have more faith to believe it could have *evolved* than is necessary to *see* (observe) and conclude it was designed. Its *design* truly shows the intelligent design, wisdom, and omniscience of God. This trochlear bone-loop-pulley- muscle system *defies evolution!* It only takes **<u>one look</u>** to see the trochlea bone-loop and superior oblique muscle system; interdependent systems - bone and muscle, *"absolutely breaks down"* Darwin's theory.[7] Interdependent Evidence of Creation is the true <u>conclusion</u>! (Not, just a "belief.")

Look (see pages 14 -22) at the intricate workings of these multiple interdependent structures. Meditate upon their structure, function, and design. Observe and conclude –that is science! The portion of the frontal bone which makes the top inner portion of the eye socket has to have a loop-shaped boney portion form *into it, the trochlea*[8,9] And **<u>at the same time</u>** the muscle has to form in a way in which it starts from the back of the eye socket (attached to the frontal bone in the *back* of the eye socket) then extends up to the *front* of the eye socket, traveling along the inside top of the frontal bone in the top of the eye socket (not attached along its length). It is attached at the ends but not attached anywhere along its course. It *must* pass into and through the loop of bone, *the trochlea,* on the frontal bone. Then, *extend* and travel *backward* to the eye where it then, *must* attach to the rear portion of the eyeball[9]. How can evolution explain a muscle evolving and **<u>at the same time</u>** a bone evolving to form a loop and a muscle which evolves forward through the loop of bone and backward to attach at the eye? "Numerous, successive, slight modifications"[10], as Darwin's challenge dictates[10], cannot explain this intentional intelligent design because ***nothing*** will work- at all, until ***everything*** is completed! Darwin's theory as Darwin himself anticipates, "absolutely breaks down[10]. The eye cannot function and will not line up straight without all the components present- and they MUST be present simultaneously! PERIOD!

This trochlear muscle system works just like a rope-and-pulley system, which we know, is an engineered design.

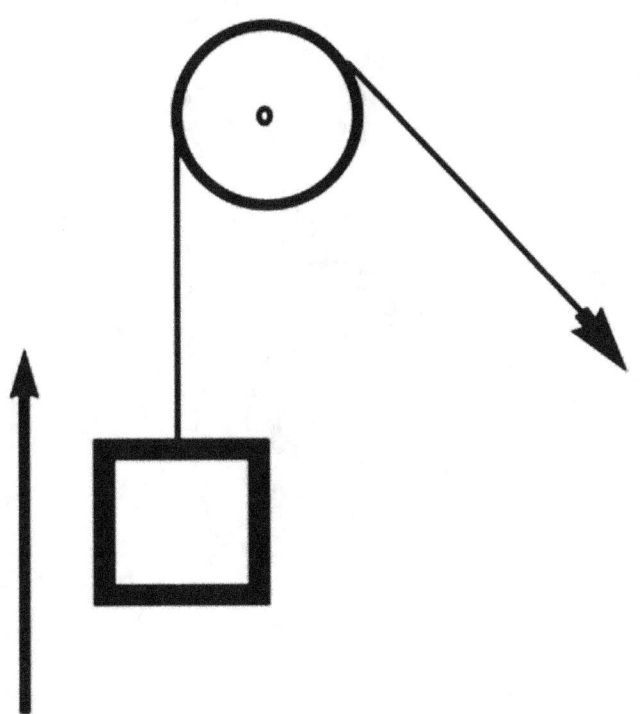

A rope and pulley system similarly set up as the trochlear muscle system.

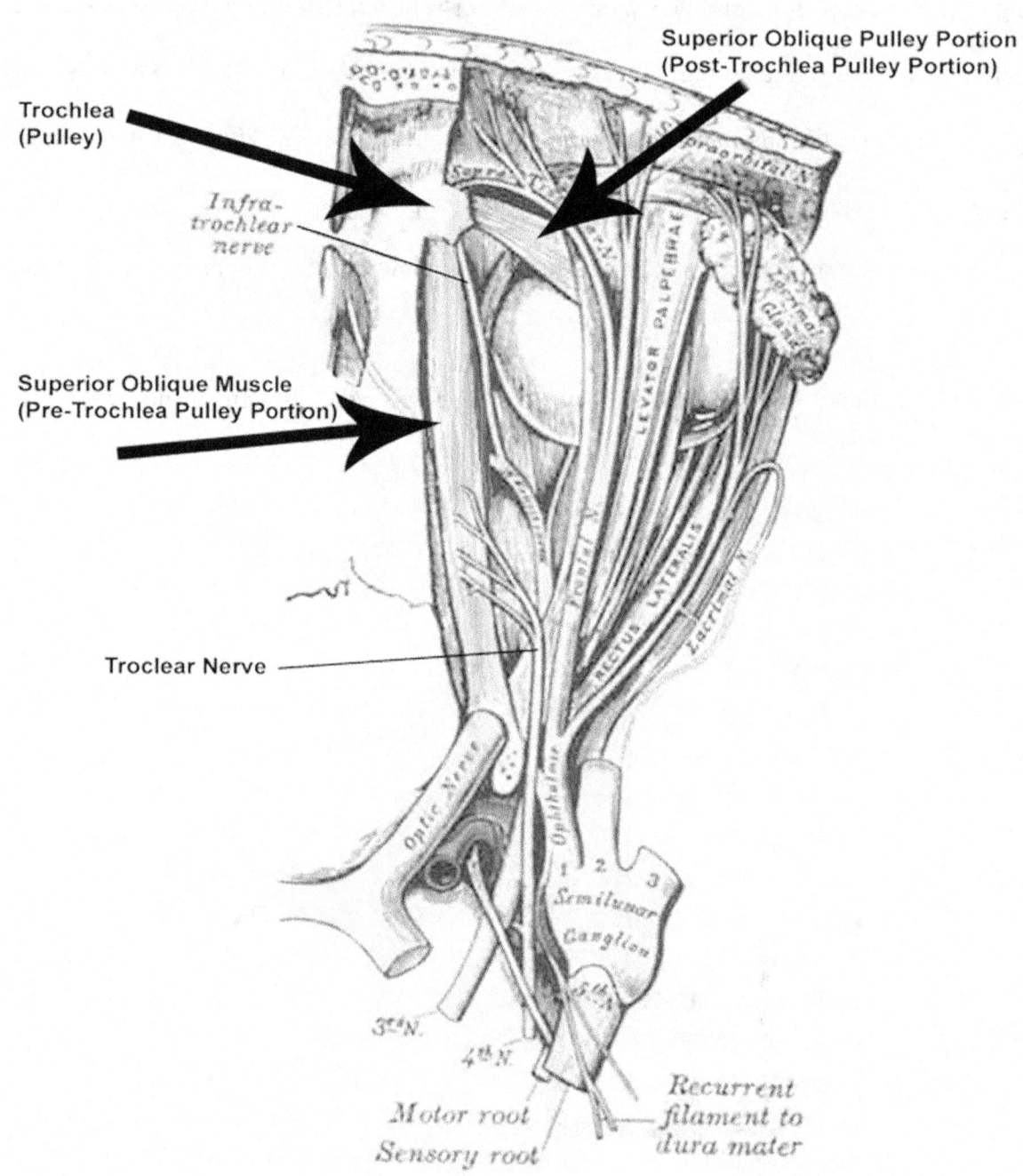

Trochlea (Pulley)

Superior Oblique Pulley Portion (Post-Trochlea Pulley Portion)

Infra-trochlear nerve

Superior Oblique Muscle (Pre-Trochlea Pulley Portion)

Troclear Nerve

From Henry Gray, F.R.S, *Anatomy of the Human Body* 20th Edition, Lea & Febiger (Philadelphia & New York, © 1918) p.885[11] with additions by Rigney

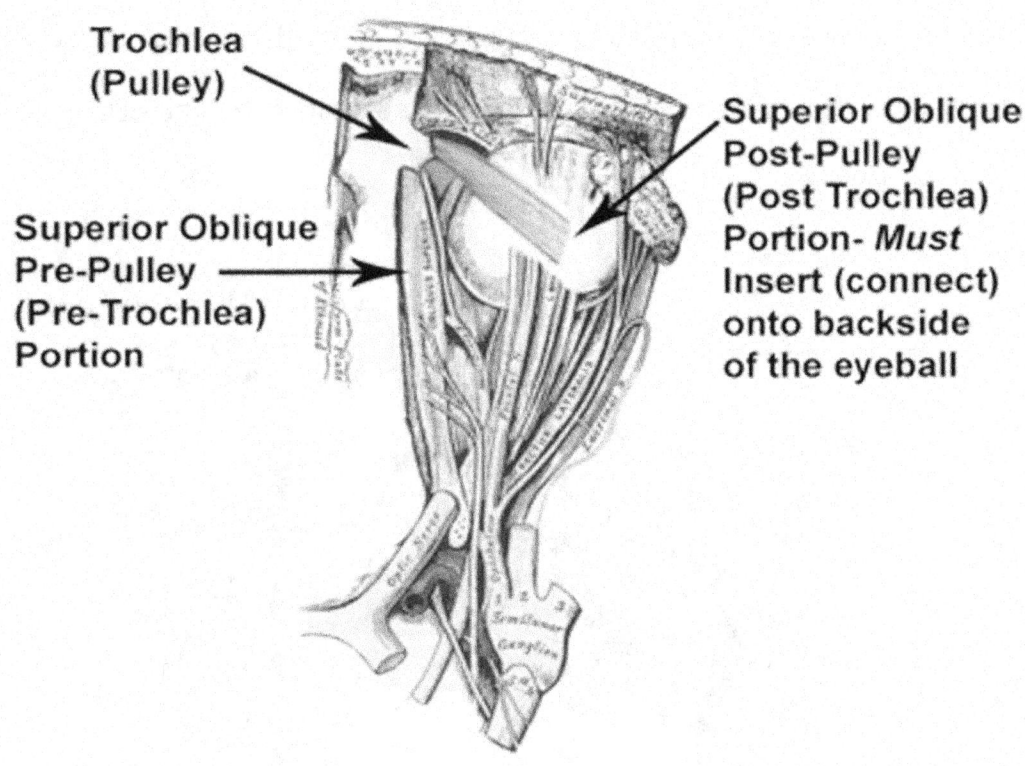

Trochlea (Pulley)

Superior Oblique Pre-Pulley (Pre-Trochlea) Portion

Superior Oblique Post-Pulley (Post Trochlea) Portion- *Must* Insert (connect) onto backside of the eyeball

From Henry Gray, F.R.S, *Anatomy of the Human Body* 20th Edition, Lea & Febiger (Philadelphia & New York © 1918) p.885[11] with modifications and additions by Rigney

The Superior Oblique muscle comes from the back, then extends through the trochlea to attach to the back of the eye. Evolution cannot explain this. The trochlea-superior oblique muscle system is **_proof_** of design and the muscle and bone system proves Interdependent Evidence of Creation!

For the trochlear muscle to function *properly*, it *must* attach to the rear of the eyeball *behind* what would be the equator of the eyeball[12]. If the muscle happened to *evolve* by chance in a way in which it attached to the eyeball in front of the equator of the eyeball, it would not rotate (or roll) the eyeball downward when looking in, which is its primary action (when you look in and then down the trochlear muscle is performing *most* of the work of all of the six eye muscles). If it was *evolved* rather than *designed* and so happened to *evolve* in a way which, by chance, it attached in front of the equator of the eyeball and not behind the equator of the eyeball it would instead pull the eye forward out of the eye socket.

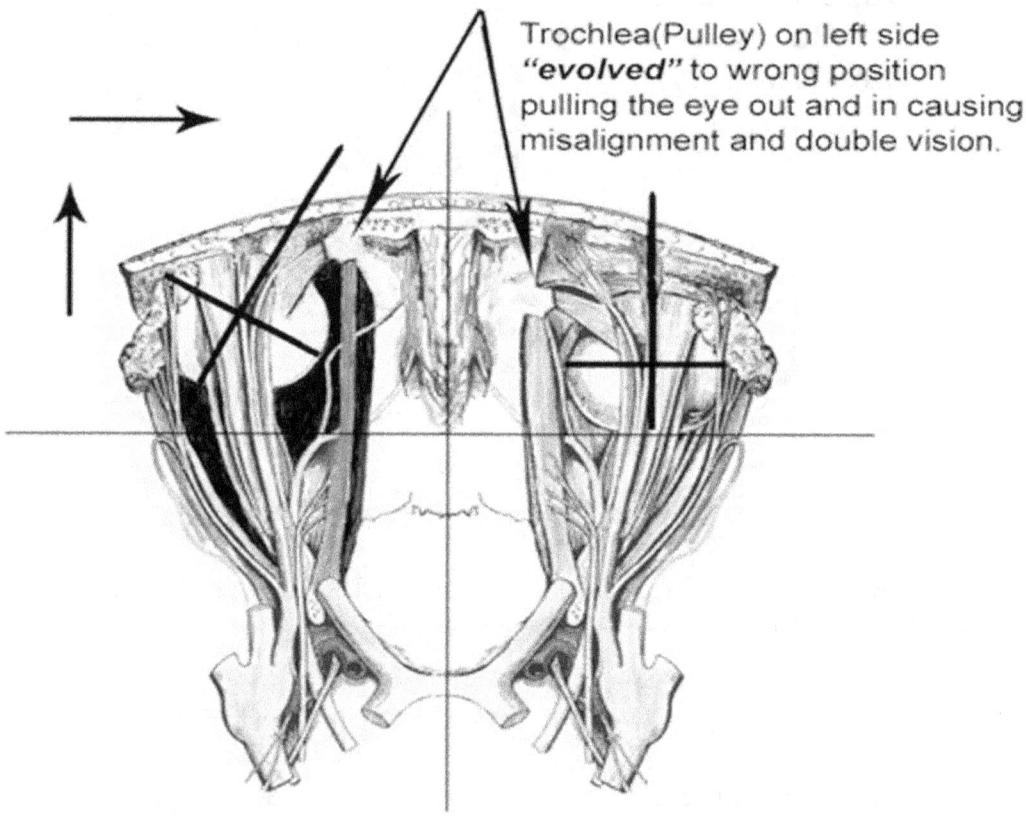

Trochlea(Pulley) on left side **"evolved"** to wrong position pulling the eye out and in causing misalignment and double vision.

From Henry Gray, F.R.S, *Anatomy of the Human Body* 20th Edition, Lea & Febiger (Philadelphia & New York, © 1918) p.885[13] with modifications and additions by Rigney

If any **_one_** part were missing from the trochlear bone-loop-pulley eye muscle mechanism, the eye could not move properly and you would see double **_constantly._**

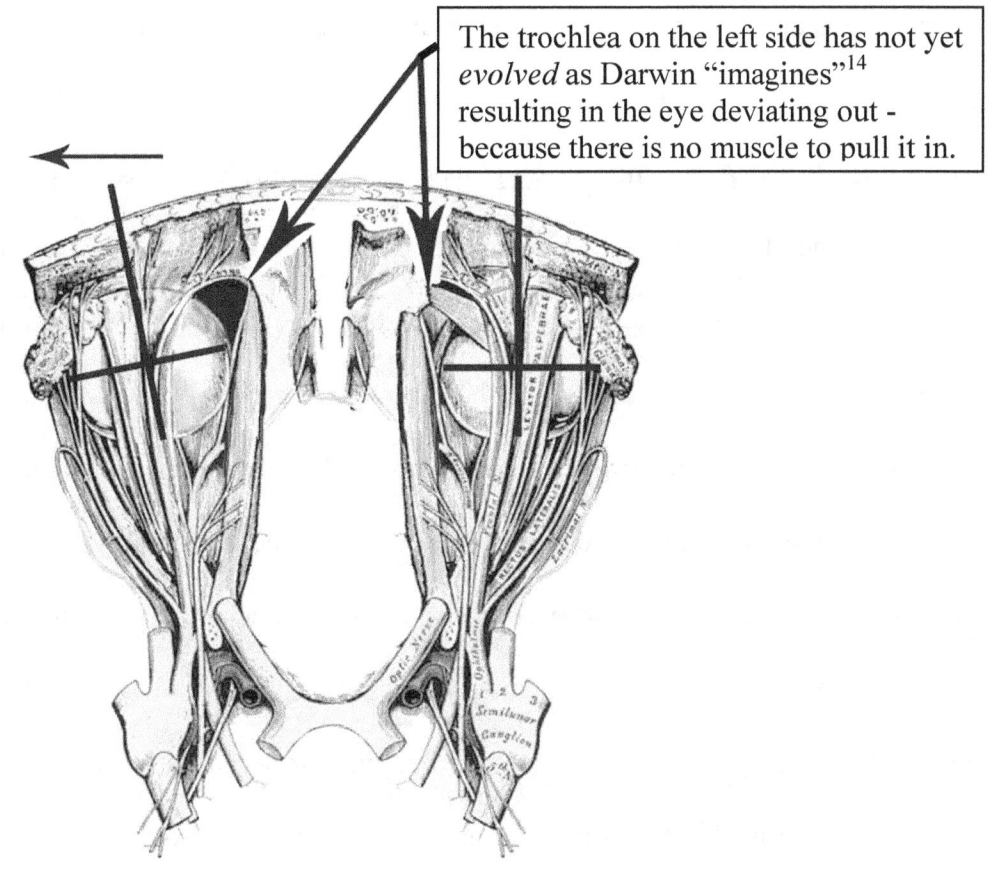

The trochlea on the left side has not yet *evolved* as Darwin "imagines"[14] resulting in the eye deviating out - because there is no muscle to pull it in.

From Henry Gray, F.R.S, *Anatomy of the Human Body* 20th Edition, Lea & Febiger (Philadelphia & New York, © 1918) p.885[13] with modifications and additions by Rigney

In this "imagined"[14] illustration, the left Trochlea Bone-Loop-Pulley has not "evolved" and the left Superior Oblique muscle has not *fully* "evolved" causing the left eye to turn out and up resulting in permanent, ***un-recoverable*** double vision.

All of the other five eye muscles, each with their respective nerves had to be simultaneously fully developed and fully functional- at the same time. If any ***one*** of the other five muscles had not developed or had *not* developed fully and simultaneously, the eye could not move properly and the person would see double constantly, and permanently. You cannot function when you see double! If the person sees double, he or she cannot survive or he or she would have to continually cover an eye to see, which would cause the covered eye to become weak and "lazy."

Additionally, *__each__* eye muscle *__must__* have an opposing eye muscle. One eye muscle could not wait for any one of the other five eye muscles to develop or evolve because they are in opposition to each other. Without the other opposing eye muscle in place and fully functional to oppose each eye muscle, there would be over action of the one which is *not* opposed, causing the eye to be out of alignment and to not work properly- resulting in double vision. For example, if the outside eye muscle of your right eye, the abduscens eye muscle (which is supplied by its own nerve), were not fully *evolved*, the right eye could not look to the right, and the eye would be out of alignment and turned in (*cross-eyed)* to the left because the function of the inside eye muscle, the medial rectus (which is supplied by a different nerve, the oculomotor nerve) is to turn the eye in (in this example turn the right eye in). If the medial rectus were to evolve without - or at a different time than the lateral rectus, because the outside eye muscle, the lateral rectus is not present or functional to pull the eye out, then the inside eye muscle, the medial rectus is unopposed. This pulls the eye in, causing the person to be cross-eyed in the right eye. Therefore, *__both__* eye muscles *__must__* be present, fully evolved, and fully functional or the eye does not line up and point straight! This applies to all six eye muscles! All six eye muscles *__MUST__* be present and fully functional *__all at the same time__* to keep the eyes straight, and for the eye to move properly in any and all directions. Furthermore, and an even greater *miracle* is; at the same time one eye and all six of its eye muscles and nerves are forming and developing *__simultaneously__*; the other eye and all six of its eye muscles must form fully, simultaneously, and perfectly *at the same time*. *All at the same time with all six eye muscles of the* *other eye!* They have to form the correct length and the proper attachment to the eye and the bones of the eye socket, at all the correct angles, in both eyes **_all_ at the very same time!** Evolution **_cannot_** be at work!

Below, both Medial Rectus muscles have not evolved causing the eyes to turn out. There is NO WAY for the eyes to turn straight if the Medial Rectus muscle has not evolved, and the person CANNOT see straight ahead **_at any time!_** If one Medial Rectus muscle does not evolve then that eye would turn out and can NOT turn in – **_ever!_**

Below, both Lateral Rectus muscles have not evolved causing the eyes to turn in. There is NO WAY for the eyes to turn straight if the Lateral Rectus muscle has not evolved, and as in this example the person CANNOT see straight ahead at any time. If one lateral rectus muscle does not evolve then that eye would turn in and can NOT turn out -ever.

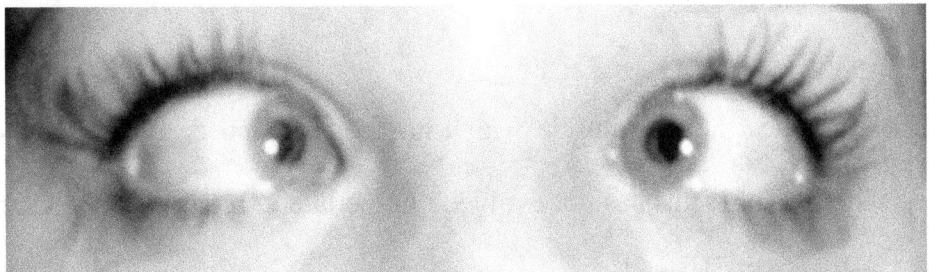

Below, the patient's right Inferior Rectus muscle has not evolved causing the right eye to pull up. There is NO WAY for the right eye to pull down straight if the Inferior Rectus muscle has not evolved, and as in this example the person's right eye CANNOT see straight ahead at any time resulting in double vision always.

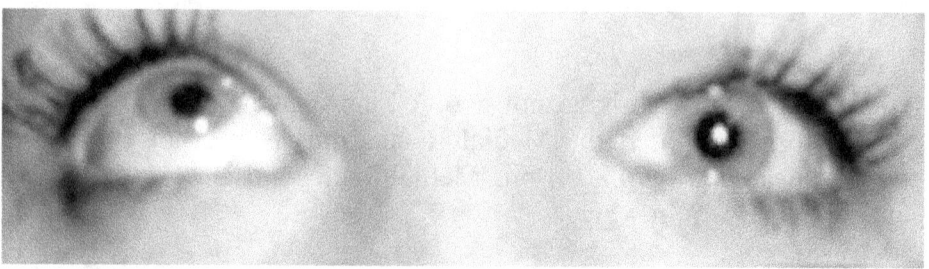

Below, the patient's right Superior Rectus muscle has not evolved causing the right eye to pull down. There is NO WAY for the right eye to pull up straight if the Superior Rectus muscle has not evolved, and as in this example the person's right eye CANNOT see straight ahead at any time resulting in double vision at all times.

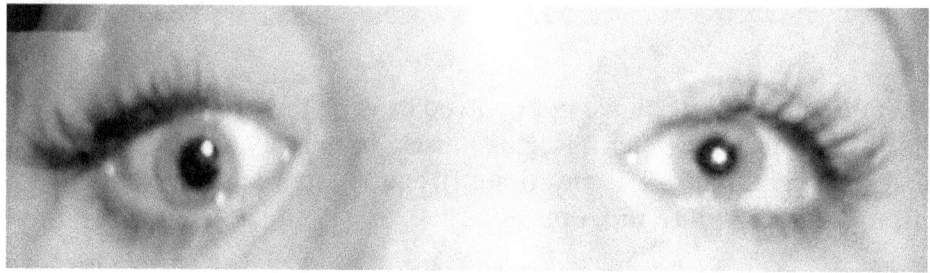

In all of these examples the eye muscle which has not evolved is not present to pull the eye straight. That is, if the medial rectus has not evolved but the lateral rectus has, the lateral rectus pulls the eye out because the medial rectus is not present to oppose the lateral rectus; the lateral rectus has full reign in pulling the eye out because it is not opposed by the medial rectus, therefore the eye turns out and there is no way for it to turn in <u>ever!</u>

Below, the patient's right Lateral Rectus muscle has not evolved causing the right eye to pull in. There is NO WAY for the right eye to pull out straight if the Lateral Rectus muscle has not evolved, and as in this example the person's right eye CANNOT see straight ahead at any time resulting in double vision.

Below, the patient's right Medial Rectus muscle has not evolved causing the right eye to pull out. There is NO WAY for the right eye to pull in straight if the Medial Rectus muscle has not evolved, and as in this example the person's right eye CANNOT see straight ahead at any time resulting in double vision.

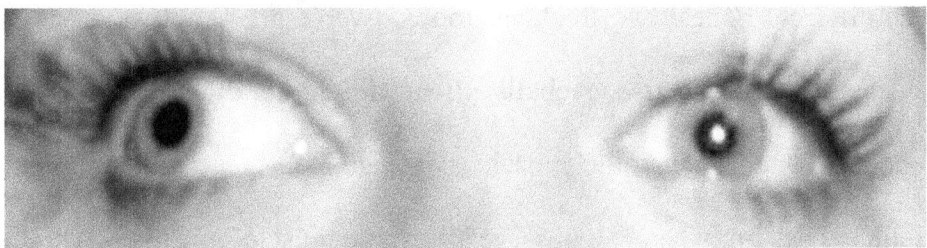

Below the patient's right inferior rectus has not evolved so the right eye cannot look down. The patient's left superior rectus has not evolved so the left eye cannot look up. At no time, will the eyes EVER work together, and the patient will see double at all times. Even if the patient were to ignore one eye and attempt to use only one eye to avoid seeing double, the patient still could not EVER look straight ahead…EVER!

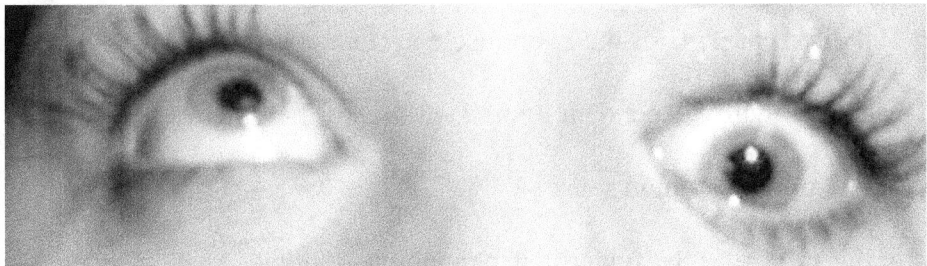

Below the patient's right superior rectus has not evolved so the right eye cannot look up at any time <u>ever</u>. And, the patient's left inferior rectus has not evolved so the left eye cannot look down at any time <u>ever</u>. At no time, will the eyes EVER work together, and the patient will see double at all times. Even if the patient were to ignore one eye and attempt to use only one eye to avoid seeing double, the patient still could not EVER look straight ahead…**EVER!**

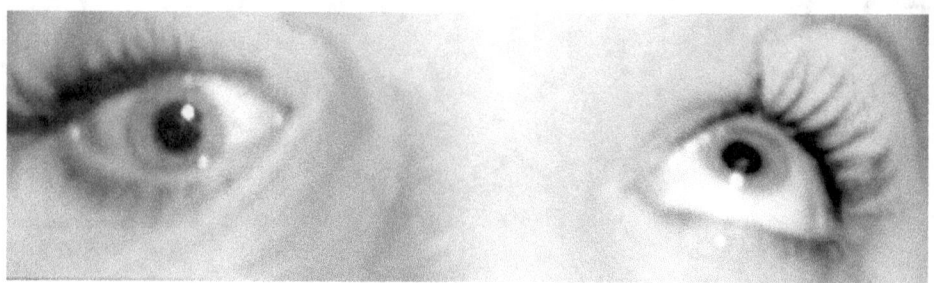

There is *no way* for the six main nerves, twelve different muscle nerves, twelve different eye muscles, fourteen different bones, two trochlear loops, twelve arteries, twelve veins and thirty-six foramen to happen to evolve on two eyeballs **all at the same time!** At this time we are only discussing the bones, foramen, and the muscle systems! Consider an even greater evidence of creation and intelligent design; when you observe all twelve muscles must attach at the correct place, and be of correct length, and attach at the exact angle at each end of each muscle. Again, the only way for this to happen is by planned design—creation! When it comes to the twelve eye muscles, Darwin's theory "breaks down[15]"!

To think *all* 108 different parts of the eye muscle system *evolved* and hooked up *properly, <u>all</u>* at the same time—each one totally dependent upon each one of the other parts—and, all of it occurred by *chance* is to utterly ignore common sense.

I *intentionally* did not say, "the odds of it happening *by chance,* given enough time, is on the order of 1 in 1 trillion trillion trillion trillion trillion trillion trillion trillion" (or 1 in 10^{108}) because that would imply it possibly could happen, "by chance" given enough time. It would be saying, "Well, the odds are *extremely extremely, extremely* slim, but there is an extremely, extremely, extremely slim *chance* it *could* happen." However, that is *not* the case. It is clearly evident they were engineered and designed—which only happens by *creation!* There is no **_chance_** of evolution!

For the eye muscles to be at all functional, let alone *perfectly and completely functional, all* the parts *must* be present *simultaneously* and *all* the parts *must* be fully functional and *completely* developed *all* at the same time.

If any *ONE* eye muscle was missing or *not* fully *evolved,* the eye could not line up straight **_at any time_** - EVER, **_PERIOD!_** No ifs ands or buts; it cannot line up straight and it also could not move in all directions. Additionally, remember the trochlea-bone-loop-superior oblique muscle-pulley system demonstrates two interdependent systems, bone and muscle that would have to evolve simultaneously. Then too remember the origin of the inferior oblique is cokpletely separate, isolated, from all the other eye muscles! It is clearly evident evolutionary processes; "numerous successive slight modifactions"[17] as Darwin's theory would have you "believe" will not work when it comes to the external eye muscles. Darwin's theory as Darwin anticipated "**_ABSOLUTELY_** breaks down."[17]

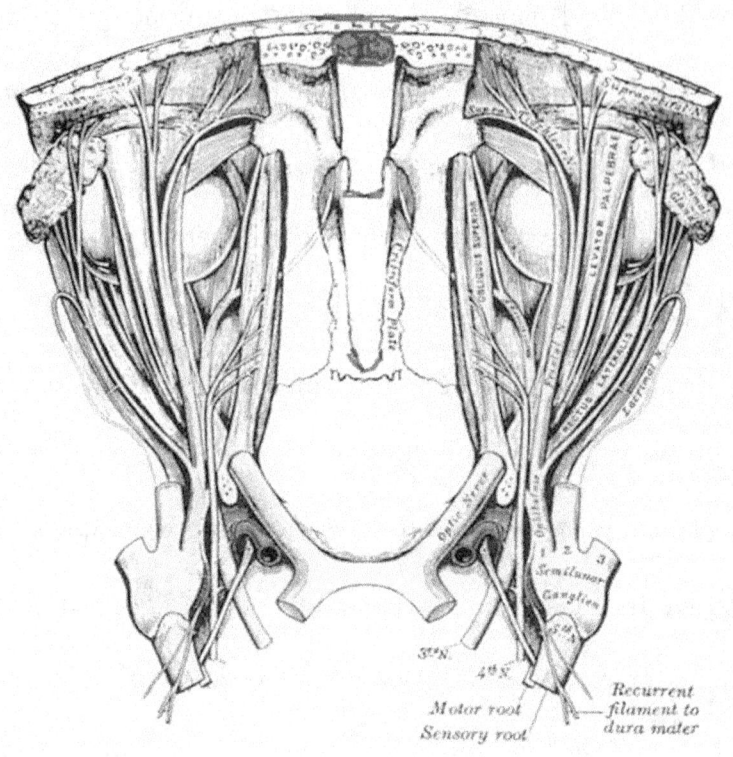

From Henry Gray, F.R.S, *Anatomy of the Human Body* 20th Edition, Lea &Febiger (Philadelphia & New York, © 1918) p.885[16] with modifications and additions by Rigney

Look at the complexity of the design, structure and function. Observe and conclude, it is clearly evident the eye could not have evolved. All the muscles, all the nerves, all the arteries, all the veins, all the bones, all the foramen had to form simultaneously because the eye cannot function if they are not ***all*** present at the same time. We have totally interdependent parts and systems, all with totally interdependent functions and **ALL *MUST*** occur simultaneously. Thus, we observe ***Interdependent Evidence of Creation.*** The only way these conditions can be met is by creation. Looking at ***ONLY*** the eye muscle system; evolution does not work when it comes to the function of the eye muscles, Darwin's theory as he himself stated, "absolutely breaks down."[17]

2. Cornea

The specific way the cornea is made _causes_ it to be clear. The cornea doesn't just happen to be clear. "Numerous successive slight modifications" as Darwin imagines[1] cannot explain this intentional, intricate, multi-dimensional design. When you examine the evidence, the only logical conclusion is Creation. Again, the evidence observed in the cornea overwhelmingly supports the theory Interdependent Evidence of Creation.

Think about the miracle of the cornea; living tissue which is clear. At first glance someone might say the word miracle is an inappropriate use here - because a miracle is something which happens which cannot be explained in human terms. A miracle is something which occurred because of direct intervention by God and could not have occurred in any other way unless God caused it to happen. Well, after examining the cornea, I think you will agree; it could not have occurred in any other way unless God caused it to happen. The cornea therefore truly is a miracle.

The cornea is made up of tissue and cells which require oxygen and nutrients to survive just as every other cell in the body, yet they do not have any direct blood vessel supply. There are no arteries or veins present within the cornea to provide the oxygen and nutrients necessary for the tissue to stay alive. It is **clearly** evident; God's creative hand is in the miraculous design of the cornea. There are at least five different distinct layers of the cornea.[2, 3, 4]

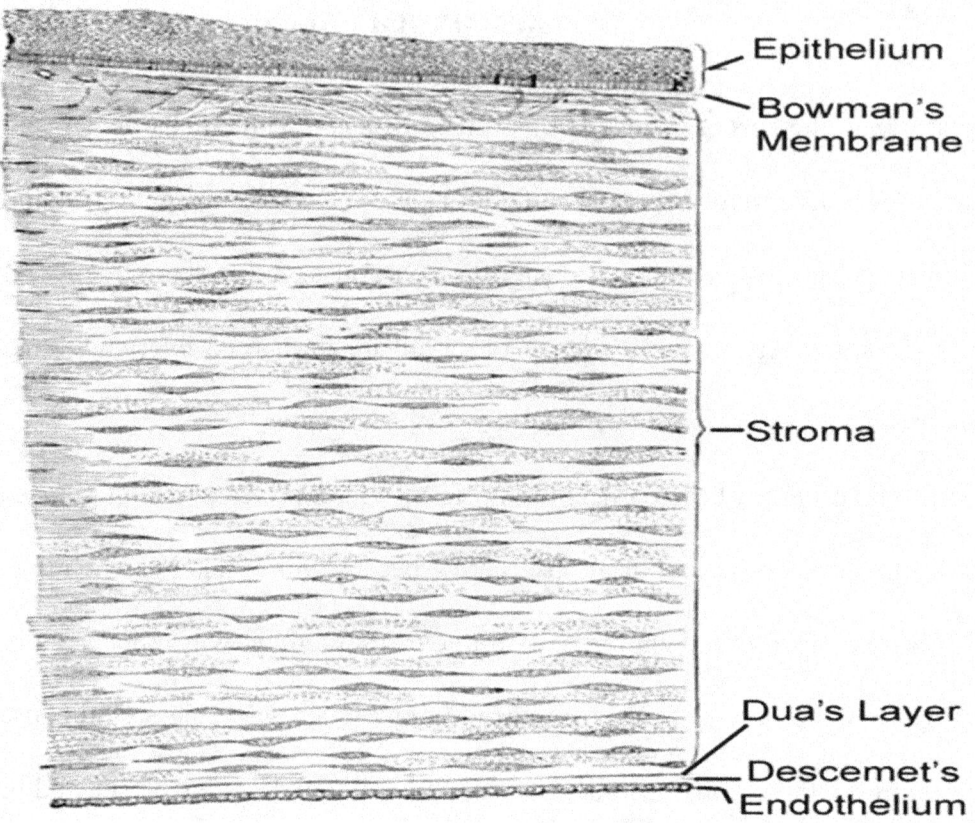

From Henry Gray, F.R.S, *Anatomy of the Human Body* 20th Edition, Lea & Febiger (Philadelphia & New York, © 1918) Page 1008[4] with modifications and additions by Rigney

From outer to inner, the five layers are; the epithelium (surface skin layer), Bowman's Membrane (outer limiting lamina or outer protective layer), Stroma (the meat or body of the cornea), Descemet's membrane (inner limiting lamina or inner protective layer), and the endothelium. A sixth layer between the stroma and Descemet's Membrane, "Dua's layer" has recently been suggested.[5] (Some would say there is a seventh layer if one considers the basement membrane layer of the surface epithelium.[6]) Some of these layers have distinct intercellular layers within themselves also.

The surface layer, the epithelium, is made up of skin cells like that of the skin covering the surface of the body except the surface skin cells of the cornea, the epithelium, lack a substance called keratin. Because they lack keratin, the skin cells covering the eye are ***clear, transparent*** skin.[6,7] If the eye were to have evolved rather than been designed, you would expect at one point the cells would have had keratin and then through the evolutionary process became *entirely* non-keratinized. The problem with this is keratin in the epithelium causes it to be OPAQUE! If the skins' cells were at one time opaque the eye cannot see. The creature cannot survive if it cannot see. How did they *"evolve"* to become transparent, yet they are skin?

The surface layer, epithelium, of the cornea is made up of skin cells (non-keratinized epithelium) which are five to eight cells thick - thinner in the center (about 5 cells thick), and thicker in the periphery (about 8 cells thick). The epithelium surface-skin- layer is approximately 50 microns thick[6,7]. (A human hair is about 72 microns thick.) The epithelial cells have a constant replenishing function, they are continually forming new cells from the basement layer which is pushed up toward the surface by the next forming basement cell. These two newly formed cells are then pushed up toward the surface by the next forming basement cell. The surface layer of old cells are continually being sloughed off the surface as new basement cells are pushing toward the surface. As a result, the entire corneal surface epithelium layer (all 5-8 cells of thickness) is replaced about every seven days,[8] yet the thickness is constantly maintained at 5 - 8 cells, and 50 microns thick. (How would it *happen* to work that way? Again, "numerous, successive slight modifications"[9] as Darwin imagines cannot explain it.)

These surface, non-keratinized skin cells (epithelium) are comprised mostly of water (about 70 % water by weight[10]) therefore they are rather soft. They must be soft to absorb oxygen and transport the oxygen into the cornea because there is no direct blood vessel supply within the corneal tissue.[10] Additionally, the epithelium cells transport some of the nutrition to the cornea via the tears.[10, 11] Stop for a moment and contemplate that! The skin cells must absorb the oxygen and absorb the nutrients in order to transport them *INTO* the cornea. Thus, the tears and the epithelium work *together* to provide the cornea with oxygen, nutrients, healing, repair, and defense against infection. If the tears were not fully, completely, simultaneously evolved at the same time the epithelium is evolved - the eye is opaque and is blind. A blind being cannot survive. (We see this problem currently (21st Century) in very severe dry eye syndrome in our day to day routine of eye care.) The cornea is totally interdependent on the tears.

Darwin states, "If it could be demonstrated that any complex organ existed, which could not possibly be formed by numerous, successive, slight modifications, my theory would absolutely break down."[12] The cornea - the entire eye for that matter, is absolutely dependent upon tears. They both must be present; the tears and the cornea, at the same time. The clarity of the cornea and the oxygen supply to the cornea and some of the nutrient supply to the cornea must come from tears. If the tears have not evolved at the same time the cornea is evolved, the eye cannot see and the being cannot survive. The tear production itself has interdependent functions which require simultaneous, fully developed, fully functional, fully completed, interdependent parts: Blood vessel supply, a provision for tear drain ducts through the bones of the eye socket into the back of the nose. Blinking of the lids and formation of the eyelid muscles, three different nerve supplies from the brain through the bones to the eyelid, the tear gland, and the cornea to sense dryness and tell the eyelid when to blink. What we see with the eye, and what this book will further illustrate is; systems, and subsystems are totally interdependent and require fully functional completed functions at the same time - because they are completely and absolutely dependent upon each other for the eye to function and for the eye to remain alive. Again, when we examine the cornea, we observe ***Interdependent Evidence of Creation***. We observe *19 interdependent systems, required* just for the cornea to function: 1.The Cornea, 2.The Tears, 3.The Eyelids, The eyelid muscle's nerves- (two different nerve supplies); 4.The Oculomotor Nerve for 5.The Levator Palpebrae Superioris eyelid muscle. 6.The Facial Nerve for the 7.Orbicularis Oculi eyelid muscle. The sensation of pain and dryness for the cornea by the 8.Trigeminal Nerve. The 9.Lacrimal Nerve to provide the stimulus to the 10.Lacrimal Gland to make tears,. The 11.Opnthalmic Artery to provide the blood supply to the 12.Lacrimal Artery for the Lacrimal Gland to create tears. The 13.Lacrimal Vein and 14.Ophthalmic Vein to carry away the blood from the Lacrimal Artery and Ophthalmic Artery, The formation of the hole in the bones of the back of the eye socket; the 15.Superior Orbital Fissure, which must be present for the Oculomotor Nerve, Trochlear Nerve, Abduscens Nerve, Trigeminal Nerve, Ophthalmic Artery, and Ophthalmic Vein to Pass through from the Brain and Internal Carotid Arteries in the skull, into the eye socket to supply the eye. The formation of the two 16.Puncta (drains) on the upper and lower lids. The two drain tubes 17.Canalliculi (drain tubes to the nose) to carry the tears to drain into the nose. The formation of the 18.Lacrimal Fossa; the foramen (holes in the bones through which the tear ducts must pass in order to drain the tears into the nose.) 19.The Aqueous Humor (the fluid that fills the inside of the eye) also provides nutrients to the cornea, and provides defense (immunity) to fight infection for the cornea, and some oxygen supply to the cornea. All 19 Systems are interdependent. Remove any **_one_** of these 19 systems and the eye will not function and oftentimes the eye will die. They are **_ALL – required;_** and each of their functions must be fully developed and fully functional because they are interdependent. Wow! To think we are only discussing what is *required* for the cornea! Again, "successive, slight modifications," as Darwin imagines[12] will not work when it comes to the cornea. *All* must be present, fully functional and all MUST occur simultaneously. One system cannot wait on another system to "evolve" because they are all absolutely and totally interdependent.

The second layer of the Cornea is called "The Anterior Limiting Lamina."[13] (It is most often referred to as Bowman's membrane.) It too is transparent clear tissue. It provides the majority of protection to foreign objects entering the deeper tissues of the eye, whereas the clear, non-keratinized, mostly water, oxygen and nutrient transporting surface epithelial cells are soft, they are easily penetrated and easily moved or manipulated. Thus, a thick tough elastic barrier is needed to protect the delicate inner layers of the cornea. This protection is provided in the design of "The Anterior Limiting Lamina" (Bowman's membrane). It is the layer the clear, non-keratinized, mostly water, oxygen and nutrient transporting epithelium cells sit upon. The anterior limiting lamina (Bowman's membrane) functions much like the foundation of a house. It is made up of a completely different embryonic developmental tissue type, mesoderm, whereas the surface epithelium is ectoderm.[14, 15] Bowman's membrane is embryologicaly derived from the same tissue type (mesoderm) as is the tissue which makes up tendons and ligaments for muscles and bones![16] During development of the eye the epithelium (the first, outermost layer) is formed as a layer of cells derived from ectodermal cells. A wave of mesodermal cells form beneath this ectodermal layer. This first wave of mesodermal cells form the endothelium (the sixth innermost layer) of the cornea. These two layers are lying adjacent to each other at about the fifth week of fetal development. Then, a second wave of mesodermal tissue invades exactly in between these first two layers at about the sixth week. This second wave of mesodermal tissue then forms the stroma of the cornea. Bowman's membrane is derived from the superficial layer of the stroma,[17] the exact site, and precise path needed to provide necessary protection.Bowman's membrane is derived from a completely different germ layer and of a completely different tissue type than the epithelium! Then at about the 3rd month Descemet's membrane (the fifth and next innermost layer) is secreted by the endothelial layer.[17] Again, evolution cannot explain these embryological processes and this embryological fact! We observe it is the DNA which DICTATES the placement, thickness, and arrangement of these six layers, and the DNA *causes* it to happen during embryological development. Evolution is NOT at work- the instruction and information in the DNA is at work. Instruction and information dictating how the cornea is built does not just happen, instruction and information is evidence of Creation. Information conveyed, instructions dictated prove Creation and cannot happen by chance!

The anterior limiting lamina (Bowman's membrane) is VERY tough but at the same time pliable. This gives the cornea strength but prevents the cornea from being brittle. If the cornea were brittle it would crack or shatter like glass. If the cornea were elastin it would be too soft and it would be difficult for it to maintain a constant shape - which is necessary for the eye to have a sharp focus. The composition of the cornea is just right - not too hard and not too soft. Without the anterior limiting lamina (Bowman's membrane) injuries to the eye would be common and infection would

27

have a free reign into the eye. In fact, when we see corneal abrasions (significant scratches to the cornea) almost always they are limited to the layer of Bowman's membrane. When the cornea is scratched often the epithelium is cleanly removed to, but not through this tough layer. (Much like using a rubber spatula to remove icing off the surface of a cake.) Injuries which fail to penetrate Bowman's membrane (the Anterior Limiting Lamina) do not leave a scar, and most injuries, thankfully, do not penetrate Bowman's membrane (because it is so tough). Additionally, the epithelial cells rest upon a basement membrane which is attached to Bowman's membrane.[17] The basement membrane holds the epithelial cells to Bowman's membrane.[18] Without the firm foundation provided by Bowman's membrane the clear, soft, non-keratinized, mostly water, oxygen and nutrient transporting epithelial cells would easily slough off the surface of the eye and be pushed around on the surface. Without the firm foundation provided by Bowman's membrane the surface epithelium cells would behave as a piece of carpet on a tile floor, or, again as would icing on a cake. (We see this condition in a disease of the cornea called *Basement Membrane Dystrophy.*[19] In this condition a person can simply rub their eyes and cause sloughing off of the surface clear, non-keratinized, mostly water, oxygen and nutrient transporting epithelial layer. People with Basement Membrane dystrophy constantly have problems with painful irritated eyes. Without the Anterior Limiting Lamina (Bowman's Membrane), injuries, infection, and recurrent surface cell sloughing would be a constant problem. Constant discomfort and regular infections would be the rule rather than the exception. Again, we observe intentional, intelligent, preventive design, each of which is evidence of a Creator.

Beneath the clear, transparent surface skin cells and tough, elastic, clear, transparent Anterior Limiting Lamina (Bowman's Membrane) is the "Stroma". It is difficult to describe in words the engineering marvel of this layer. If you can speak any one word at all "WOW!" might be the best one. The creator/designer's hand is *clearly* evident (pun intended) when you observe the design of the stroma. If one truly contemplates the complexity and minutia involved within this layer, total awe for the Creator should be the logical conclusion and response. Darwin with the best technology available at his time, (a compound lens light microscope), could not KNOW about these small fibrils and how complex of an arrangement they are in because they are beyond resolution with light microscopy.[20]

The tissue type of the stroma is a connective tissue (collagen) and like the tendons and ligaments, it is derived embryologicaly from mesodermal tissue.[20] The stroma is comprised of very small collagen *microfibrils* which are taken together to create *fibrils*. These *microfibrils* and *fibrils* are so small they **cannot** be detected with a light microscope[20] (which is all Darwin had at the time of the writing of his theory). These *microfibrils* and *fibrils* can only be appreciated with the aid of an electron microscope at a magnification of 48,000 times.[20] Again, Darwin in his day, with the best technology available at the time, could not know about these small *microfibrils* and *fibrils* and how complex of arrangement they are in.[20] The corneal stroma consists of individual *microfibrils* which are about 4 nanometers in diameter.[21, 22] (A human hair is about 72,000 nanometers thick so it would take 18,000 of the microfibrils to equal the size of a human hair. Think about that, 18,000 of them together would equal the thickness of a human hair. Stop and meditate on how complex and minute the details of the arrangement of the fibers of the cornea are. Five of these microfibrils are taken together to form a 36-nanometer *fibril*.[21, 22] (The *microfibril* itself is only 4 nanometers thick and there are five present -which would make one assume that the thickness of a *fibril* would be about 20 nanometers rather than 36 nanometers, but the *microfibril* has a twisted helical shape, and while the *microfibril* is only 4 nanometers thick, because it has a twisted shape and because there are five of them in a fibril, the twisted shape along with the presence of other proteins and proteoglycans makes each *fibril* to be about 36 nanometers thick.[22, 23, 24] So the total *fibril* thickness, comprised of five 4-nanometer microfibrils is about 36 nanometers rather than the anticipated 20 nanometers. (It would take 2000 fibrils to equal the thickness of a human hair- and remember each fibril is composed of 5 micofibrils.) Then, about 200-250 of these 36 nanometer *fibrils,* each composed of 5, four nanometer *microfibrils-* are laid down together to form a *lamella* (the word lamella is Latin for *layer*) or sheet which runs across the entire surface of the cornea.[21, 22, 23, 24] Each individual *lamella* (layer) or sheet, is about 2 microns thick (2,000 nanometers thick)[21,22, 23, 24] (a human hair is about 72 microns thick, or 72,000 nanometers thick). If you were to divide a human hair into 36 equal threads each one of the 36 threads would be 2 microns thick. One of those threads 1/36th of a hair thick would be the same thickness as one of these lamella or sheets. Another way to say this is; it would take 36 of these two-micron lamella (or sheets) to equal the thickness of a human hair. These two-micron thick *lamella* (sheets) are layered on top of each other but they are arranged in alternating directions 90 degrees perpendicular to each other. For example, one *lamella* or sheet is layered with the fibers running vertical, or up and down. Then the next *lamella* or sheet is layered on top of the preceding sheet but with the *microfibrils* and *fibrils* running horizontal, or left and right. The lamella (or sheets) are stacked one upon the other and placed 90 degree orthogonal (perpendicular) in alternating layers, about 10-25 microns wide by two microns thick, that then runs from edge to edge across the entire surface of the cornea in *lamella* or sheets[21, 22, 23, 24] (see the picture on page 32). The corneal stroma is about .5 mm or 500 microns thick[23] or 500,000 nanometers thick (remember. the microfibril is 4 nanometers thick) so since the lamella or sheets are 2 microns thick each, there would then be about 250 of these 2 micron thick *lamella* or sheets laying or stacked upon each other in opposite, orthogonal, 90 degree directions to each other[22, 23, 24, 25, 26, 27] in order to make up the corneal stroma.[25]

In summary, five 4-nanometer *microfibrils (about 1/18,000 of a hair)* are bundled together to form a 36-nanometer *fibril (1/2000 of a hair)*. Then about 200-250 of these *fibrils* are laid together in a *lamella* (layer) or sheet which spans the entire front surface of the cornea and is 2000 nanometers thick (two microns thick or 1/36 of a hair thick) by 10,000-15,000 nanometers (10-15 microns or about 1/5 of a hair) wide. Then about 250 of these two-micron thick *lamella* (layers) or sheets are laid on top of each other in alternating 90 degree (perpendicular) orthogonal sheets or layers (lamella) to make up the stroma.[19, 20, 21, 22, 23, 24, 25, 26, 27]

5, 4 nm Microfibrils twisted together make one 36 nm fibril.
250 fibrils make up alternating 2 micron lamella sheets.

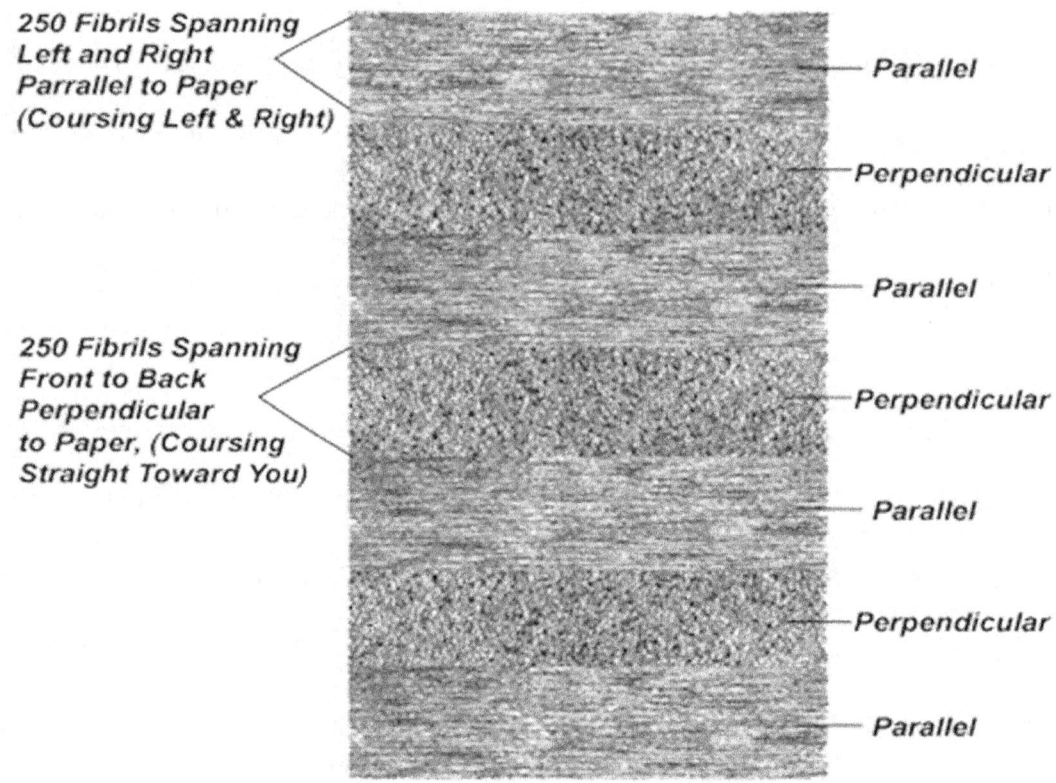

250 Fibrils Spanning
Left and Right
Parrallel to Paper
(Coursing Left & Right)

250 Fibrils Spanning
Front to Back
Perpendicular
to Paper, (Coursing
Straight Toward You)

Parallel

Perpendicular

Parallel

Perpendicular

Parallel

Perpendicular

Parallel

250 Fibrils laid together to form a 2 micron Lamella (sheet).
These 2 micron lamella are laid 90 degrees perpendicular to
each other in alternating arrangement. 250 Lamella make up
the Corneal Stroma, here we see 7 of the 250 lamella (sheets).

If you could, visualize the individual strands of nylon in a nylon rope and see how the strands consist of microscopic threads. The strands which are made up of microscopic threads are then placed into groups which are woven together; this would be similar to the make up of corneal microfibrils and fibrils making up the lamella (layers) of the stroma, except that the strands are not interwoven, they are stacked and layered in alternating directions. Here again intelligent design is clearly evident. Stacked fibers give the cornea more rigidity and strength to forces coming toward the eye. This design of the lamella (layers) of the stroma are similar to stacking boards in a manner in which the grain of the wood is running in alternating perpendicular directions, 90 degrees, from each consecutively stacked board. This design provides the greatest strength to forces coming directly toward the board,[29] or in this case, as we apply the same principle to the stacked layers of the cornea, it provides maximum protection to objects flying at, or coming directly toward the eye.

When we look at the *design* of the layers of the cornea it is evident *the layers are <u>arranged</u> in alternating 90 degree crossed angles* in order to provide the most structural support and integrity. Plus, this arrangement of multiple *microfibrils* used to make multiple *fibrils* used to make sheets or *lamella* (layers) is a common engineering principle. Multiple strands running in the same direction provides increased strength but at the same time provides flexibility.[30] An everyday common example of this would be copper wire. There are two types of wires; there is a solid single copper wire, and there is a multi-strand copper wire. For example, you can have a solid single copper wire which is 1/8 inch thick, or you can have a 1/8 inch thick multi-strand wire which is made up of 8, 1/64 inch strands. In both instances, they are 1/8 inch thick. However, a single solid copper wire 1/8 inch in diameter is stiffer than a 1/8 inch diameter copper wire composed of 8, 1/64 inch strands even though they are both 1/8 inch thick. If you were to bend the 1/8 inch single solid copper wire many times over, a crease will develop in the wire and eventually it will break. Whereas it would take a lot more bending to break the 8 strand 1/8 thick multi-strand wire.[30]

A similar example is seen with a single 2-inch piece of lumber (a board that is two inches thick) in comparison to a 2-inch sheet of plywood. The two inch piece of lumber is stronger in resistence to compression than is the plywood, but the two inch piece of lumber is more likely to crack or break when exposed to blunt force. Additionally, as we see with plywood, as the layers of lamination increases the strength increases. For example, a 4-layer plywood sheet is 24 % stronger than a 3-layer plywood sheet, and a 5- layer plywood sheet is 13% stronger than a 4-layer plywood sheet.[31] Multiple layers provide added strength, *and* multiple layers are less rigid and less likely to crack or break than a single sheet of equal thickness.[31] If you make multiple strands into layers, then arrange the layers 90 degrees from each other you have increased the strength even more. This clearly intentional design makes the cornea strong but not brittle, so it will not crack or break. The cornea must be strong but at the same time the cornea cannot be too rigid or brittle, otherwise we would have to worry about our cornea chipping or cracking; as when a window chips or cracks. Crack the cornea and the fluid cannot stay in the eye. If the cornea were to crack, germs could then enter the eye. If the cornea could be chipped, a chip would cause us to feel like something was in our eye every time we blink. What a disaster! A chip would also allow debris and germs to lodge or settle into the depression of the chip and increase the likelihood of infection. The cornea must be strong-but not too rigid or brittle. At the same time, if the cornea were too soft the eye could not focus a consistently sharp image. Think about the mirrors at the funhouse at a circus; how they distort and stretch your reflection. If the cornea were too soft our vision would be distorted just as the reflection in those mirrors are. Intentional design is clearly evident, just as it is seen in a sheet of plywood or multi-strand wire.

The corneal design is collagen microfibrils, combined into fibrils, laid into a lamella (layer) or sheets, then these lamella or sheets are laid one on top of each other at exactly 90 degree angles perpendicular to each other in layers to form the stroma of the cornea. Evolution cannot explain this intentional design. To say they evolved to form in this specific pattern would be similar to walking in the woods, and just happening to find a copper multi-strand wire wrapped in plastic insulation. After closely observing these wires wrapped in plastic insulation, you conclude the small wires somehow combined together on their own and became wrapped in plastic insulation completely by chance with no outside influence. It would be similar to walking in the woods and finding a sheet of plywood and observing the *many layers* within the piece of plywood are *arranged* with the grain laying in 90 degree, perpendicular, alternating layers. You also observe the layers are *glued* each one to the other. After this close inspection which demonstrates multiple layers, laid in a perpendicular 90 degree alternating arrangement, you then conclude the plywood taken as a whole, combined itself to form in this arrangement, completely by chance, with no outside influence. You conclude it *evolved* by chance.

If you were walking in the woods and happened to encounter a multi-strand wire wrapped in plastic insulation; you would *know* that it was created. If you were walking in the woods and happened to encounter a sheet of plywood; you would *know* it was created. You may ask, "I did not *SEE* someone create or make it, so how can I *KNOW* it was created?" You still can logically determine and *KNOW* they were made - by someone or something at some time - just by examination and deduction; observation and conclusion. You can *SEE* someone's creative hand had done the creating, you observe Interdependent Evidence of Creation, I.E.C. The point is, you don't have to *see* someone make it, to *know* it was created. You may say to yourself, "well, I did not *SEE* someone make it so I do not *KNOW* with absolute *100% CERTAINTY* someone created it; and until I *SEE* someone create it and thereby KNOW with 100 % *CERTAINTY* someone created it; I will *believe* it happened on it's own by chance. As a result, *YOU DECIEVE YOURSELF! Or, YOU CHOOSE TO BE DECEIVED! YOU CAN **KNOW**, AND YOU DO **KNOW** WITH 100% CERTAINTY – EVEN THOUGH YOU DID NOT SEE SOMEONE CREATE THEM. BE LOGICAL! BE REAL! BE TRUE TO YOURSELF! DON'T BE DECEIVED! YOU OBSERVED INTERDEPENDENT EVIDENCE OF CREATION! YOU <u>KNOW</u> IT WAS CREATED! You KNOW with 100% certainty – even though you didn't see it be created. Even though you didn't see the creator; you 100% know they were created.*

Just as with a multi-strand wire wrapped in plastic insulation; just as with a piece of plywood; the cornea demonstrates Interdependent Evidence of Creation. However, the cornea is infinitely more powerful in exhibiting Interdependent Evidence of Creation than a multi-strand plastic insulated wire, or piece of plywood!

What we *SEE*, that is, <u>what is observed</u> with the cornea, is the cornea must be strong but not too rigid. It must have *just right rigidity*, and at the same time *just right flexibility* so the cornea does not shatter or break. It is *LIVING*, **CLEAR** tissue like glass, but **better** than glass in that it is not easily broken, chipped, or shattered. And, if it develops a scratch, new cells are constantly growing to repair itself. A scratched lens causes ghost images, aberrations, halos, and glare. In fact, due to the unique design and make-up of the cornea it ***cannot*** break, chip, or shatter. WOW! THAT WE *KNOW!* What a design! We take for granted that our cornea, the main WINDOW to the eye, can be submitted to significant trauma and maintain its function- ***BECAUSE*** of its' specific design. We don't have to continually worry about damage or injury to our cornea. In fact, the sad thing is we take it for granted and never give a thought to the possibility our cornea may chip or crack. Have you *ever* thought or been concerned that you may crack or chip your cornea? Whereas, if you are carrying around a camera, cell phone, or computer tablet of any type with a glass or plastic lens, surface, or window; you are constantly concerned with its well-being and you consciously take steps to protect the window, lens, or glass cover. The cornea is a miraculous intelligently engineered design you should thank God for-daily. Stop and thank Him right now! (Thank you God for designing the cornea just right.) The cornea is marvelously (fearfully and wonderfully) *MADE*! (WE CAN *KNOW* IT WAS MADE - EVEN THOUGH WE DID NOT *SEE* IT *BE* MADE!) What we *SEE* with the cornea is a window, like glass - but better than glass. Again, God's design is better than anything man can make.

Additionally, these *microfibrils*, arranged into *fibrils*, taken together into *lamella* (layers) or sheets, stacked one on the other at alternating right angles are not only important for <u>structural reasons,</u> this design is absolutely necessary for transparency. And, while the right balance of strength and rigidity is a consideration for the design of the *microfibrils*, *fibrils*, and *lamella*, it is also ***the reason*** for the clarity of the cornea. This alternating lamella (layers) design is the ***only*** design which allows the cornea to be clear. We can clearly *SEE* and it is plainly evident these *alternating layers -*

of just the right thickness and design <u>*causes*</u> the cornea to be clear. Think about that; the layers need to be *just right* - the cornea cannot be too hard or too soft – and at the same time the lamella or layers must be of the *exact* thickness to <u>*cause*</u> the tissue of the cornea to be clear. These two requirements must be met precisely, <u>***and***</u> simultaneously. Do you really think it is just a coincidence or it just happened that way? Do you really think evolution can explain this? Can you not *SEE* the design of the corneal stroma – and, like a multi-strand wire, like a sheet of plywood *KNOW* it too was Created? Here again we observe Interdependent Evidence of Creation.

Living tissue… that is transparent; Living cells which are clear. Stop and consider the ***<u>miracle</u>*** of this statement! (Use of the word ***<u>miracle</u>*** is with intention.)

The ***reason*** the cornea is clear is because of the <u>precise</u> and ***<u>EXACT</u>*** thickness of the lamella (layers). These *microfibrils*, *fibrils*, and 90 degree alternating *lamella* (layers) or sheets are so precisely arranged – (precise to the level of 200 nanometers, or about 1/36 of a human hair) it causes internal reflection of light waves back onto themselves so they cancel each other out - *making* the cornea clear. Francis Heed Adler said, "As long as the fibrils are regularly arranged in a lattice and separated by less than a wavelength of light, the cornea will remain transparent."[32]

Here is how it works; a light wave encounters a lamella (or sheet) and is reflected backward, the reflected light wave goes backward to encounter an incoming light wave. The spacing and thickness of the lamella is *precisely the exact thickness* so that the peak of the reflected light wave encounters the valley of the next incoming light wave so they exactly cancel each other out. (The peak of the next incoming light wave encounters the valley of the reflected light wave.) This precise design causes the light rays to cancel each other out as they enter the eye *making* the cornea clear. The thickness of the *lamella* (or sheets), *fibrils* and *microfibrils* and their 90 degree alternating arrangement has to be *just right* and precisely *exact* for this to occur, otherwise the cornea is opaque.[32, 33, 34] Maurice in 1957 stated, "The cornea is clear because the uniform collagen fibrils are arranged in a regular lattice so scattered light is destroyed by mutual interference. Loss of transparency would result if this regular arrangement were altered by hydration or mechanical forces."[33, 34] The lamella are arranged precisely in this manner to create the clarity of the cornea. That is, the arrangement and spacing is **the reason** the cornea is clear. The fibers are "arranged in a regular lattice."[33, 34] "*Arranged*" in a "regular" lattice implies purposeful intentional design.

Conversely, Darwin states on page 158 of his book *The Origin of Species*, "we ought **in imagination** to take a thick layer of transparent tissue, with a nerve sensitive to light beneath, and then **suppose** every part of this layer to be continually changing slowly in density, so as to separate into layers of different densities and thicknesses, placed at different distances from each other, and with the surfaces of each layer slowly changing in form."[35] Darwin goes on to state, "let this process go on for millions on millions of years; and during each year on millions of individuals of many kinds; and may we not **believe** that a living optical instrument might thus be formed as superior to one of glass, as the works of the Creator are to those of man?"[35, 36] **Note:** Darwin uses the word **"Creator"**[36] (underlining by Rigney)

You can "imagine" it from now through eternity, but the *reality* is: the eye cannot see, the individual cannot function, the being will not survive while it is waiting for its' eye to **evolve** to a point of clarity.

Darwin says, (page 158 *The Origin of Species)* "we ought in ***imagination*** to take a thick layer of *transparent tissue*, with a nerve sensitive to light beneath, and then ***suppose*** every part of this layer to be continually changing slowly in density, to separate into layers of different densities and thicknesses, placed at different distances from each other, and with the surfaces of each layer slowly changing in form. Further we must ***suppose*** there is a power always intently watching each slight accidental alteration in the transparent layers; and carefully selecting each alteration which, under varied circumstances, may in any way, or in any degree, tend to produce a distincter image. We must ***suppose*** each new state of the instrument to be multiplied to the million; and each to be preserved till a better be produced, and then the old ones to be destroyed."[37]

He says, "may we not believe"[37]. He says, "we *must* suppose"[37.] Why ***must*** we suppose? Is it just because he says we should? He says, "we ought to imagine.[37]" Why ought we imagine? Is it just because he *says* we should? *Imagination* and *supposition* without supportive evidence is a belief. He thus says, "may we not believe."[37] A "belief" "supposed", a "belief" "imagined"[37] - without evidence requires faith. Faith in something not seen, just *IMAGINED, SUPPOSED* OR ***BELIEVED*** is a religion. Evolution is really a religion. If *evolutionism* is allowed to be taught in schools then so should *creationism.*

It is obvious, (I think), with this rather *flippant, haphazard* statement[37] that he did not know or even contemplate the requirements; (requirements much, *much* greater than *"imagination"*, *"supposition", and "belief"*) which ***must*** be met (not ***"ought"*** to be met; not ***"supposed"*** to be met; not ***"imagined"*** to be met, not ***"believed"*** to be met)[37] in order for living tissue to be *transparent.* Furthermore, he is in absolute error when he says, "with a nerve sensitive to light beneath.[37]" We

KNOW, this is an incorrect "supposition."[37] He did not know the "nerve" is **NOT** what is "sensitive to light," it is instead the retina. His "supposition" and therefore his "imagined" theory and his "belief " is false! (We will discuss this errant statement, "nerve sensitive to light beneath"[37] more in chapters 16, 17, 18 on the retina, the optic nerve, and the visual cortex later in this book.)

Just because you can *"imagine"* it; *"imagining" it* does not make it so. Then, just because after you *"imagine"* it, you then *say* it is so, does not *mean* it *is* so. *Imagination* is *NOT* science. Where is the evidence? *Imagination* and *supposition* without evidence is, as Darwin himself says - "a belief". A "belief" requires faith and therefore his theory, as he *"believes"* it, is a religion.

The *reality* is; if you can't see - you can't survive. Think about that. *Imagine*[37] *THAT;* a being living, functioning, surviving while it is waiting for its eye to develop to a point of seeing. Now, *IMAGINE*[37] each species trying to live, function, and survive while it is waiting for its' eye to develop to a point of seeing; IMAGINE[37] *THAT*, for all the different seeing species which exist on earth or have ever existed on earth.

Why is it we *NEVER* find *ANY* species ANYWHERE either in the past or present- NO FOSSIL evidence ANYWHERE, no OBSERVED SPECIES PRESENT TODAY ANYWHERE with a partially evolved eye. *AND*, *why* is it we always observe two eyes, not one, not three, not five, not ten? {I am aware there is a two-eyed fish that splits it's superior (top) field of view and its' inferior (bottom) field of view into above-the-water field of vision and below-the-water field of vision. Some call it "a four-eyed fish" but it only has two eyes. It uses its superior field of vision for above-the-water vision and its inferior field of vision for under-the-water vision} Even the insects with compound eyes have two eyes, not one, not three, not five, not ten. We see insects, fish, reptiles, spiders, crabs, mammals; all with two eyes- but no partially evolved eyes, anywhere!

Could it be the plan of the Creator is <u>evidenced</u> in consistency of design. Could it be the genetic code, the DNA, supplied by the Creator follows the same instructions, plan, and blueprint for all the different species, and this is why we *see* the similarities? We see it when we find fossil evidence and when we observe life on earth today. What we observe is; the eye is fully developed -fully developed for what a particular species is doing, and how the species is living. I am not aware of ANY FOSSIL, ANYWHERE, which has EVER been discovered with a partially evolved eye. And to my knowledge (which admittedly is very limited), the fossils which have been discovered, of any seeing species, *all* have two - fully developed eyes. Also, the fossil record of non-seeing species all show no eyes. We either observe no-eyes for non-seeing species, (bacteria, fungi, archaea) or two fully-developed eyes for seeing species. We observe NO partially evolved eyes and NO evolving eyes, anywhere!

Even when we observe a species which suggest some form of evolution by logical observation and deduction, such as for example in my observation, the Halibut fish – it still evidences two eyes.

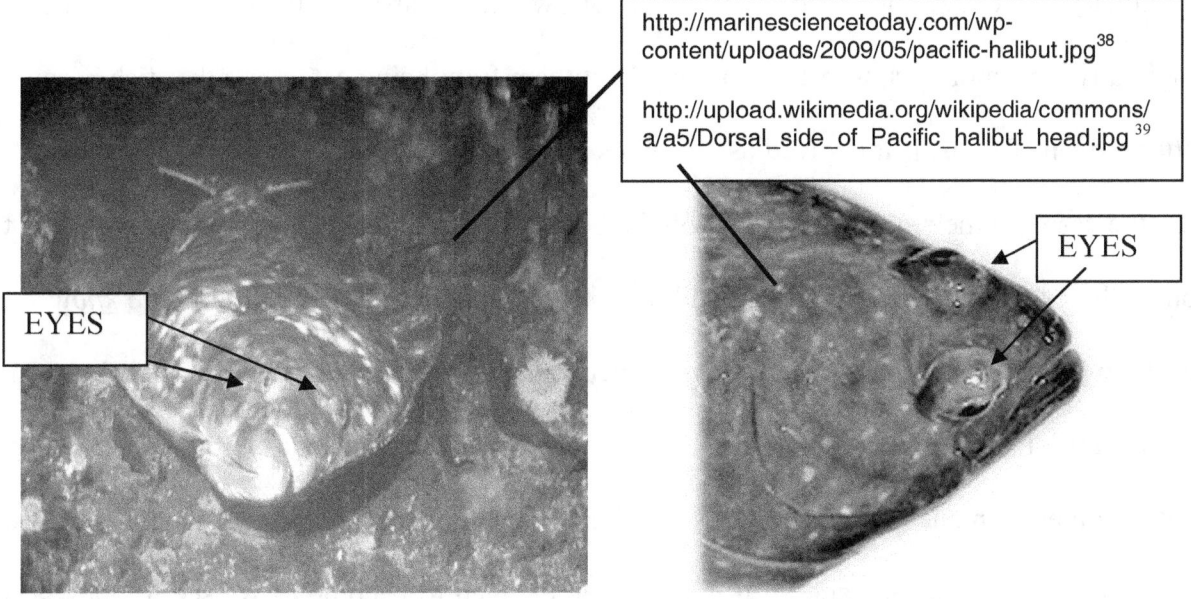

http://marinesciencetoday.com/wp-content/uploads/2009/05/pacific-halibut.jpg[38]

http://upload.wikimedia.org/wikipedia/commons/a/a5/Dorsal_side_of_Pacific_halibut_head.jpg[39]

EYES

EYES

This fish has two eyes on one side of its head. The Halibut body appears to be like that of a regular fish except both of its eyes are on one side of its body. It lies on its side on the ocean floor much like a manta ray or stingray. It looks as if it's a regular fish that swims with its dorsal fin up and its belly down, but instead it lives on the ocean floor laying on what you would normally think is its side; as if it were a normal fish that had died and is laying on the bottom on its side. (I am not aware if it swims vertically or horizontally.) When I first saw this fish I thought it actually did evolve to have both eyes on one side of its' head. However, knowing what we know about vision and an unused eye, we know the Halibut's eyes could not have evolved. Furthermore, if we follow the contrivances of evolution; evolution cannot explain the Halibut's eyes (both on one side) either.

If the Halibut's eyes *evolved* due to lying on its side on the bottom, as a competitive advantage for survival (as the evolution theory would suggest), and if it once was like a regular fish with two eyes, one eye on each side of its head, and if the Halibut lying on one side flat on the bottom gave it an evolutionary advantage for survival over Halibut swimming upright; possibly it gave them an advantage in finding more fish to eat. It possibly kept them from being eaten by other fish because it was better hidden. That is, if it evolved as Darwin's theory suggests, as a competitive advantage. IF this were the case and the reason why it laid on the bottom on its side; we would expect it to lose the function of the eye facing the bottom and end up with one functioning eye. At least we *know*, (that is, I don't "imagine" or "suppose" or "believe" when I make this statement) we lose the function of an eye when after birth it is continually not used. The unused eye becomes lazy and then, because of disuse, it becomes permanently blurred. This is called deprivation amblyopia.[40] If the Halibut at one time had two eyes; one on each side of it's head, then evolved to become as we see it today, a fish which lays on its side on the bottom - so only one eye is continually used; instead of two eyes both on one side of its head we would expect, by evolution, the Halibut would have only one eye on the top of its head. (Because of the reality of what we know happens to an eye when it is not used.)

But, if we give *more credence* to the evolutionary process than would generally be expected, if the halibut did *evolve* two eyes on one side of its' head – from once having one on each side of its' head. That is, if the one eye evolving from the under side of the head, evolved to become with the other eye, on one side of what is now the top of the head, and thus we now see two eyes on the top of its head. The problem is we find no evidence anywhere, of any kind, of the Halibut's eye *becoming relocated* to an evolving location *somewhere*, or the Halibut's eye becoming relocated to an evolving location *anywhere*, on any Halibut, or any fossil record of the Halibut. There is no evidence the second eye evolved from one side "by numerous successive slight modifications" [41] as Darwin's theory suggests; and then began to function with the other, as two eyes - but now through evolution both are on what is now the top of its head. There is no evidence of "numerous, successive, slight modifications."[41] as Darwin's theory would suggest. None!

So: there is no evolutionary evidence of numerous, slight, modifications.[41] *And*, there is no evidence of evolution based upon the Halibut's habits changing from that of a normal fish swimming upright then evolving to a fish laying on the bottom on its side; because if it was to have happened in this manner it would be one eyed. The evidence suggests its DNA makes it that way. This would then obey the principle observed; it is the overall plan of the Creator when he wrote the program for the DNA of a life. We see two eyes on top of the halibut rather than the expected one eye if evolutionary processes were actually at work. What I am saying is; the fact we see two eyes on the top of one side of the head of a Halibut is more evidence of creative design, and this fits the overall plan we see in all of nature; TWO EYES, because the written instructions of the Creator in the overall plan of the DNA is preserved and maintained; that plan being; two eyes.

I would have to concede and admit there were evolutionary processes at work if today I observed the Halibut only had one eye on what is now the top of its head. That is what we would expect to see if evolutionary processes were at work. The fact it has two eyes on the top of its head is *evidence* (not I "imagine", not I "suppose", not I "believe") of creation. The Halibut having two eyes on the top of its head is due to its DNA demonstrating the Creators' instructional blueprint and plan in the DNA, to have two eyes on the top of its head as demonstrated in all *seeing* species. I will discuss more about this in Chapter 22, *Vision,* page 208.

Then - after writing about all of this, as I was proofreading the entire book for errors, I wondered, "does a Halibut fish swim vertically or horizontally?" It was then I was overjoyed to discover the halibut fish is <u>*NOT*</u> as I previously wrote of on page 45 of this book, "*Even when we today see a species that does <u>suggest some form of evolution</u> by logical observation and deduction, such as for example in my observation, the Halibut fish*" <u>I was flat out wrong!</u> The halibut fish does not even **remotely** "suggest some form of evolution." Let's look at why the Halibut fish does not **<u>remotely suggest</u>** evolution.

After searching for information on whether the Halibut fish swims vertically or horizontally I found that, THERE IS NO EVOLUTINARY CONTRIVANCES AT WORK **<u>WHAT-SO-EVER</u>**! AGAIN, IT IS THE DNA AT WORK!!! THERE IS NOT EVEN **<u>*THE HINT*</u>** OF EVOLUTION AT WORK AS I PREVIOUSLY THOUGHT MIGHT BE EVIDENT! The halibut fish is born with eyes on <u>*each*</u> side of its head and swims vertically as a regular fish throughout its pre-juvenile life. Then, during juvenile growth, one eye <u>*MIGRATES*</u> from one side, to become both-on-one side. Again, it is the DNA which dictates the eye migrating from one side to become both on one side. AS A RESULT WE <u>OBSERVE</u> THAT THERE IS *NO EVOLUTIONARY PROCESS AT WORK* **<u>*WHAT-SO-EVER.*</u>** THE DNA DICTATES THE MIGRATION – NOT EVOLUTION! IT IS THE DNA AND **<u>*THE DNA ALONE*</u>** **WHICH DICTATES THE DIFFERENCE! WOW! EVEN FURTHER EVIDENCE OF CREATION!** Even when at first glance there appears to be "some form of evolution" we find; **no evolution**. The DNA dictates this process. THE INFORMATION PROVIDED BY THE CREATOR IN THE DNA **<u>*INSTRUCTS*</u>** THE EYE TO MIGRATE TO BECOME BOTH ON ONE SIDE. THE EYE MIGRATING TO BECOME (NOT EVOLVE) BOTH ON ONE SIDE IS DUE **<u>*SOLEY*</u>** TO THE WRITTEN INSTRUCTIONS IN THE GENETIC CODE WITHIN THE DNA. EVOLUTION IS NOT **<u>*REMOTELY*</u>** INVOLVED!

I am aware there are *non-seeing* species, such as bacteria, molds, fungi, archaea, cave dwelling, or gas vent inhabiting, and deep sea creatures which do not have sight, and cannot see at all - yet survive. When I say, "a being cannot survive without seeing" I am referring to the millions of seeing species we currently observe on earth or in the fossil record. Those *seeing* species require vision to survive and could not have survived while waiting on their eyes to evolve to a point of *seeing*. I agree there are some species which can survive without *seeing*. However, we do not observe ANY evidence of those *non-seeing* species then going on to become *seeing* species, they remain a *non-seeing* species. There is no evidence they become a *seeing* species even if you "imagine"[39] or suppose"[41] or "believe"[41] they can go from *non-seeing* to *seeing*. *Evolving* from *non-seeing* to *seeing* *cannot happen* for the numerous reasons outlined *throughout* this entire book. (In fact, it is exactly what this entire book is about.) What we do know; that is, *what the evidence demonstrates*, is neither *seeing* or *non-seeing* species **EVER** had **_partially evolved eyes_**. **_And_**, the seeing species we observe; both alive and in the fossil records, always have two eyes. Why is this? It is because of the blue print of design in the DNA. Furthermore, as outlined in *this* chapter; the clarity of the cornea *requires* very specific, in fact multiples of very specific conditions to be met. Multiple conditions which are specific to the level of *1/18,000 of a human hair!* If you want to "imagine"[41], or "suppose"[41], or "believe"[41] these non-seeing creatures evolved into seeing beings - you choose to *believe*. However, the ***evidence*** does not support it, and belief without evidence is faith not science.

What we do observe; for all the seeing beings, the cornea *evolving* to a point of clarity as Darwin states "by numerous, successive, slight modifications"[41] does not, will not, and neurologically, (as we will see in chapters 16-19 and chapter 22 of this book) CANNOT work! The being could not survive if it could not see. Close your eyes for the next five minutes and *"imagine"* what life would be like. Then with your eyes closed for the next five minutes *"imagine"* closing your eyes for the rest of the day; "imagine" if you could survive and provide and protect yourself. "Imagine" surviving the rest of your life while not being able to see. Do you suppose you could survive on your own if you could not see? *"Imagine"* that! Close them now, just for five minutes and "imagine" living, while waiting for your eye, in fact two eyes, to evolve!

One of the tools evolutionists use to help cause a person to believe in evolution and not consider the miraculous design of the body is over-simplification - which is a form of a lie. For example, when doctors, scientists, researchers, evolutionists, etc. refer to the cornea they use the term "fibers" to describe its composition. For example, "The fibers of the cornea are arranged thus and such......" or, "the corneal fibers are this or that...." When you over simplify this truly complex design, you cause the person to not consider the reality of the intelligent design. In doing so, that is, by oversimplifying, you purposefully are causing, (and thereby *deceiving*) people to be much less impressed by, and to not understand or consider, the complexity of the intelligent design of the cornea. The truth is, the stroma of the cornea is **_very_** complex and MUST be made exactly and precisely, in order to be clear and remain clear. When you look at the way in which the cornea is *made* – five, 4 nanometer "microfibrils" combined together into a 36 nanometer "fibril"; then combine 200-250 fibrils together to form a "lamella"; which is 10-15 microns wide by 2 microns thick; then "arrange" 250 of these lamella laid into "sheets"; that are stacked in **_alternating_** 90 degree right angles. Then 250 of them laid at alternating right angles 90 degrees apart, on top of each other so they are of just the right thickness- 2 microns (or $1/36^{th}$ of a hair) the exact and perfect thickness, to cause internal cancellation of light waves in order to make the cornea clear. When we observe the *way* in which the cornea is *made* it is *easy* to appreciate the miraculous intelligent design of the cornea; microfibrils, fibrils, lamella, arranged in alternating perpendicular sheets. Illustrated here are six alternating sheets. About 250 of these alternating sheets make up the stroma of the cornea. Each one is 1/36th of a hair thick and there are 250 of them. One layer is running left and right. The next is running toward and away from you.

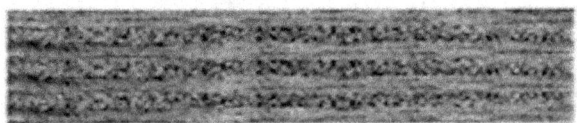

Source: Corneal Dystrophies:https://upload.wikimedia.org/wikipedia/commons /b/b9/Congenital_ stromal_dystrophy_2.jpg
Website: http://ojrd.biomedcentral.com/articles/10.1186/1750-1172-4-7
Author: Klintworth GK[42]

The fourth layer of the cornea is the posterior limiting lamina (most often referred to as Descemet's membrane. It is 10-12 microns thick[43] (about 1/6th of a human hair). This is an elastic membrane made up of fine elastic-like fibers banded together. The make up of this layer – elastin-like fibers (but not true elastin[44]) causes it to be very resistant to chemical agents and more importantly invasion by infection. In the eye, this final barrier is extremely important. Once, bacteria, viruses, fungal, or amoebic organisms pass through the cornea into the warm fluid which fills the inside of the eye they grow extremely rapidly and loss of the eye is highly possible, something almost certain to occur if medical intervention is not undertaken. Eugene Wolff has said, "When the entire cornea has broken down into pus, we often see the thin posterior limiting lamina offering resistance and remaining unimpaired for days."[45]

The fifth layer of the cornea is the endothelium. These cells are one layer thick and line the very inner surface of the cornea. They have no regenerative ability. The number you are born with is all you will ever have. In fact, their number is constantly declining through cell death and mechanical sloughing as we age. Fortunately, God was gracious and gave us MORE than we needed. There is an *overabundance* of these cells.

The function of the endothelial cells is the same as a sump pump in a basement. In a basement, when the moisture in the dirt, or water in the dirt after heavy rains encounters the concrete walls of the basement, the water seeps in through the pores in the concrete. This causes water to accumulate. To get rid of the water in the basement you have to continually mop it up or put in a pump to pump it out. The same situation is present with the cornea, but <u>to a much greater and problematic extent</u> *because* the fluid which fills the eye must have a high enough *pressure* to keep the eye round and firm. This internal pressure of the eye causes the fluid to be *forced* into the body of the cornea causing the cornea to fill with water and swell. Swelling thickens the cornea which then changes the prescription of the cornea and causes it to be out of focus. More importantly, swelling causes the spacing of the 250 alternating sheets of the stroma to be too thick and then the cancellation of light rays by internal refection which occurs in the stroma (the peak of the incoming wave encounters the valley of the reflected wave) does not occur and the cornea becomes cloudy, opaque, and you cannot see. The endothelial cells constantly pump the water attempting to seep into the cornea. Every moment of every day they are pumping the water out. This keeps the water out of the cornea, maintaining the proper thickness of the microfibrils, fibrils, and lamellar sheets, and keeps the cornea clear. Without the endothelial cells the cornea would be cloudy and opaque. The being cannot wait on the endothelium to evolve to have a clear cornea. Everything, all the different interdependent parts must be present - simultaniously, from the very beginning. Without the tears the cornea could not get nutrients. Without the proper thickness and arrangement of the microfibrils, and fibrils, and lamellar sheets, the cornea would not be clear. If the cornea were too thick or too thin it would be out of focus. If the cornea were too steep or too flat it would be out of focus. If the fibers were too brittle the cornea would chip, or crack, or break. If the fibers were too soft the eye would easily be distorted and could not produce a consistently clear image. Without the endothelium, the cornea would be swollen and watery. They **all** must be together, and together properly, fully developed, fully functional. The only way to meet all these criteria and meet them simultaneously (which is an <u>absolute</u> requirement for the cornea to function) is through creation. We observe Interdependent Evidence of Creation, I.E.C.

All these requirements which *MUST* be met to have tissue which is transparent and clear; make it is easy to conclude, that is – it is evidence of Creation. But there is even more amazing evidence yet; consider as tissue, living cells need oxygen and nutrients to survive. And, all tissues, in their function, have an accumulation of waste products. Yet there are no blood vessels in the cornea to provide nutrients, oxygen, and removal of waste products. If blood vessels were in the cornea they would disturb the intricate and precise spacing of the alternating microfibrils, fibrils, and lamellar sheets arranged in perpendicular layers. As a result the cornea could NOT be transparent and clear. Additionally, the blood vessels themselves would scatter the light, obscure the light waves, and interfere with the ability to see clearly. Consider that the cornea is made up of living tissue - with no blood supply. Now *that* is a miracle. Meditate on the miracle of living, breathing, nutrient-needing, waste-producing cells, arranged in tissue which is transparent. How can that be? Living tissue with no blood supply! How can that happen? The answer is; oxygen is absorbed directly from the air into the cornea. Is that true? Can that really happen? Yes, it does, and it happens in the eye for exactly the same reasons it happens in the lungs - and everyone knows it happens in the lungs (this too is a miracle). The oxygen in the air is absorbed directly into the blood through the lungs, and the oxygen in the air is absorbed directly into the cornea. The reason this happens has to do with the atmospheric (air) pressure and the concentration of oxygen in the air (the partial pressure of oxygen in the air). Oxygen makes up about 21% of the air. Nitrogen is about 78%. So 99% of the air is Nitrogen and Oxygen, the rest is Carbon dioxide and other gases. The atmospheric pressure at sea level is 760mmHg[46]. (You could think of it as the weight of air at sea level) So, of the total atmospheric pressure; which is 760mmHg, the pressure concentration of oxygen is about 21% of that, or about 159.6mmHg, this is the partial pressure of oxygen in the air. The partial pressure of oxygen in the blood has a normal range of 75 -100mmHg[47] or about an average of 87.5mmHg. Since the pressure concentration of oxygen in the air is about 159.6mmHg but that of blood is only 87.5mmHg, the oxygen diffuses from the area of high concentration - 159.6mmHg of air, to the area of lower concentration, 87.5mmHg of blood, *into* the blood by diffusion. (The gas diffusion law or Fick's Law states, "The rate of transfer of a gas through a sheet of tissue is proportional to the tissue area and the difference in gas partial pressure between the two sides and inversely proportional to the tissue thickness."[48])

Several studies [49, 50, 51, 52] show the concentration of oxygen in the cornea (or the partial pressure of oxygen in the cornea) to be about 24mmHg. Again, the pressure concentration of oxygen in the air (the partial pressure of oxygen) is 21% of 760mmHg or about 159.6. Therefore, it is *very easy* for oxygen to diffuse into the cornea since the partial pressure of oxygen in the air is 159.6mmHg while oxygen in the cornea is about 24mmHg. Another way to say thi is; there is *more than 6 times* the pressure of oxygen in the air compared to that in the cornea so the oxygen is forced or diffuses easily into the cornea and the cornea thereby *breathes* directly from the air by diffusion as is observed by Fick's Gas Diffusion Law. (There is also some provision of oxygen to the cornea from the blood vessels immediately adjacent to the cornea at the sclera {the white part of the eye} and there is also a small amount of oxygen provided to the cornea from the aqueous {the fluid which fills the front part of the eye}, because the aqueous is derived directly from the blood).

This is what happens when your eyes are open, but, what happens when you close your eyes at night? Your eyes cannot receive the oxygen in the air by diffusion when your eyes are closed. So how can the cornea *breathe* without blood vessels or access to the air - while you sleep? The answer is, when you close your eyes at night, the oxygen is absorbed from the tiny blood vessels (capillaries) which line the inside of the upper eyelid.

Is this really true? Can this really happen? YES! And it does! At night when the eyelid is closed there is no direct access to the oxygen from the air *and* there are no blood vessels within the cornea either. This presents a BIG problem. However, studies have shown the partial pressure of oxygen at the surface of the blood vessels which line the inside of the upper lid is about 56.7mmHg,[50,51,52] again that of the cornea is 24mmHg, a difference of more than 2 times - so oxygen easily diffuses into the cornea from the upper eyelid blood vessels - again due to the gas diffusion principle of Fick's Law. Amazing!

Now, ***even more amazing yet!*** Not only is the design, thickness, and arrangement of the microfibrils, fibrils, and lamella (sheets) precisely arranged in 90-degree perpendicular layers *THE* precise and exact thickness required for cancellation of lightwaves the reason the cornea is transparent living tissue. Plus, the tissues of the cornea can breathe without a blood supply! And, the tissues of the cornea get the necessary nutrients required to live and function from the tears and the aqueous! And, the endothelial cells which line the inner surface of the cornea keeps the Aqueous Fluid which fills the eye from seeping into the cornea causing swelling and clouding of the cornea. Furtermore, not only does the cornea breathe from the tiny blood vessels (capillaries) that line the inside of the eyelid while the eye is closed during sleep, but in addition to all of this, taken as a whole; ***all*** of these ***interdependent parts***; must then be all taken together, and ***all*** of them together… must also be <u>of the exact thickness</u> and also of the <u>exact curvature</u> that is ***required*** to focus the light rays on the retina. WOW! *"IMAGINE"* <u>THAT</u> Darwin![53] Imagine ***ALL*** that ***simultaneously***! To think it ***ALL*** happened by chance is to purposefully ignore all logical and rational thought. You must ***choose*** to suppress the truth and believe a lie!

Consider now the ***exact thickness*** required for the cornea to provide a distinct and clear focus on the retina. The cornea is normally about 560 microns thick.[54] If the cornea were instead only 18 microns thicker, which is about 1/4th of a human hair too thick, the eye would be about 20% blurred for far vision.[55] If the cornea were 36 microns too thick, which is about 1/2 of a human hair too thick, the eye would be about 50% blurred for far vision.[55] If the eye were 72 microns too thick, which is about one hair thickness too thick, the eye would be about 80% blurred for far vision.[55] (This level of vision is considered Legally Blind.[55]) The same thing would happen if the cornea were

1/4th, 1/2, and 1 hair too thin (except when the cornea is too thin it causes near vision to be out of focus). If the cornea were only 18 microns thinner, which is about 1/4th of a human hair too thin, the eye would be about 20% blurred for near vision. If the cornea were 36 microns too thin, which is about 1/2 of a human hair too thin, the eye would be about 50% blurred for near vision. If the eye were 72 microns too thin, which is about one hair thickness too thin, the eye would be about 80% blurred for near vision. (Again, this level of vision is considered Legally Blind.[55]) If the thickness is incorrect ***one hair*** the eye is legally blind!

This same focus problem applies to the steepness and flatness of the *curvature* of the cornea. Consider now the ***exact steepness of curvature*** required for the cornea to provide a distinct and clear focus on the retina. The average curvature of the cornea when the eye is in proper focus and seeing 100 % clearly is a radius of curvature of about 7.85 mm (see 43.00 diopter curve [B] on page 65).[56] If the radius of curvature of the cornea were 1/2 of 1 mm (less than the thickness of a credit card) of radius ***too steep*** or a 7.35mm radius (see 46.00 diopter curve [A] on page 64 below) the eye is about 50% out of focus for far vision (50% blurry). If the radius of curvature of the cornea were 1/2 of 1mm (less than the thickness of a credit card) of radius ***too flat*** or an 8.35 mm radius (see 40.00 diopter curve [C] below on page 65) the eye is about 50% out of focus for close vision (50% blurry). Illustrated below are the three examples of curvature I gave as examples above. As you can see the naked eye cannot detect a difference in these three different curves, (I drew them and I know which is which, and even then, I cannot see the difference!) Yet, this amount of difference makes the difference between seeing clearly and seeing very blurry- 50% blurry. Can you tell the difference in curvature between these three examples? In these illustrations both A and C; the vision is 50% blurry. You could definitely tell the difference if it were your eye!

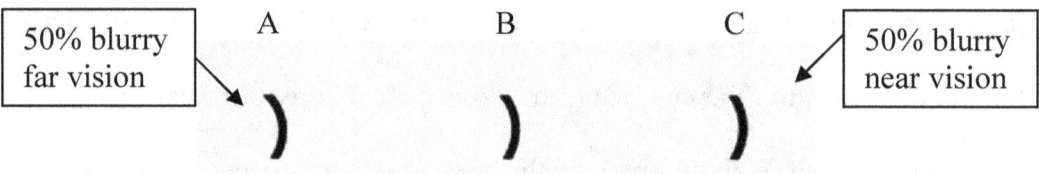

These are each the actual radius of curvature of the cornea: A. Too Flat Curvature 40.00, B. Normal Curvature 43.00, and C. Too Steep Curvature 46.00. A & C are 50% blurry!

Again, if the cornea curvature is only 1mm radius of curvature too flat or too steep the person is legally blind! (Additionally, when we discuss the sclera later in this book, we will see the ***length*** of the eye is also critical to about 1mm. If the length of the eye were 1mm too long or 1mm too short the person would see about 50 % blurry.)

We see *multiple* criteria ***must*** be met and must be met ***simultaneously*** because they are all dependent and interdependent functions. For the cornea to live, to function, to be transparent, and to be in focus they all must be met ***simultaneously***, and they all must be ***completed*** functions. The other systems cannot wait on any one system to evolve. If we just consider the cornea, (that is ignoring the twenty-four other major systems of the eye listed in this book), we see the cornea alone has at least 17 specific criteria which ALL MUST be met in order for it to function properly.

1. Tear film must be present to supply nutrients, oxygen, and immune response (to fight infection). 2. Epithelium must be lacking keratin. 3. The mirofibrils, fibrils, and lamellar sheets of the stroma must be of exact thickness (2 microns) and is critical to about 1/100 of a human hair. The reflected light ray from a lamella must exactly cancel the incoming light ray and is critical to 1/100th

of a human hair. 4. The endothelium must be present to keep the cornea pumped free of the fluid which seeps in due to the pressure of the eyeball. 5. The oxygen must be able to be absorbed directly into the cornea from the air. 6. The oxygen must be able to diffuse from the capillaries (which must also be present) which line the inside of the upper lid into the cornea while we sleep. 7. The atmospheric pressure must be high enough to force the oxygen into the cornea by diffusion. 8. The concentration of oxygen in the atmosphere must be high enough to diffuse into the cornea. 9. The aqueous must be present to help supply nutrients, oxygen, and some of the immune defenses. 10. The thickness of the cornea cannot be *too thick* and is critical to ¼ of the thickness of a human hair. 11. The thickness of the cornea cannot be *too thin* and is critical to ¼ of the thickness of a human hair. 12. The curvature of the cornea cannot be *too steep* and is critical to about ¾ of a human hair. 13. The curvature of the cornea cannot be *too flat* and is critical to about ¾ of a human hair. 14. The pressure of the eye must be high enough to keep the cornea round. 15. The cornea cannot be too firm or brittle and cannot chip or crack. 16. The cornea cannot be too soft and must be firm enough to provide a consistent non-distorted curvature to provide a clear focus on the retina. 17. The surface cells of the cornea must be scratch resistant, and scratch "self-repairing" and provide continual replenishing of the surface with new cells.

Now, remember in addition to these 17 **_criteria_** required to be met in order for the cornea to live, work and be clear, there are also the *19 interdependent **systems** **required*** for the cornea to live work and be clear **_also_**: 1. The Cornea, 2. the Tears, 3. the Eyelids, the eyelid muscles nerves- (two different nerve supplies); 4. the Oculomotor Nerve for the 5. Levator Palpebrae Superioris eyelid muscle. 6. The Facial Nerve for the 7. Orbicularis Oculi eyelid muscle. The sensation of pain and dryness for the cornea by the 8. Trigeminal Nerve. The 9. Lacrimal Nerve to provide the stimulus to the 10. Lacrimal Gland to make tears, The 11. Opnthalmic Artery to provide the blood supply to the 12. Lacrimal Artery for the Lacrimal Gland to create tears. The 13. Lacrimal Vein and 14. Ophthalmic Vein to carry away the blood from the Lacrimal Artery and Ophthalmic Artery, The formation of the hole in the bones of the back of the eye socket; the 15. Superior Orbital Fissure, which must be present for the Oculomotor Nerve, Trochlear Nerve, Abduscens Nerve, Trigeminal Nerve, Ophthalmic Artery, and Ophthalmic Vein to pass through from the Brain and Internal Carotid Arteries in the skull, into the eye socket to supply the eye. The formation of the two 16. Puncta (drains) on the upper and lower lids. The two 17. Canaliculi (drain tubes to the nose) to carry the tears to drain into the nose. The formation of the 18. Lacrimal Fossa; the foramen (holes in the bones through which the tear ducts must pass in order to drain the tears into the nose.) 19. The Aqueous Humor (the fluid that fills the inside of the eye) also provides nutrients to the cornea, and provides defense (immunity) to fight infection for the cornea, and some oxygen supply to the cornea. All 19 ***systems*** must be present, ***and*** all 17 required ***criteria*** must be met. Thirty-six (36) different things must happen ***all at the same time*** for the cornea to function properly… **36 all at the same time!** And remember also, we saw 108 things which must be met- just for the eye muscles to work. (Plus, we will later see there ***MUST*** be present the other 24 different interdependent systems. *All* are required for the function of the eye itself also - all, at the same time also!) Observe and conclude, 108 and 36 equals 144 simultaneous requirements which must be met; and so far we are just through Chapter 1 Bones of the eye and Chapter 2 Cornea. Does the ***evidence*** support evolution? What we observe is Interdependent Evidence of Creation, I.E.C.

Each and every single component must be present and fully functional. Each and every one of these critical criteria must be met - because they are all interdependent. Again, we observe Interdependent Evidence of Creation. **_ALL_** these things mut be present at the same time and occur completely and simultaneously, exactly and precisely, otherwise the cornea would not be clear; the eye could not see and the creature would be blind. A blind creature cannot survive.

On page 158 of his book *"The Origin of Species"* Darwin said, "we **ought** in imagination to take a thick layer of transparent tissue, with a nerve sensitive to light beneath, and then **suppose** every part of this layer to be continually changing slowly in density, to separate into layers of different densities and thicknesses, placed at different distances from each other, and with the surfaces of each layer slowly changing in form"[53]. Darwin further states, "let this process go on for millions on millions of years; and during each year on millions of individuals of many kinds; and may we not **believe** a living optical instrument might thus be formed as superior to one of glass, as the works of the Creator are to those of man?"[57] (Bold portions by Rigney not Darwin)

We **ought _NOT_** to **_imagine_** but instead look at the reality. Look at the evidence. The reality is; the eye cannot see, the being cannot function, and the individual cannot survive while it is waiting on any one part to evolve. Looking only at the cornea, (without considering the other twenty-four systems of the eye are **_all_** absolutely interdependent.) *evolving* to a point of clarity "by numerous, successive, slight modifications"[58] does not work when it comes to seeing. (You will also see there are also neurological reasons why evolution does not work for the eye and vision, in chapters 16, 17, and 18 of this book.) The being could not survive if it could not see.

The criteria which *must* be met in order for the eye to see and for the being to survive is; each and *every* single component, had to be present at the same time, and happen completely and simultaneously, exactly and precisely- all of them all at the same time. The requirement is fully developed, interdependent functions. The only way for this criteria to be met, the only way *all* this can happen ***simultaneously*** is by (in fact, this describes, and is the definition of) Creation. Darwin may "***believe*** a living optical instrument might thus be formed as superior to one of glass, as the works of the Creator are to those of man."[59] But what we instead ***observe*** is Interdependent Evidence of Creation. The evidence proves the eye was Created. The theory Interdependent evidence of Creation is proven. The theory of evolution is "believed"[59]. Where is the proof of the "theory"?

3. Iris

The iris is made up of two circular muscles that open and close the pupil to regulate how much light gets into the eye. One muscle, the pupillae dilator, opens the pupil, and the other muscle, the sphincter pupillae, closes the pupil.[1] There are two different nerves to the two muscles of the pupil. One nerve stimulates the muscle which opens the pupil, the other nerve stimulates the muscle which closes the pupil. These two nerves come from two completely different places and from two entirely different routes from the brain. Even more amazing is they are directed by two completely different neurological divisions of the brain, the sympathetic nervous system and the parasympathetic nervous system. It is these two vastly different routes and nerve supply of the iris which *PROVE* the eye "could not have been formed by numerous successive slight modifications"[2] as Darwin's theory suggests. These two different nerve routes <u>alone</u> "absolutely break down"[2] Darwin's theory.

The nerve which stimulates the Iris Dilator muscle (the muscle that *opens* the Iris larger) has a long and indirect 10-step route: 1.Impulses for the iris Dilator muscle originate in the optic nerve. 2.They exit at the midbrain and travel within and down the spinal cord. 3.The Iris Dilator nerve exits from the spinal cord just below the level of where the cervical and thoracic vertebrae meet (the area where the neck turns into the back) in the upper part of the back at the 1st 2nd and 3rd thoracic vertebrae. Here the Iris Dilator nerve exits the spinal cord and 4.proceeds across the tip of the lung then, over 5. to the heart and the cardiac plexus. The Iris Dilator nerve 6. proceeds up with the upper blood vessels of the heart, and up to the eye, traveling along and with the carotid arteries.[3] Again, the Iris Dilator nerve is controlled by the *sympathetic division* of the central nervous system. All the sympathetic nerves of the sympathetic nervous system are interconnected by a network of ganglion immediately adjacent to, and along each side of the spinal cord called the sympathetic chain. (Ganglion house the nerve cell bodies. The sympathetic chain is a network of ganglion interconnections of sympathetic fibers which interconnect into a series of sympathetic ganglion that run along each side of the spinal cord and receive input from nerves of the spinal cord, and relay connections throughout the nervous system.[4,5]) The Iris Dilator nerve 7.links up to and connects into the sympathetic chain in what is called the superior cervical ganglion (in your neck right about the jaw line). After this interconnection into the sympathetic chain, the Iris Dilator nerve exits the superior cervical ganglion and 8.proceeds up to the eye, traveling with and wrapped around the internal carotid arteries within the network of nerves called the internal carotid artery plexus. These sympathetic fibers of the Iris Dilator nerve branches off from the internal carotid artery plexus at about the level of the optic nerve into two branches. A branch connects into the middle ear, and a branch 9. passes through the *ciliary ganglion* of the eye.[6] (The ciliary ganglion lies between the optic nerve and the lateral rectus muscle.) One branch then passes into the ophthalmic division of the trigeminal nerve and the long ciliary nerve. Then from the ciliary ganglion and the long ciliary nerve, it travels 10. to the Iris Dilator muscle of the iris.[7] This Iris Dilator nerve opens the pupil, allowing the eye to get more light when in the dark, or when excited.

In short, just for ease of thinking, we could say the path goes from the retina, to the midbrain, down the spinal cord, to the heart, then back up to the eye.

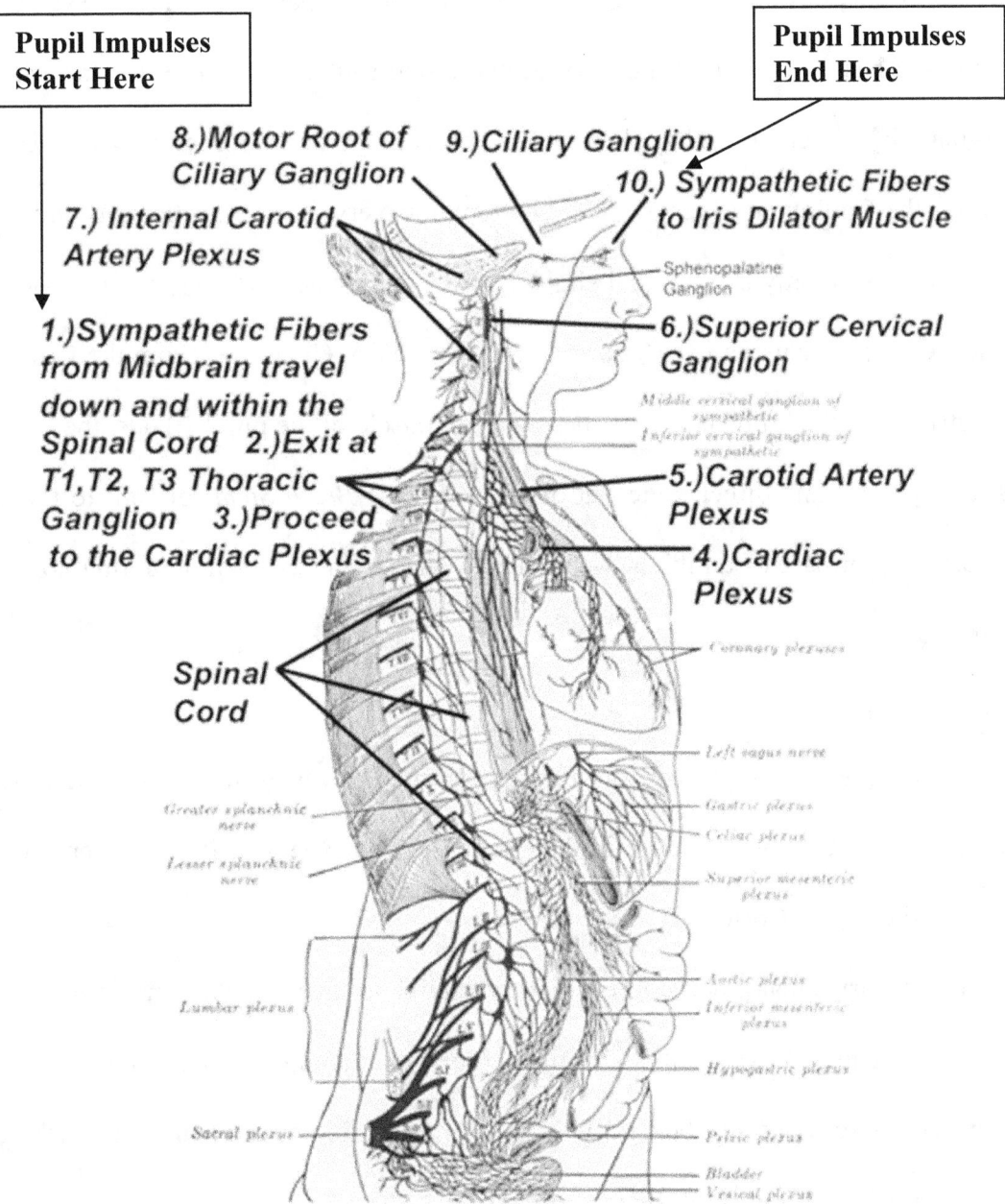

10 Step Route for Pupil Dilating Nerve

Pupil Impulses Start Here

Pupil Impulses End Here

8.)Motor Root of Ciliary Ganglion

9.)Ciliary Ganglion

7.) Internal Carotid Artery Plexus

10.) Sympathetic Fibers to Iris Dilator Muscle

1.)Sympathetic Fibers from Midbrain travel down and within the Spinal Cord   2.)Exit at T1,T2, T3 Thoracic Ganglion   3.)Proceed to the Cardiac Plexus

6.)Superior Cervical Ganglion

5.)Carotid Artery Plexus

4.)Cardiac Plexus

Spinal Cord

Sphenopalatine Ganglion

Middle cervical ganglion of sympathetic

Inferior cervical ganglion of sympathetic

Coronary plexuses

Left vagus nerve

Gastric plexus

Celiac plexus

Superior mesenteric plexus

Greater splanchnic nerve

Lesser splanchnic nerve

Aortic plexus

Inferior mesenteric plexus

Hypogastric plexus

Lumbar plexus

Sacral plexus

Pelvic plexus

Bladder

Vesical plexus

Sympathetic Fiber Pathway from Midbrain to Iris Dilator Muscle

From Henry Gray, F.R.S, *Anatomy of the Human Body* 20th Edition, Lea & Febiger (Philadelphia & New York, © 1918) p.969[8]

63

The other nerve which stimulates the Iris sphincter muscle, the muscle which *Closes* the iris smaller when in excessive light is regulated by the parasympathetic nervous system.[8] The Iris Constrictor nerve has a very short and very direct 4-step route: 1.Impulses for the Iris constrictor muscle originate in the retina. 2.They then exit the optic nerve at the midbrain and travel directly to the eye within the Oculomotor Nerve. (The Oculomotor Nerve comes to the eye directly from the brain stem, which is directly behind, and slightly lower than the eye (two to three inches behind the eye). The Iris Constrictor nerve then 3.connects to the Ciliary Ganglion. From the Ciliary Ganglion the Iris Constrictor nerve 4.travels to the iris sphincter muscle of the pupil.[8] Thus, the Iris Constrictor nerve has a very short and direct route out of and from the *Brain Stem* forward to the front of the skull. In short, just for ease of thinking, we could say the path goes from the retina, to the midbrain, then back to the eye.

In the illustration below;

V = the route of the **V**ision impulses from the retina which travel from the eye to the brain through the Optic Nerve. (The portion of the Brain that processes your vision is in the lower-back portion of the brain called the Visual Cortex.)

D = the route of the iris **D**ilator nerve fibers from the retina to the iris **D**ilator muscle.

O = Sympathetic Chain. The sympathetic chain is a series of ganglion (ganglion behave somewhat as a nerve hub that house nerve cell bodies) which lie on each side of the spinal cord where the nerve fibers exit the spinal cord at each of the 31vertebrae.

C = the route of the iris **C**onstrictor nerve fibers from the retina to the iris **C**onstrictor muscle. The route of retina-through-V-to-C-to-Constrictor of iris is very short and direct with 4 total steps. The route of retina-through-V-to-D-to-Dilator of iris is very different and very long with 10 total steps. Evolution CANNOT explain these vastly different routes. Here again we observe Interdependent Evidence of Creation, I.E.C.

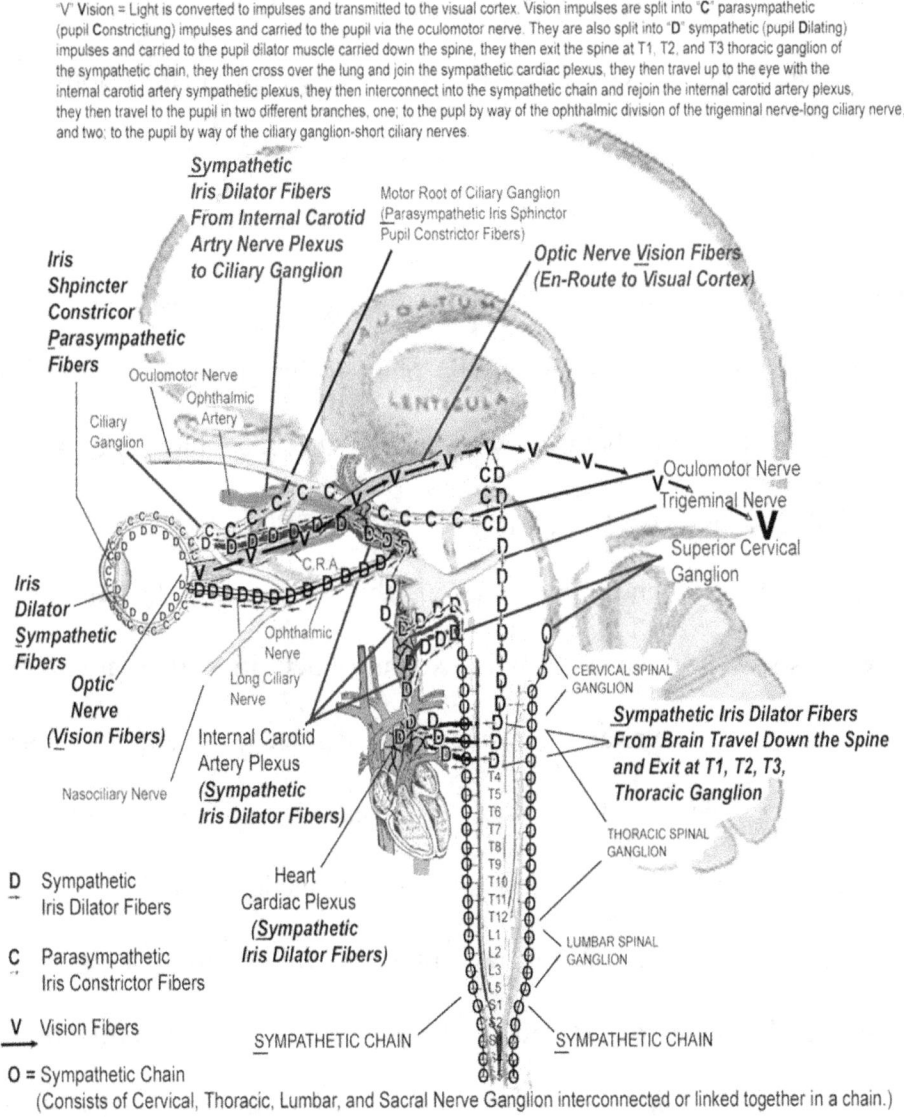

"V" Vision = Light is converted to impulses and transmitted to the visual cortex. Vision impulses are split into "C" parasympathetic (pupil Constrictiung) impulses and carried to the pupil via the oculomotor nerve. They are also split into "D" sympathetic (pupil Dilating) impulses and carried to the pupil dilator muscle carried down the spine, they then exit the spine at T1, T2, and T3 thoracic ganglion of the sympathetic chain, they then cross over the lung and join the sympathetic cardiac plexus, they then travel up to the eye with the internal carotid artery sympathetic plexus, they then interconnect into the sympathetic chain and rejoin the internal carotid artery plexus, they then travel to the pupil in two different branches, one: to the pupl by way of the ophthalmic division of the trigeminal nerve-long ciliary nerve, and two; to the pupil by way of the ciliary ganglion-short ciliary nerves.

O = Sympathetic Chain
(Consists of Cervical, Thoracic, Lumbar, and Sacral Nerve Ganglion interconnected or linked together in a chain.)

From Henry Gray, F.R.S, *Anatomy of the Human Body* 20th Edition, Lea & Febiger (Philadelphia & New York © 1918) p.776, 1021, 541, 887[9] combined with modifications and additions by Rigney **V-C = Constrictor path, V-D = Dilator path**

These two pupil muscles and their nerves must be formed at ***exactly*** the same time. How could these two muscles "***evolve***" into two completely different nerves with two completely different routes at exactly the same time? How could a nerve, for that matter why *would* a nerve *evolve* a pathway from the brain, ***down*** the neck, ***over*** across the lung to the heart, then join with upper blood vessels of the heart, and back ***up*** to the superior cervical ganglion, ***up*** (along with) the internal carotid arteries, ***over*** to the ciliary ganglion, then on to the eye? Again, Darwin's theory *absolutely breaks down!* [10]

Why?... The evidence concludes God *designed* these Iris Dilator sympathetic fibers to interconnect into the sympathetic chain; as is His plan of design for *ALL* the sympathetic nerve fibers. We observe ***ALL*** the sympathetic nerve fibers of the body have interconnections into the sympathetic chain.[11] In this manner the sympathetic fibers are all *interconnected*. This is true for ***ALL*** sympathetic nerves. The evidence observed, (and this is my, the author, opinion) suggests since the sympathetic nervous system is involved with excitatory activities, they are *all* interconnected to facilitate rapid readiness when excitation is necessary - such as when one fears for their life. Whatever the reason, we observe: the design and plan is; all sympathetic nerve fibers are interconnected; and the sympathetic Iris Dilator fibers obey this design and plan.

It is clearly evident *evolution* could not have played a role in these *vastly different routes* of development, supply, and innervation. Based on the observed consistency of *the plan* and evident intentional design, the conclusion is Creation. Again, we observe Interdependent Evidence of Creation when we observe the two iris nerves.

The Iris Pupil Dilator nerve fibers travel down, over, up and over to the iris dilator muscle of the eye. The Iris Pupil Constrictor nerve fibers travel down and over to the iris constrictor muscle of the eye. We observe two entirely different nerve routes to the iris muscles. Again, "numerous successive slight modifications"[12] cannot apply. Either the iris opens and closes or it doesn't. Why would there be a reason for it to gradually form, and gradually evolve - knowing it will not function until everything is present, and everything is connected? And, how can evolution explain the two entirely different routes the nerves follow? Darwin's theory "absolutely breaks down"[12] if you understand the iris. It just takes one look to see this HAD to be Created. Look at the two vastly different routes to the pupil. Even more destructive evidence as to the possibility of evolution is the fact this must happen for *each eye* **at the exact same time** (these two different routes happen on both sides of the body at exactly the same time). WOW! "Imagine" that Darwin! Again, it would require faith to "believe" it!

For the iris dilator and iris constrictor muscles of the iris to function we observe thirty-one (31) different interdependent systems: For the iris constrictor muscle there must be 1. The Brain must be present to receive the visual fibers from the 2. Eye. The 3. Optic Nerve must be present to carry the nerve fibers from the brain to the eye. 4.The Optic Foramen must form in the 5. Bones of the eye socket (seven different bones are required to form the eye socket). 6.The Oculomotor Nerve must be present to carry the 6. Parasympathetic Nerve Fibers. 7. The Ciliary Ganglion must be present to house the nerve cell bodies. 8. The Short Ciliary Nerves must be present to carry the nerves from the Ciliary Ganglion to the Eye. 9. The Rods and Cones of 10. the Retina must be present to convert the light into 11. Visual Impulses. 12.The Iris must be present to house 13 the Iris Constrictor Muscle.

14.The Cornea and 15. the Lens must be present to focus the light on the Retina. 16. The Sclera must be present to support the Iris.

At the same time, all of these interdependent structures are necessary for the Iris Dilator muscle. The Iris Dilator Muscle must have the Brain and 17. the Midbrain to branch off 18. the Sympathetic fibers down 19.the Spinal Cord. 20.The 1st 2nd and 3rd Thoracic nerve Ganglion must be present on the Spinal Cord. 21.The Cardiac Plexus must be present for the sympathetic fibers to join. The Carotid Artery Plexus, must then follow up with the 22. Carotid Arteries. The 23. Sympathetic Chain must be present for the Iris Dilator fibers to connect into the 24. Superior Cervical Ganglion. The 25. Internal Carotid Arteries must branch off to form the 26. Internal Carotid Artery Plexus. The Internal Carotid Arteries must branch off to form the 26. Ophthalmic Arteries. The Sympathetic Iris Dilator fibers must follow the Ophthalmic Artery to the Ciliary Ganglion. The Iris Dilator fibers pass through the Ciliary Ganglion and proceed to the Eye in the Short Ciliary Nerves. They travel in the Retina to the Iris and the 27. Iris Dilator muscle. The 28. Short Ciliary Arteries of the Optic Nerve must be present also, as well as the 29. Long Ciliary Arteries to supply the Iris Dilator and Iris Constrictor muscles with the necessary 30. Blood flow. The 31. Aqueous Humor must be present to keep the eye Round. All together, we see Thirty-One (31) interdependent structures necessary for the function of the pupil., **_ALL_** of this is required for **_each eye_**. Double of everything just mentioned! "Numerous, successive, slight modifications?"[12] No! Interdependent Evidence of Creation? Yes! 62 interdependent systems; I.E.C.

10-Step Pathway for Sympathetic Iris Dilator Nerve Fiber Route to Dilate and Open the Pupil

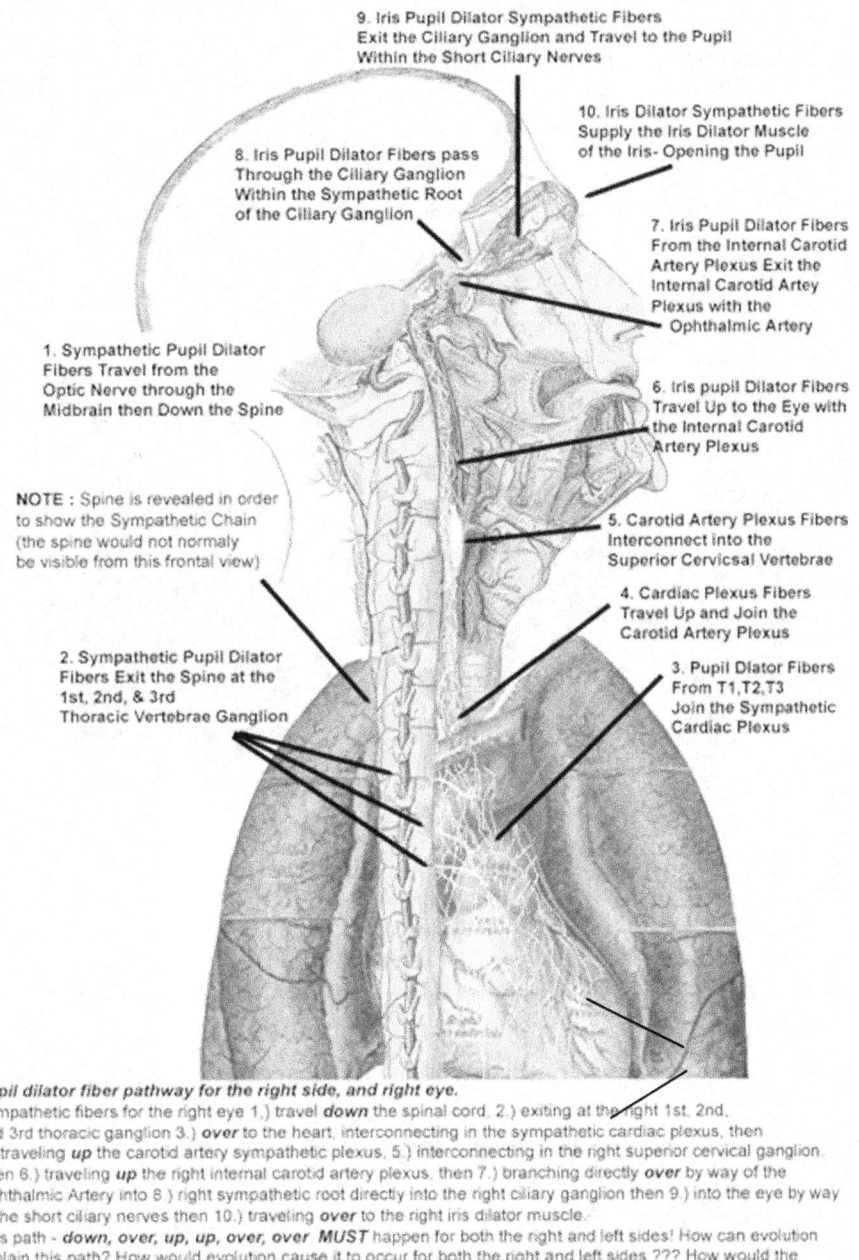

9. Iris Pupil Dilator Sympathetic Fibers Exit the Ciliary Ganglion and Travel to the Pupil Within the Short Ciliary Nerves

10. Iris Dilator Sympathetic Fibers Supply the Iris Dilator Muscle of the Iris- Opening the Pupil

8. Iris Pupil Dilator Fibers pass Through the Ciliary Ganglion Within the Sympathetic Root of the Ciliary Ganglion

7. Iris Pupil Dilator Fibers From the Internal Carotid Artery Plexus Exit the Internal Carotid Artey Plexus with the Ophthalmic Artery

1. Sympathetic Pupil Dilator Fibers Travel from the Optic Nerve through the Midbrain then Down the Spine

6. Iris pupil Dilator Fibers Travel Up to the Eye with the Internal Carotid Artery Plexus

NOTE : Spine is revealed in order to show the Sympathetic Chain (the spine would not normaly be visible from this frontal view)

5. Carotid Artery Plexus Fibers Interconnect into the Superior Cervicsal Vertebrae

4. Cardiac Plexus Fibers Travel Up and Join the Carotid Artery Plexus

2. Sympathetic Pupil Dilator Fibers Exit the Spine at the 1st, 2nd, & 3rd Thoracic Vertebrae Ganglion

3. Pupil Dilator Fibers From T1,T2,T3 Join the Sympathetic Cardiac Plexus

Pupil dilator fiber pathway for the right side, and right eye.
Sympathetic fibers for the right eye 1.) travel *down* the spinal cord, 2.) exiting at the right 1st, 2nd, and 3rd thoracic ganglion 3.) *over* to the heart, interconnecting in the sympathetic cardiac plexus, then 4.) traveling *up* the carotid artery sympathetic plexus, 5.) interconnecting in the right superior cervical ganglion. Then 6.) traveling *up* the right internal carotid artery plexus, then 7.) branching directly *over* by way of the Ophthalmic Artery into 8.) right sympathetic root directly into the right ciliary ganglion then 9.) into the eye by way of the short ciliary nerves then 10.) traveling *over* to the right iris dilator muscle.
This path - *down, over, up, up, over, over MUST* happen for both the right and left sides! How can evolution explain this path? How would evolution cause it to occur for both the right and left sides ??? How would the right and left sides evolve completely, fully, and simultaneously? ... It didn't ! *God designed it ! God created it !*

From Henry Gray, F.R.S, *Anatomy of the Human Body* 20th Edition, Lea & Febiger (Philadelphia & New York, © 1918) p.566, 569, 527, 766[13] combined, with additions and modifications by Rigney

Additionally, during embryonic development of the eye, there is no pupil. At about the sixth fetal month the iris tissue must dissolve away (exactly and precisely) in order to create the perfectly round circular opening that becomes the pupil.[14]

How can the tissue evolve to dissolve into a perfect circle? How can evolution explain that there is no pupil in the embryo yet the tissue dissolves away in just the right manner to create a pupil? It is evident the DNA dictates the design, not evolution.

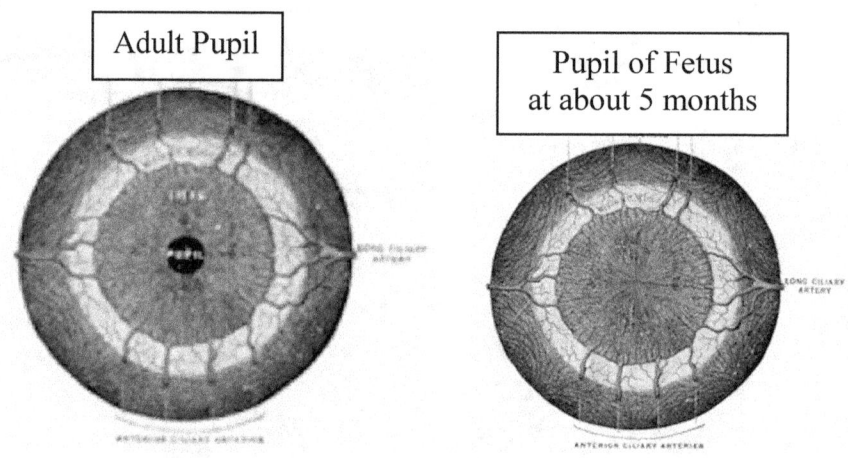

From Henry Gray, F.R.S, *Anatomy of the Human Body* 20th Edition, Lea & Febiger (Philadelphia & New York, © 1918) p.1013[14] with modifications by Rigney

If the pupil *evolved* to become a round opening - the eye is blind until it evolves the opening. When observing all the interdependent structures of the pupil, Darwin's theory of "numerous, successive, slight modifications"[14] cannot be supported. His theory as he himself states,"absolutely breaks down."[15] What we observe is two different muscles, two different nerves, two completely different routes of innervation for ***two different eyes*** all occuring ***at exactly the same time!*** We observe Interdependent Evidence of Creation! Think about this also, if the pupil did not open and close, we would only see well in the light - or only see well in the dark. It requires foresight - (foresight indicates *planned* Creation) to know that the pupil needs to be able to change size so we can see well in all different kinds of light; dim, medium, or bright.

4. Aqueous

Aqueous is the fluid which fills the front of the eye.[1] The aqueous fills from the posterior (rear portion) lens forward to the cornea. The aqueous is made by a gland that encircles the inside of the eyeball; this gland is called the Ciliary Body. There must be a continuous flow of *fresh* aqueous fluid into, and back out of, the eye for the eye to be round and maintain its shape. The aqueous flows in and out continually, at all times day and night, and keeps the eye round and firm. The aqueous is derived from the arterial blood, yet it is clear. The aqueous provides the nutrients for the cornea and the lens for those tissues to live and breathe because they do not have arteries or veins to bring them nourishment and oxygen. If arteries and veins were in the cornea and lens, they would interfere with the light rays and block your vision. God made the aqueous to bring the nutrients to the lens and the cornea and to keep the eye round and firm!

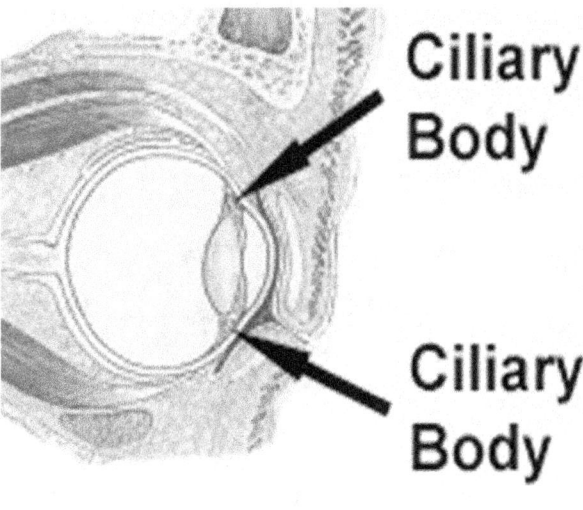

Ciliary Body

Ciliary Body

From Henry Gray, F.R.S, *Anatomy of the Human Body* 20th Edition, Lea & Febiger (Philadelphia & New York, © 1918) Page 1021[2] with modifications and additions by Rigney

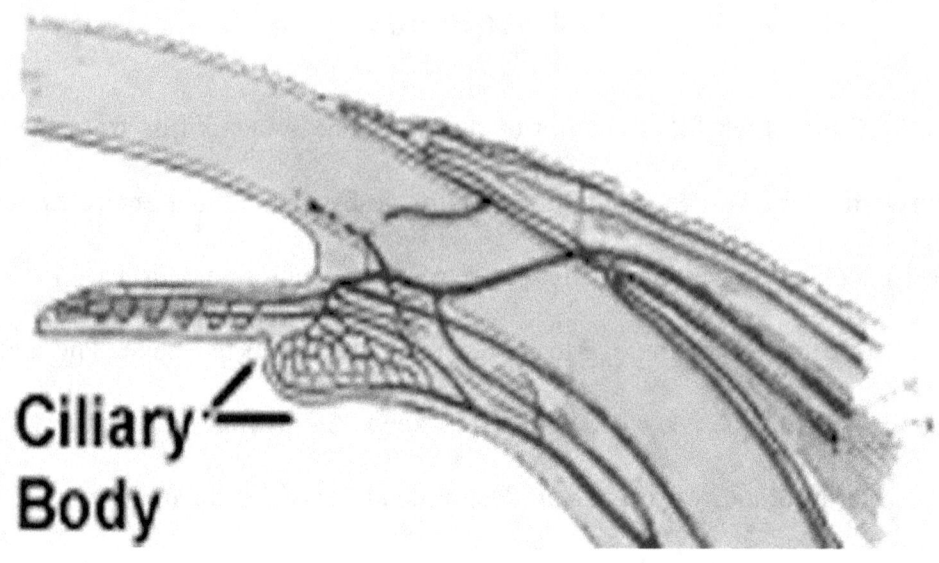

Ciliary Body

From Henry Gray, F.R.S, *Anatomy of the Human Body* 20th Edition, Lea & Febiger (Philadelphia & New York, © 1918) Page 1012[3] with modifications and additions by Rigney

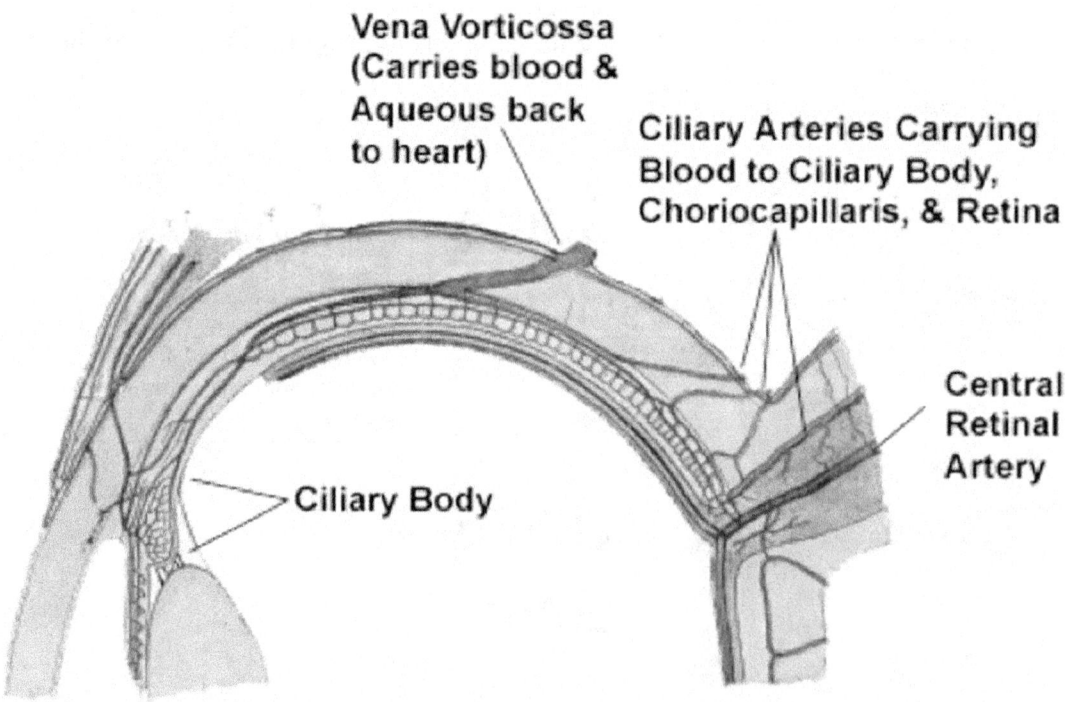

Vena Vorticossa (Carries blood & Aqueous back to heart)

Ciliary Arteries Carrying Blood to Ciliary Body, Choriocapillaris, & Retina

Central Retinal Artery

Ciliary Body

From Henry Gray, F.R.S, *Anatomy of the Human Body* 20th Edition, Lea & Febiger (Philadelphia & New York, © 1918) Page 1012[3] with modifications and additions by Rigney

Without the aqueous, the cornea and lens could not stay alive and you could not see. Without the aqueous, the eye could not stay round; the eye would be soft and mushy.

How could the eye evolve a clear fluid from the blood *through* the bones and then into the eyeball? It is clearly evident God designed and created it to all work and function together.

The Ciliary Body (where the aqueous is made), and the Trabecular Meshwork & Schlem's Canal (where the aqueous is drained) are in completely different areas of location within the eye - yet both MUST be present and fully functional <u>at the same time</u>. If the Ciliary Body evolved and began to produce fluid before the Trabecular Meshwork & Schlem's Canal drain system is evolved, the eye pressure explodes the eye. If the Trabecular Meshwork & Schlem's Canal drain system evolved before the Ciliary Body fluid making system, the eye cannot be filled or pressurized, then the eye is a soft, deformed, nonfunctional, a permanently damaged mass of "junk" tissues with no purpose or function. Evolution does not make any logical sense when you think about these two interdependent tissue systems; the Ciliary Body-fluid-making-system and the Trabecular Meshwork-Schlem's Canal-drainage system. They are located in two completely different areas yet are absolutely and totally interdependent. Their location, required simultaneous function, and absolute interdependence are evidence of Creation. We can observe and conclude they were created. We observe Interdependent Evidence of Creation. We observe 16 different interdependent systems at work. 1.Bones 2.Foramen 3.Sclera 4.Cornea 5.Internal Carotid Artery 6.Ophthalmic Artery 7.Ciliary Arteries 8.Ciliary Body 9.Aqueous 10.Trabecular Meshwork 11.Schlem's Canal 12.Ciliary Veins 13.Vena Vorticosa 14.Ophthalmic Vein 15.Facial Vein 16.Internal Jugular to Heart.

5. Aqueous Flow

Aqueous, the fluid which fills the front part of the eye is produced from the blood vessels in the Ciliary Body gland inside the eyeball. The Aqueous flows in, but at the same time, it must have a way to flow out of the eye.

If the drainage system evolved *before* the fluid producing ciliary body, there is no aqueous fluid and the eye would be soft, mushy, and deformed; and the delicate inner structures would not be supported, they would be damaged, causing blindness.

If the ciliary body- aqueous producing system evolved *before* the drainage system of the Trabecular Meshwork and Schlem's Canal there would be too much Aqueous fluid, the eye pressure would be too high and would damage the Optic Nerve causing blindness, (this is what glaucoma is).

For the Aqueous flow to proceed properly the eye would have to evolve a way for the fluid to flow in and at the ***exact same time*** evolve a place for the fluid to flow out - in two completely different parts and areas-of the eye –but at the exact same time. It is clearly evident it could not just *happen* or gradually *evolve* to happen. One part cannot function without the other part; they are absolutely, totally, and simultaneously interdependent. "Numerous successive slight modifications"[1] as Darwin imagines will not work here- it has to happen simultaneously at two different areas within the eye and at exactly the same time. Again, Darwin's "theory absolutely breaks down."[1]

God made a way for the aqueous to flow in and, *at the same time,* and in a completely *different* area, designed a way for it to flow out. We observe Interdependent Evidence of Creation. Thank you, God!

Page 84 shows the Aqueous in-flow/out-flow mechanism.

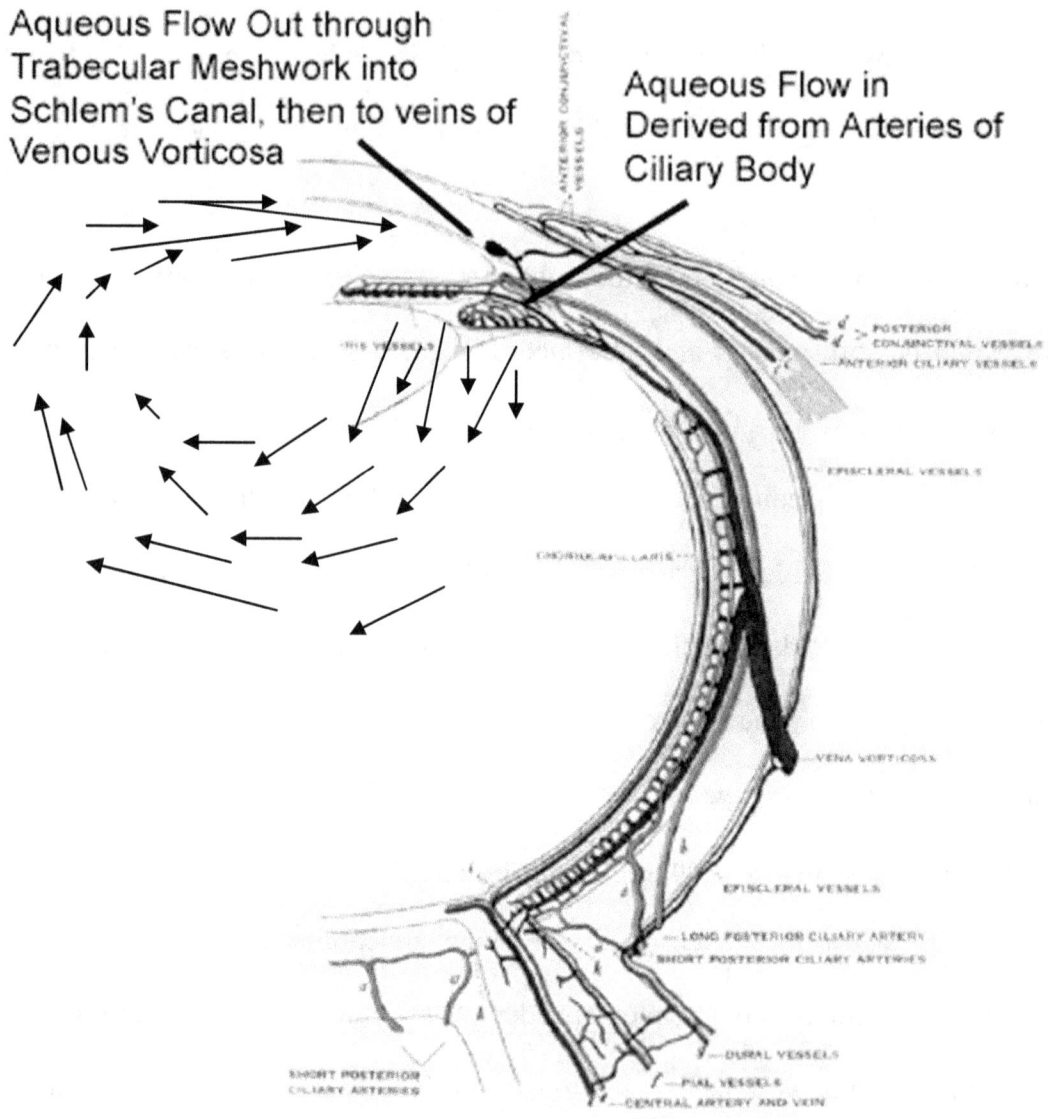

Aqueous Flow Out through Trabecular Meshwork into Schlem's Canal, then to veins of Venous Vorticosa

Aqueous Flow in Derived from Arteries of Ciliary Body

From Henry Gray, F.R.S, *Anatomy of the Human Body* 20th Edition, Lea & Febiger (Philadelphia & New York, © 1918) Page 1012[2] with modifications and additions by Rigney

Note: The aqueous remains in the front portion of the eye as the back portion of the eye is filled with a thicker fluid called the vitreous, I have drawn it this way for ease of illustration and ease of visualization. The aqueous flows primarily through the space between the front of the lens and the back of the iris, which in this illustration is not visible.

6. Trabecular Meshwork

The aqueous flows out of the eye in what is called the *angle* of the eye where the cornea, sclera, and iris meet. For the aqueous fluid to flow out, the aqueous fluid must first pass through the *trabecular meshwork,* then into *Schlem's canal,* then into the *veins* of the *episclera and venous vorticosa* and then into the bloodstream from which it was derived[1,2,3,4] (See pages 80-87).

The aqueous fluid must be present to keep the eye firm and round like a ball. And fresh flow is *continually* needed to provide nutrients to the lens and cornea. The *arteries* must be present in the *ciliary body* to provide the nutrients. The *ciliary body gland* must be present to make the fluid and at the same time the *trabecular meshwork*, *Schlem's canal,* and the *veins* mut all *be present* for the fluid to drain out.[1, 2, 3, 4] They *all* had to be created and fully functional at the same time. If they were not *all* present, the fluid could come in, but it could not get out; and the pressure would kill the eye.

God's design of the trabecular meshwork flow and drain *system* truly shows the wisdom, **_creative genius_**, and omniscient (all-knowing) hand of God. The trabecular meshwork lets fluid drain out, but at the same time, it won't let *too much* fluid drain out. It is *miraculous!* The trabecular meshwork is Created and designed like a sponge that the fluid must *seep* through in order to get out. It works like a reverse-pressure-release-valve and its design is very complex and requires foresight and preventive knowledge. It is apparent knowledge of potential problems had to be known and actions and measures implemented and compensated for by intelligent design and actual engineering in order for those potential problems _not_ to occur. And they had to happen simultaneously in two completely different locations within the eye. I.E.C.! Thank you God !

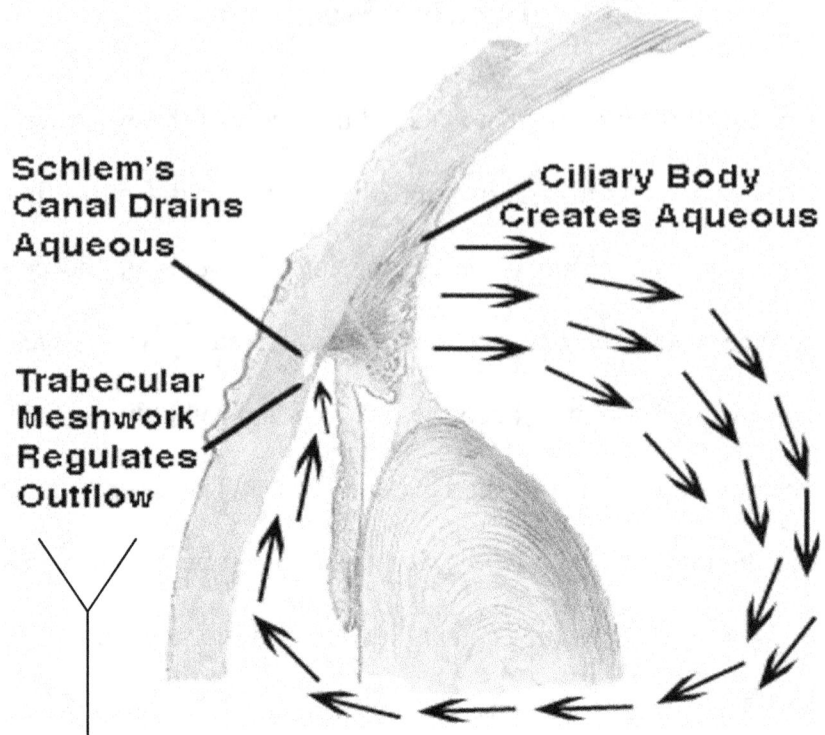

Schlem's Canal Drains Aqueous

Ciliary Body Creates Aqueous

Trabecular Meshwork Regulates Outflow

From Henry Gray, F.R.S, *Anatomy of the Human Body* 20th Edition, Lea & Febiger (Philadelphia & New York, © 1918) Page 1019[5] with modifications and additions by Rigney

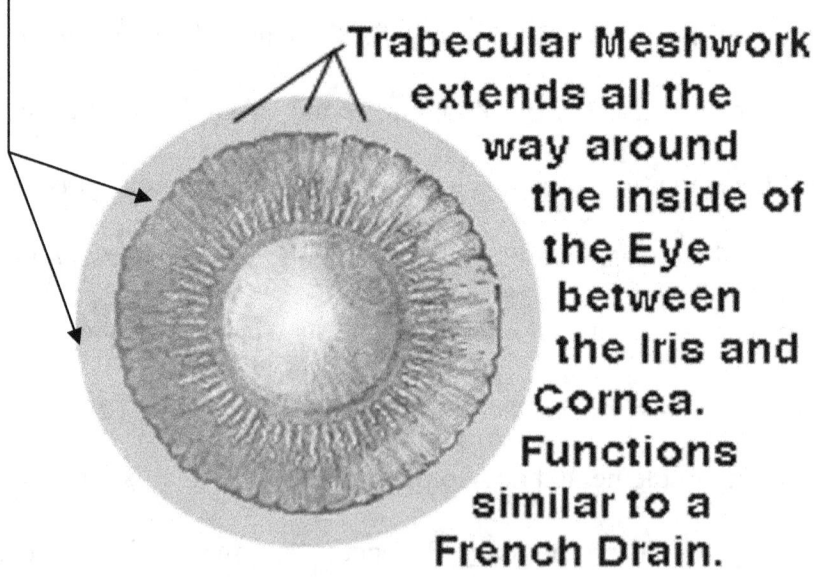

Trabecular Meshwork extends all the way around the inside of the Eye between the Iris and Cornea. Functions similar to a French Drain.

From Henry Gray, F.R.S, *Anatomy of the Human Body* 20th Edition, Lea & Febiger (Philadelphia & New York, © 1918) Page 1011[6] with modifications and additions by Rigney

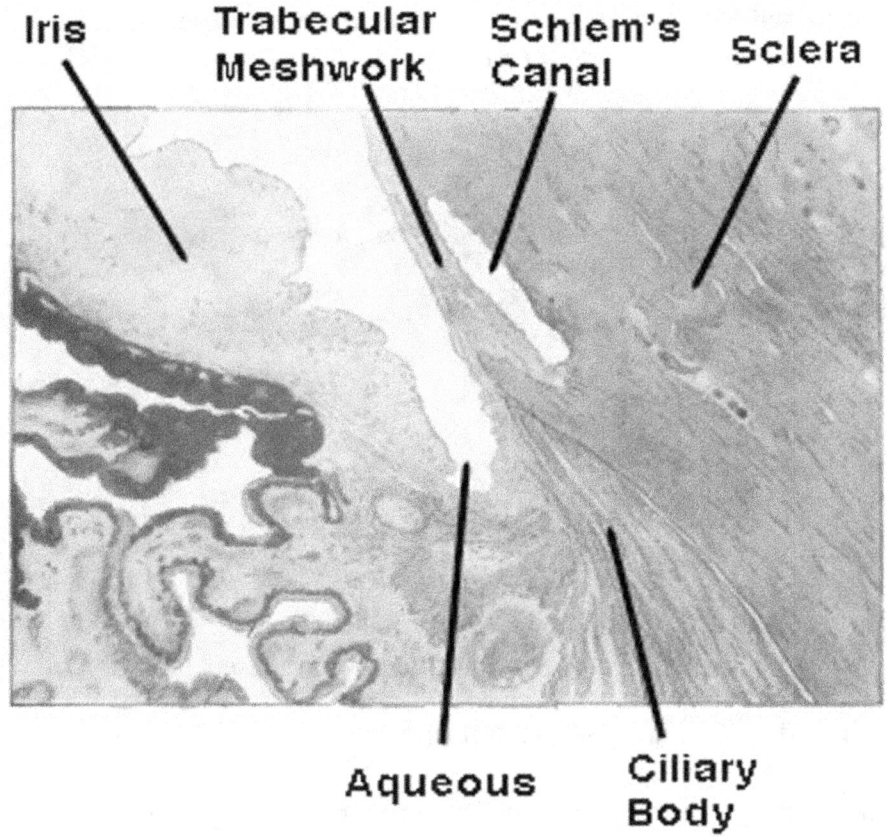

From Henry Gray, F.R.S, *Anatomy of the Human Body* 20th Edition, Lea & Febiger (Philadelphia & New York, © 1918) Page 1007[7] with additions by Rigney

If, instead of the trabecular meshwork design, the eye ***evolved*** a drain (like the drain in a sink), there would be no way to <u>keep</u> the aqueous *in* the eye; it would just drain out, leaving the eye empty, soft, mushy, and not round. The eye would be a soft mushy mess. The eye would not have pressure to keep it round. The lens inside the eye would not be supported, and the retina would detach, among other things. This type of damage does not heal in a way that would allow the eye to see after it has healed. The damage would be *permanent,* irrecoverable, and the eye would not see. A simple open-drain-design will not work.

**Distorted Eye due to evolved-open-drain-eye, rather than the;
Designed-trabecular meshwork-drain-system-eye.**

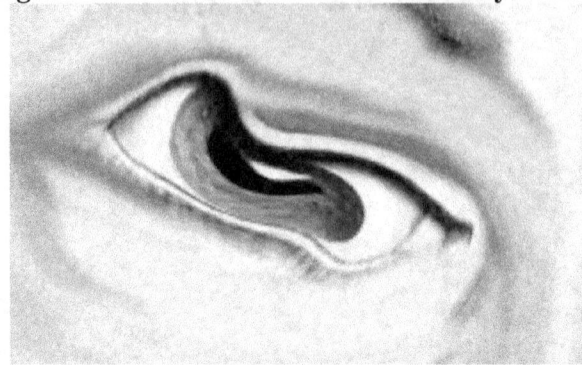

Or, if instead of the trabecular meshwork being **_designed_**, a simple open-drain pressure-release-valve **_evolved_**. A pressure-release-valve would cause a problem when you are simply rubbing your eye. As you press on your eye as you are rubbing it, the increased pressure from rubbing would cause the fluid to be forced to drain out, leaving the eye soft and mushy. Or, if you were to have pressure exerted on your eye (as may happen if you were in a fight and someone had you in a headlock and your eye was being *squished*). As the eye is being *squished;* the pressure-release-valve drain would allow the fluid to be forced (or squirt) out, leaving the eye a soft mushy mess. A third example is when you are sleeping and you are in a position in which you are somewhat laying your head sideways into the pillow and the pillow is pressing in on your eye. In each of these examples, if your eye had evolved a pressure-release-valve-drain rather than the designed Trabecular Meshwork drain system, as the pressure on the eye from the outside increases, it would cause the fluid to be forced out while the eye is being pressed upon. Then, when the pressing onto the eye is stopped, most of the fluid would have been released through the pressure-release-valve and the eye would be a soft mushy mess and it would *not* be round. And again, permanent damage to the internal structures, especially the lens and the retina would happen. We know today when this type of damage occurs to the eye, the eye cannot see and will not heal in a way which would enable it to *ever* see again. All this just from sleeping with some pressure on the eye!

God knew ahead of time these potential problems so He designed a miraculous spongy design: *the Trabecular Meshwork.* The Trabecular Meshwork works like the pores of a sponge and causes the openings of the spongy design to close when the pressure goes up, just as when you compress a sponge it becomes less spongy and less porous. This keeps the fluid from draining out <u>when</u> the eye is pressed upon but at the same time allows the fluid to flow out under normal conditions.

God's design is a *reverse* pressure release valve. The Trabecular Meshwork closes up *more* preventing the fluid from draining out *when* the eye is pressed upon, which enables the eye to maintain its pressure especially when it is pressed upon, preventing loss of the fluid which would cause a soft mushy eye - resulting in significant permanent internal damage. This *design* requires foresight and knowledge of what **<u>could</u>** occur and is engineered to **<u>prevent</u>** the problem from occurring. Prevention of a problem before it occurs requires foresight and *knowledge* which cannot be explained by *any* evolutionary process.

In God's unique *design* of the trabecular meshwork, the fluid pressure is not too high and not too low—it is just right! —and the flow of fresh fluid in and out of the eye can happen without the pressure going up (again, the cornea and lens *must* have fresh aqueous fluid to stay alive). God provided a way to keep the fluid from draining out during an accident or trauma or when pressure is exerted on the eye from day to day activities like eye rubbing and sleep position pressure forces. Wow! Thank you, God! This design of the trabecular meshwork requires foresight and *preventive planning* (taking steps to stop something from possibly happening). Preventive planning *is* intelligent design and requires engineering. The foresight and preventive planning necessary for the design and function of the Trabecular Meshwork drain system defies evolution. It either works or it doesn't. It can't gradually begin to work. The eye is either a soft mushy mess or it will explode, **_PERIOD !_** The Trabecular Meshwork drain system all by itself "absolutely breaks down" Darwin's theory.[5] How could the eye *evolve* a place for the fluid to be made and at the same time and in a different place *evolve* a place for the fluid to drain out? If they did not *evolve* at the exact same time and if it did not *evolve* to a fully completed and functional state, the eye would either be too soft or too hard - it could not function and it would die. Fully functional and simultaneously fully completed *interdependent*- but *different* parts- in completely different yet interdependent locations cannot happen from any evolutionary process. It can only happen with *Creation*! We observe Interdependent Evidence of Creation. Thank you, God!

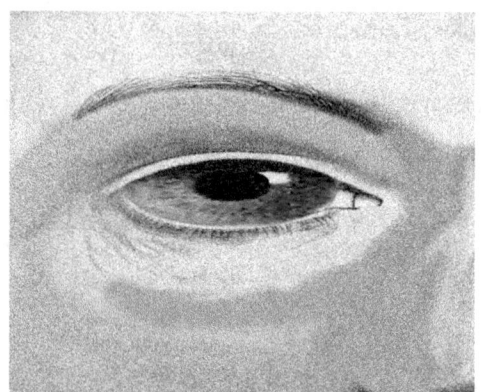

Developmentally High Pressure
Resulting in Megalocornea

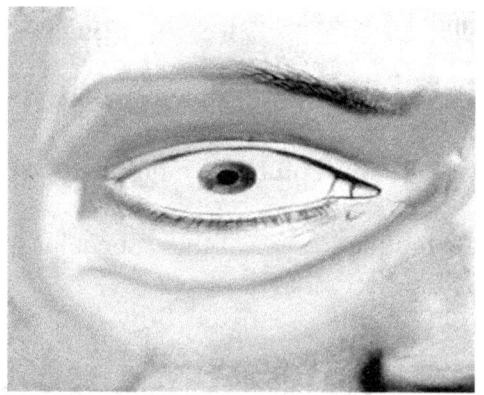

Developmentally Low Pressure
Resulting in Microcornea

During embryonic development of the eye, if the Trabecualr Meshwork has not *evolved* properly the pressure is too high and results in megalocornea. If the Trabecular Meshwork is *evolved* too much the pressure is too low resulting in microcornea. (Actual disease states)

We observe 11 interdependent systems when we observe the aqueous in-flow/ outflow system for both eyes. 1.Ciliary Body 2.Long Ciliary Artery 3.Ophthalmic Artery 4.Internal Carotid Artery. 4.Trabecular Meshwork 5.Schlem's Canal 6.Ciliary Veins 7.Vena Vorticosa 8.Ophthalmic Vein 9.Internal Jugular Vein 10.Sclera 11.Cornea. Again, this must all happen for each eye at the exact same time. Plus, it is clearly evident these two systems are absolutely and fatally interdependent. If the inflow system were to evolve before the outflow system, the eye would explode. If the outflow system evolves before the inflow system the eye is a soft mushy mess and none of the internal structures would be supported and the eye would die.

Darwin states on page 145 of his book *The Origin of Species*, "Long before having arrived at this part of my work, a crowd of difficulties will have occurred to the reader. Some of them are so grave that to this day I can never reflect on them without being staggered; but, to the best of my judgment, the greater number are only apparent, and those that are real are not, I think, fatal to my theory."[6]

The aqueous inflow/outflow system is '***absolutely***' fatal to Darwin's theory. The only means by which this system could not be fatal is for the *entire inflow system* and *entire outflow system:* to fully evolve; exactly; simultaneously; for both eyes; at the exact same time. Again, this can only happen by Creation. Everything must be fully present, fully developed, and fully functional at the same time. This can only happen by Creation. The Trabecular Meshwork, Ciliary Body, and Aqueous Flow ***must*** all occur **simultaneously**. The theory of *Interdependent Evidence of Creation* is proven. The theory of *Evolution* is false. Darwin evidently was not aware of these two systems and their fatal interdependence. I think he would agree they are fatal to his "theory".

7. Eyelids

The eyelids protect the eye from injury and infection. They help keep the eye moist. They also help keep out excessive light when needed.

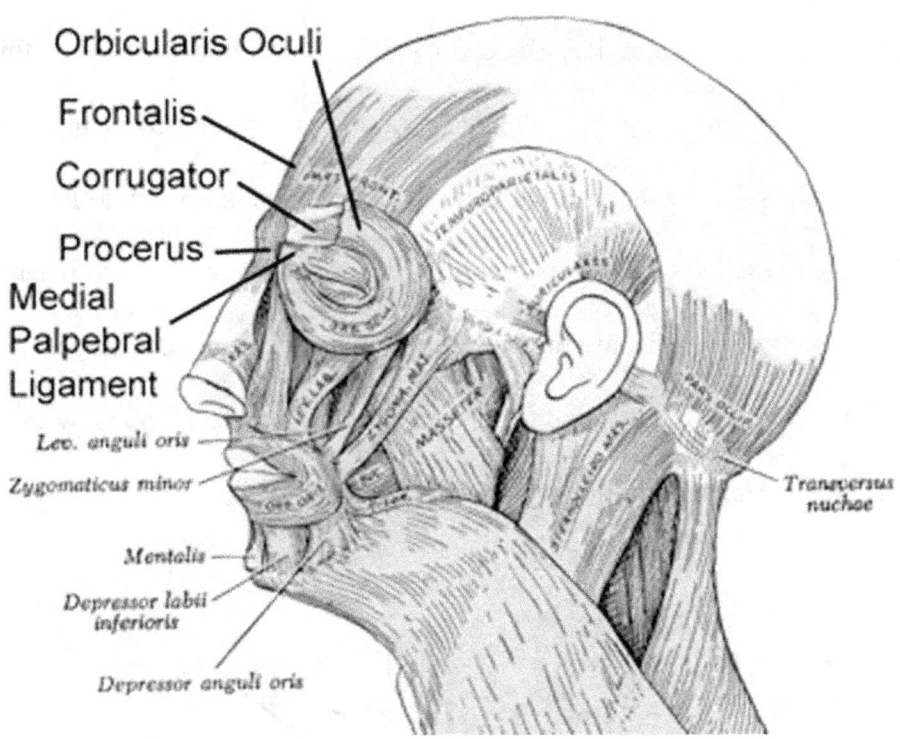

From: Henry Gray, F.R.S, *Anatomy of the Human Body* 20th Edition, Lea & Febiger (Philadelphia & New York © 1918) p.379[1] with Modifications by Rigney

Without the eyelids, the eye would dry out and become cloudy and opaque. If the eyelids did not form properly to create an opening, the eyes would not open and you could not see.

There are three different muscles which make up the eyelid. They are supplied by three different nerves.[2] The orbicularis oculi eyelid muscle, above, is supplied by the Facial Nerve, Cranial nerve #7[2] It is the muscle that wraps around the eye like a purse string. When you squeeze the eye shut (very hard) it is the muscle most affected. It does not lift the eyelid. Lifting of the lid to open the eye is provided by a second eyelid muscle the Levator Palpebrae Superioris. The Levator Palpebrae Superioris is supplied by Cranial Nerve #3, the Oculomotor Nerve.[3] Cranial Nerve #3, the Oculomotor Nerve has two branches, a superior branch and an inferior branch. The superior branch supplies motor (movement) impulses to the Levator Palpebrae Superioris.[3] The third muscle within the eyelid is Mueller's muscle. The nerve that supplies Mueller's muscle is supplied by Sympathetic Fibers from the Superior Cervical Ganglion of the Internal Carotid Artery plexus, the same fibers associated with the Iris Dilator Muscle. Mueller's Muscle is what helps to open the eyes wide when we are excited. When we are sleepy, lack of significant stimulation of it causes our eyelids to get droopy.[4] Thus, we observe three different muscles of the eyelid –supplied by three different nerves, with three completely different routes of supply. Again, evolution cannot explain this process!

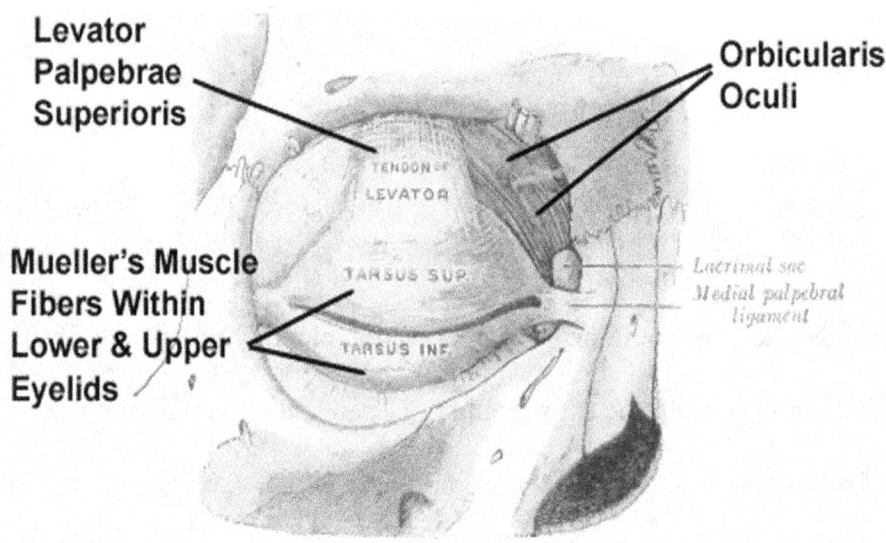

From Henry Gray, F.R.S, *Anatomy of the Human Body* 20th Edition, Lea & Febiger (Philadelphia & New York, © 1918) Page 1027[5] with modifications and additions by Rigney

If the nerve which supplies the Levator Palpebrae Superioris did not evolve at the same time as *any* of the other parts of the eye, the eye could not open at all. If the eye doesn't open, the being cannot see. Here again we observe Interdependent Evidence of Creation. All 25 interdependent systems must be fully functional, **_ALL_** at the same time!

The inside pink skin of the eyelid is designed to extend backward along the globe of the eyeball—but not adherent or connected to the globe of the eyeball. Then about three fourths of the way back on the backside of the eyeball, the pink skin of the eyelid joins and connects and becomes the skin on the outside of the white globe of the eye at what is termed the Fornix. This prevents any foreign matter, dirt, debris, mucus, germs, or infection to make its way behind the eyeball where it could become permanently *lodged,* creating pain and promoting infection. This warm moist area would be a perfect breeding ground for bacteria, fungus, virus and amoeba; infection would be a constant problem. Wisdom, foresight, knowledge, and understanding are necessary to *prevent* problems. To *prevent* something from happening, one must understand the potential problem then take steps to keep it from happening. Again, here also, we observe intelligent design and further evidence of Creation. "Imagine[6]" as Darwin says we "must[6]", how the skin of the *eyelid* evolved to become joined to the skin of the *eyeball*, thereby preventing foreign matter, debris, germs, etc. from getting behind the eye where a perfect breeding ground for germs would be present. If we are to "imagine[6]" how that could occur by "numerous, successive, slight modifications[6]", the problem is two-fold: 1. If the skin of the inside of the eyelid evolved, before the skin of the lid became as we actually observe it– the likelihood of the eye surviving and therefore the being surviving is unlikely; constant infection with ultimate loss of the eye would be the rule; so the species could not propagate and no "numerous, successive slight modifications[6]" could actually happen. 2. I can't *"imagine[6]"* a means by which it could have occurred by "numerous, successive, slight modifications[6]" - can you? (I do understand that simply because *I* can't *imagine* [6] it, doesn't mean that it couldn't happen. But, at the same time - just because you can *imagine* [6] it, (**_IF_** you actually can) doesn't mean it *could* happen either!) You can, *"believe[6]"* it just happened that way by chance. However, this requires faith. Rather than "believe[6]" - observe the evidence and *conclude:* "it was **_created_**!" Again, we observe Interdependent evidence of Creation, I.E.C.

The skin (conjunctiva) of the eyelid (Palpebral Conjunctiva) turns without a break and becomes the skin (Bulbar Conjunctiva) of the eyeball. This design prevents debris from becoming lodged behind the eye causing infection.

Superior Fornix
Pink eyelid skin
(Palpebral Conjunctiva)
becomes clear
eyeball skin
(Bulbar Conjunctiva)

Bulbar
Conjunctiva
Clear Skin
on Eyeball

Inferior Fornix
Pink eyelid skin
(Palpebral Conjunctiva)
becomes clear
eyeball skin
(Bulbar Conjunctiva)

From Henry Gray, F.R.S, *Anatomy of the Human Body* 20th Edition, Lea & Febiger (Philadelphia & New York, © 1918) page 1021[7] with modifications and additions by Rigney

The inside of both the upper and lower eyelid contains oil glands; they are called Meibomian Glands. These oil glands within the eyelids extend the entire length of the eyelid from the margin of the lid all the way to the crease of the upper lid and to the edge of the rim of the orbit of the lower lid. There are about 23 of them within each eyelid. Each one has its' own pore at the eyelid margin. Their presence and placement is further evidence of Creation. "Why?" you might ask. Oil floats on top of water. The Lacrimal gland produces the *watery* component of the tears. The Meibomian glands produce *oil*, which then creates an oily layer on top of the tears. This placement of the Meibomian glands on the margins of the eyelids enables the oil layer to be placed specifically where it is needed – on the surface of the tears. The oily layer on top of the water layer prevents the tears from evaporating. It also increases the surface tension of the tears and helps prevent the tears from spilling over the lids onto the face and helps to make sure the tears drain out the puncta. This oily layer also helps to make the eyelids slick and prevents the eyelids themselves from scratching the cornea as the lids blink.

In addition to the Meibomian glands, there are four other glands within the eyelid; the Tarsal Glands of Moll, the Glands of Zeiss, the Glands of Krause, and the Glands of Wolfring.[8] The glands of Krause and Wolfring contribute to the watery layer of the tears. The Glands of Moll and Ziess are associated with the eyelashes and prevent the lashes from being brittle. Within the Bulbar Conjunctiva (the skin that covers the white globe of the eye, the Sclera) there are Goblet cells. Their function is to secrete mucus into the tears. The mucin layer forms the bottom layer of the tears and functions to help keep the tears adherent to the eye.[9] The mucin layer also has anti-infective properties, so it is a final barrier to infection.[9] The design of the tears is specifically created so the tears are made up of three layers; the superficial oily layer, the middle watery layer, and the inner mucin layer. Removal of any one of these layers creates a dry eye problem.

The placement and function, of each of these layers is further evidence of planned intentional design, and again demonstrates Interdependent Evidence of Creation, I.E.C.

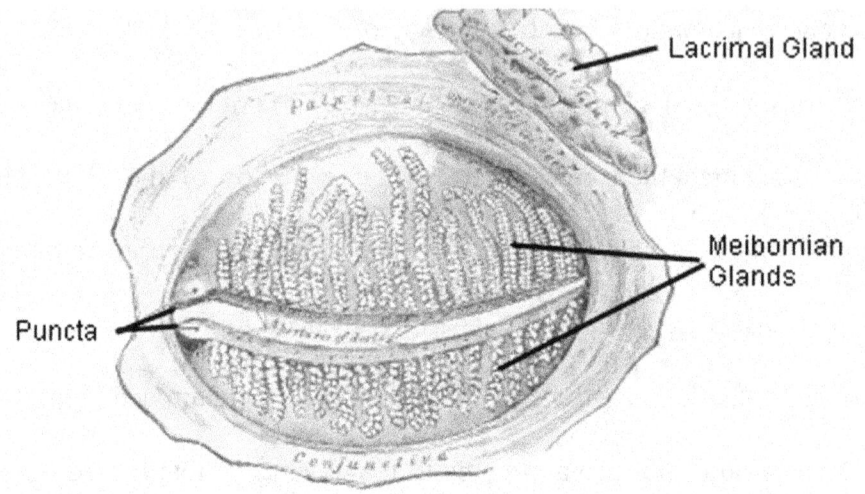

From Henry Gray, F.R.S, Anatomy of the Human Body 20th Edition, Lea & Febiger (Philadelphia & New York, © 1918) p.1027[10]

Without the Meibomian Gland oily layer the tears evaporate too quickly and the eye dries out. Without the Meibomian Gland oily layer the eyelid itself would irritate the eye. Without the Lacrimal Gland watery layer the cornea is dry and cloudy. Without the mucin layer the tears don't adhere to the cornea and roll off like water on a waxed car. All three layers need to be present otherwise the cornea is dry, cloudy and painful. Do you really think evolution can explain these specific problems; and somehow by chance evolve the correct solution which answers the specific requirements? Again, we observe planned intentional design and further evidence of Creation. When we observe the lids, we observe 11 different Interdependent Evidences of Creation, I.E.C.

8. Tears

Tears are made by the Lacrimal Gland, which is located under the upper-outer part of the eyebrow (exposed below). Tears keep the delicate front surface of the eye, the Cornea, moist. They also provide some nutrients to the Cornea. The tears help transport oxygen to the Cornea, and they provide what is needed to help prevent the eye from getting infections.[1]

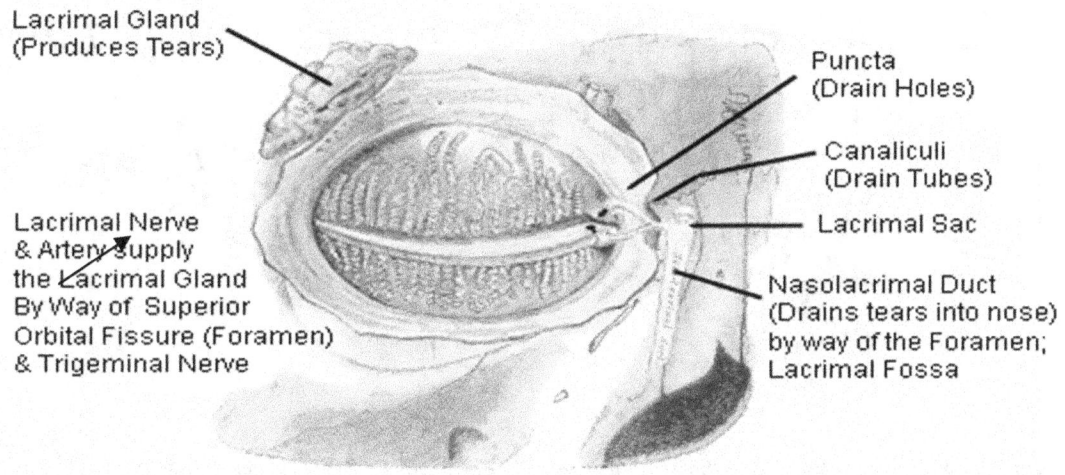

From Henry Gray, F.R.S, *Anatomy of the Human Body* 20th Edition, Lea & Febiger (Philadelphia & New York © 1918) Page 1027 & 1029[2] merged with modifications and additions by Rigney

The Lacrimal Gland makes the tears. It is in the upper outer portion of the eye under the upper eyelid. As a result, tears flow across the eye to drain out the tear drain ducts located in the inner corner of the eyelid. The Puncta are the openings of the drain in each lid. There are two Puncta, one on the upper and lower lid. The Puncta connects to a drain tube, the Cannaliculus, which carries tears into the side of the nose through a hole (or foramen) the Lacrimal Fossa on the inside bone of the eye socket. The tears then drain from the back of the nose into the throat.

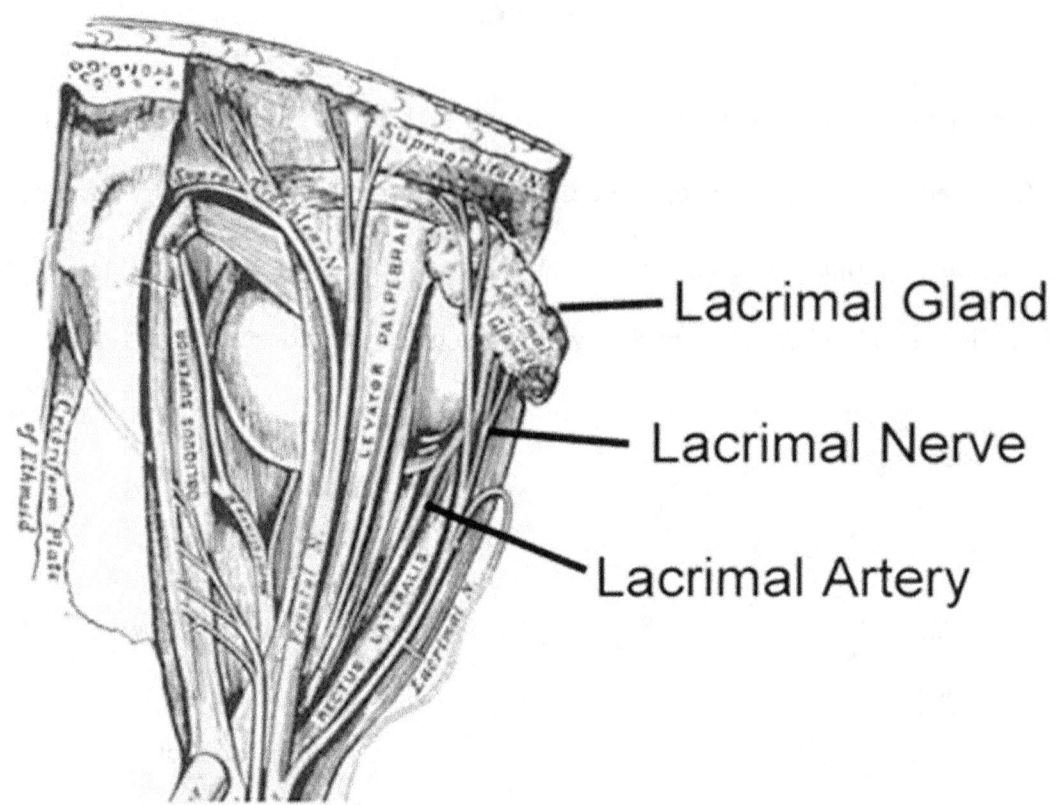

From Henry Gray, F.R.S, *Anatomy of the Human Body* 20th Edition, Lea & Febiger (Philadelphia & New York, © 1918) Page 885[3] with modifications and additions by Rigney

Thinking about the process of evolution, how could hard bone *evolve* to become a soft tear gland? How can the soft Lacrimal Artery, Lacrimal Vein, and Lacrimal Nerve -necessary for the Lacrimal Gland to function *evolve* through hard bone? How can the Lacrimal Gland *evolve* to begin to one-day make tears? What process would one-day cause the tear gland to start to form? How can a tear gland form from nothing into something- and at the same time form the required artery, vein, and nerve needed to make it function and work properly? Again, we observe Interdependent Evidence of Creation.

The watery component which makes up tears is derived from blood from the artery which supplies the Lacrimal Gland. As a result, tears contain the antibodies from the blood necessary to fight infection.[4]

Without tears your eye would feel dry and painful, and become cloudy and not clear. Your eyes would hurt and you could not see. Without tears, your eye would easily get infected.

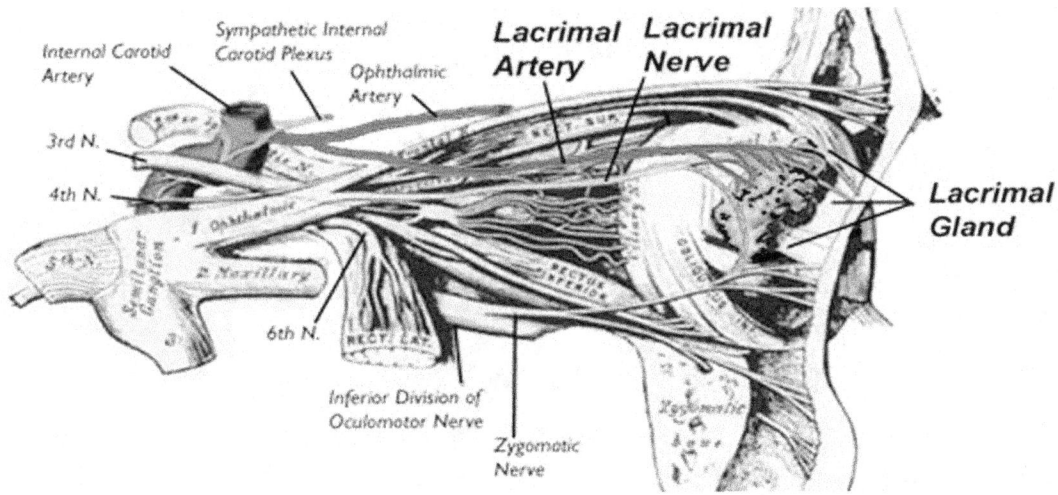

From Henry Gray, F.R.S, *Anatomy of the Human Body* 20th Edition, Lea & Febiger (Philadelphia & New York, © 1918) Page 887[5] modifications and additions by Rigney

Either the tear gland works or it doesn't. If it doesn't the eye cannot live, it will become cloudy and opaque. The tear gland "could not possibly have been formed by numerous, successive, slight modifications"[6] as Darwin suggests. The tear gland is either present and the eye functions and sees, or it is not present and the eye does not function and does not see, this we know –we don't have to "imagine"[7] it. Again, Darwin's "theory absolutely breaks down."[8] Interdependent Evidence of Creation is observed. We observe 13 interdependent systems for the tears; the 1.Foramen of the

Superior Orbital Fissure must be present for the arteries and nerves to supply the Lacrimal Gland. The 2.Ophthalmic Artery must be present to supply blood for the 3.Lacrimal Artery so the 4.Lacrimal Gland can make the 5.Tears. 6.The Meibomian glands 7.The Goblet Cells 8.The eyelids to distribute the tears. 9.The Levator Palpebrae Superioris, 10.Obicularis Oculi, 11.Cranial nerve #3, 12.Cranial nerve #7 for those muscles, and 13.Cranial nerve #5 to sense dryness.

9. Tear Flow

The watery portion of the tears is made by the tear gland, the Lacrimal Gland, located under your upper outer eyebrow. The tears drain out from two drains (Puncta): one on the inside lower lid and one on the inside upper lid; both are near the nose. The tear drain tubes (Canaliculli) pass through the foramen in the bone, the Lacrimal Fossa, on the inside corner of the nose next to the eye, and after passing through the bone, it connects into the inside back of the nose. Your tears drain into the back of your nose and then into your throat.[1]

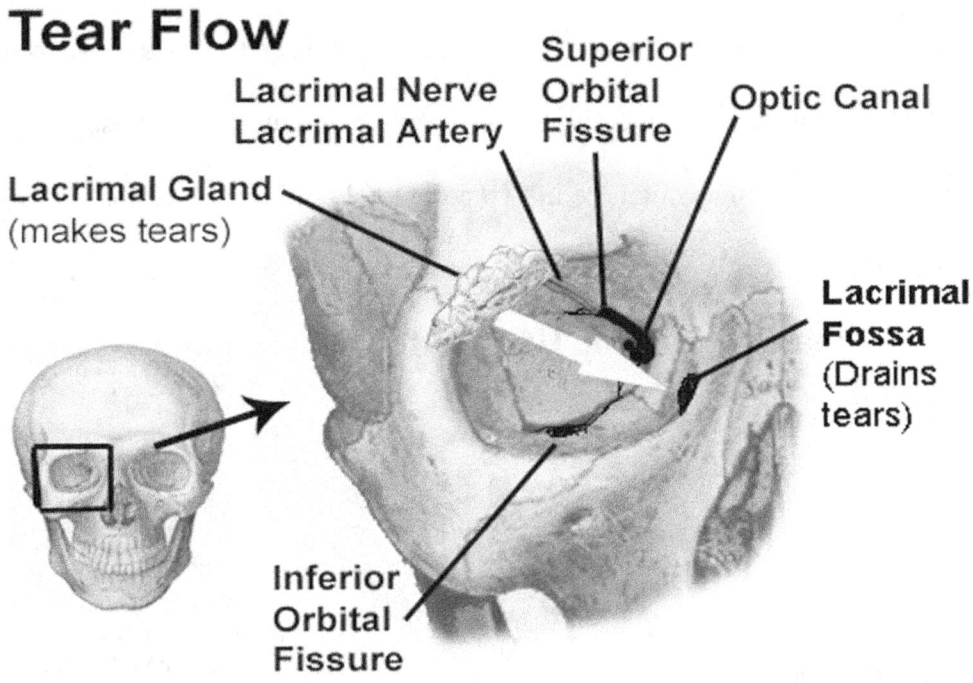

From Henry Gray, F.R.S, *Anatomy of the Human Body* 20th Edition, Lea & Febiger (Philadelphia & New York, © 1918) Page 186[2] with modifications and additions by Rigney

How could soft tear drain duct tissue evolve through a hard bone so the tears can drain into the back of the nose?

Without the Tear *Gland,* your eye would be dry and painful. Without the Tear *Drain,* the eye would be runny and gunky, and you would be weeping or crying all the time. Without the *Foramen* (openings) in the bones, the Tear Drain could not drain the tears. They *all* had to be created and work together *all* at the same time!

Look at the location and placement of the tear-flow <u>system.</u> The Tear *Gland System,* which makes the tears, is in the outer upper portion of the eye socket. The Tear *Drain System,* which drains the Tears, is in a totally different area, in the inner portion of the eye socket. They must be formed at exactly the same time. If the Tear Gland formed before the Tear Drain ducts, your eyes would constantly water. If the Tear Drain ducts evolved before the Tear Gland the eye would be super, super dry – and why would a Tear Duct evolve when a Tear Gland is not present? It is clear these two completely different systems which are in two completely different locations are interdependent and were designed and created. We observe Interdependent Evidence of Creation, I.E.C.!

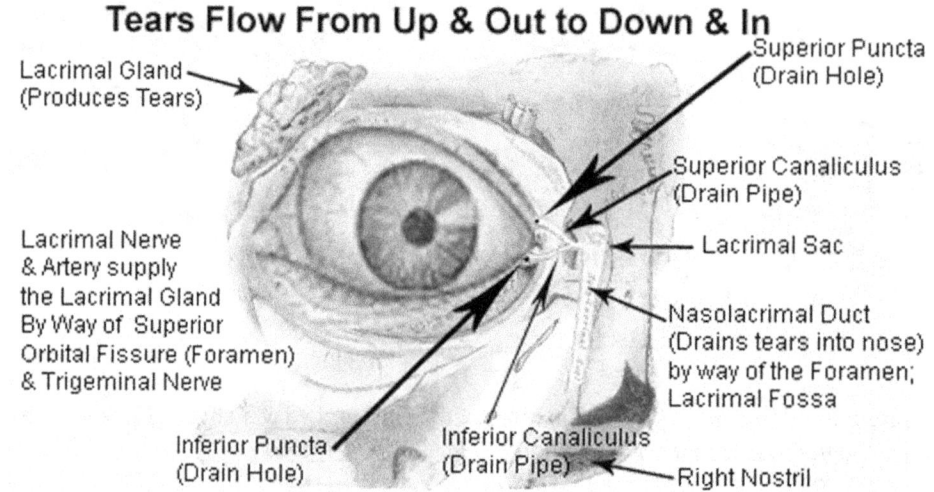

Tears Flow From Up & Out to Down & In

From Henry Gray, F.R.S, *Anatomy of the Human Body* 20th Edition, Lea & Febiger (Philadelphia & New York © 1918) Page 1026 & 1027[3] with modifications and additions by Rigney.

How could these have evolved totally, and simultaneously - in two such different areas of location within the eye? They didn't! It is *clearly evident* God *designed* and *created* them to work together. He created them to work together to be fully functional at exactly the same time. "Numerous successive slight modifications"[4] as Darwin's theory "supposes"[5] cannot explain this. The Lacrimal Gland, located **up and out** and the Lacrimal Drain (which is made up of multi-tissue-systems; bone, foramen, ducts, canals), located in a completely different area, **down and in**, "absolutely break down" Darwin's theory.[6] We observe Interdependent Evidence of Creation, I.E.C.

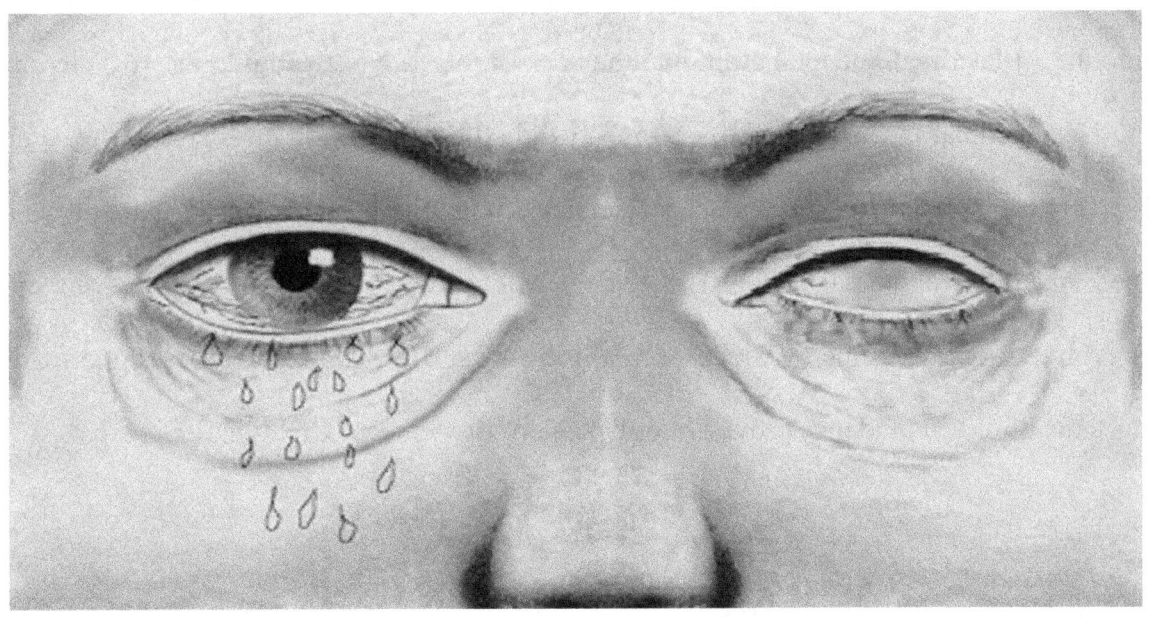

In this illustration, I am attempting to demonstrate what would happen if evolution were at play and the lacrimal gland and drainage <u>system</u> did not evolve simultaneously. The person's right eye is weeping due to the <u>tear duct</u> NOT having "evolved" while the lacrimal gland HAS evolved.

Alternatively, the left eye is dry, cloudy, red, irritated, and squinting from the <u>tear gland</u> NOT having evolved, but the drainage <u>system</u> HAS evolved. As a result the eye is totally DRY and opaque.

What is observed is Interdependent Evidence of Creation. The 1.Lacrimal Gland makes the 2.Tears. The 3.Lacrimal Nerve and 4.Lacrimal Artery must be present for the Lacrimal Gland to produce the tears. The 5.Superior Orbital Fissure (Foramen) must be present in the back of the eye socket for the Lacrimal Artery and Lacrimal Nerve to supply the Lacrimal Gland. Simultaneously, there must be the two Puncta (6&7), and the two Canaliculi (8&9) present in the 10.Upper and 11.Lower Eyelids. 12.The Foramen: the Lacrimal Fossa must be present in the 13.Maxillary Bone to drain into the 14.Lacrimal Sac then drain into the back of the 15.Nose and into the 16.Throat by way of the 17.Nasolacrimal Duct. This all must happen and all be present on *both eyes* <u>at the same time.</u> We observe 17 additional Interdependent Evidences of Creation! Again, all must be present, fully developed and fully functional AT THE SAME TIME – FOR BOTH EYES! Does the evidence support evolution? Or, does the evidence - *simultaneous interdependent evidence*, support Creation? What does the observed evidence prove? Interdependent Evidence of Creation, I.E.C.

10. Lens

The lens is created in a way to allow the eye to change focus from far to near and from near to far. The lens, like the cornea, is made of clear tissue (living cells which are *clear)*. Stop and think about this. Meditate on the miracle of the fact living breathing, functioning cells need nourishment and have waste products- yet are clear!

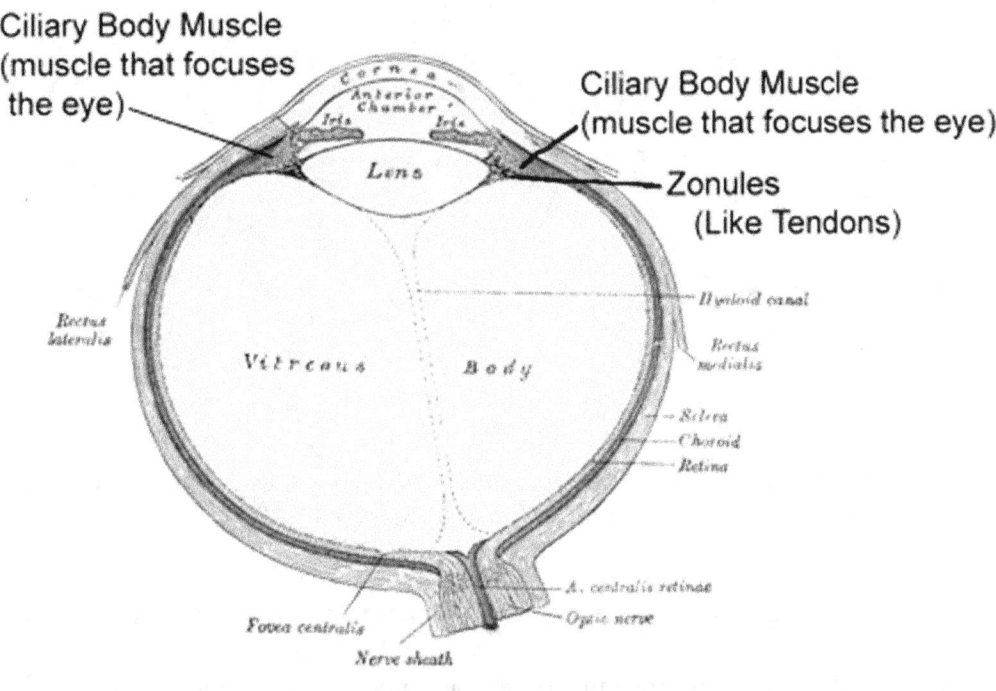

From Henry Gray, F.R.S, *Anatomy of the Human Body* 20th Edition, Lea & Febiger (Philadelphia &New York, © 1918) Page 1006[1] with modifications and additions by Rigney

The lens receives nerve impulses from the brain by way of the parasympathetic nerve fibers of the Oculomotor Nerve Cranial Nerve #3. We observe three additional Interdependent Evidences of Creation 1. Oculomotor Nerve, 2. Lens and 3. Ciliary Muscle.

Plus, the lens has to be the precise shape and thickness to fine-tune the focus of the light on the retina. If the lens is too thick, the eye would not be able to see distance. If the lens were too thin, the eye would not be able to see close.

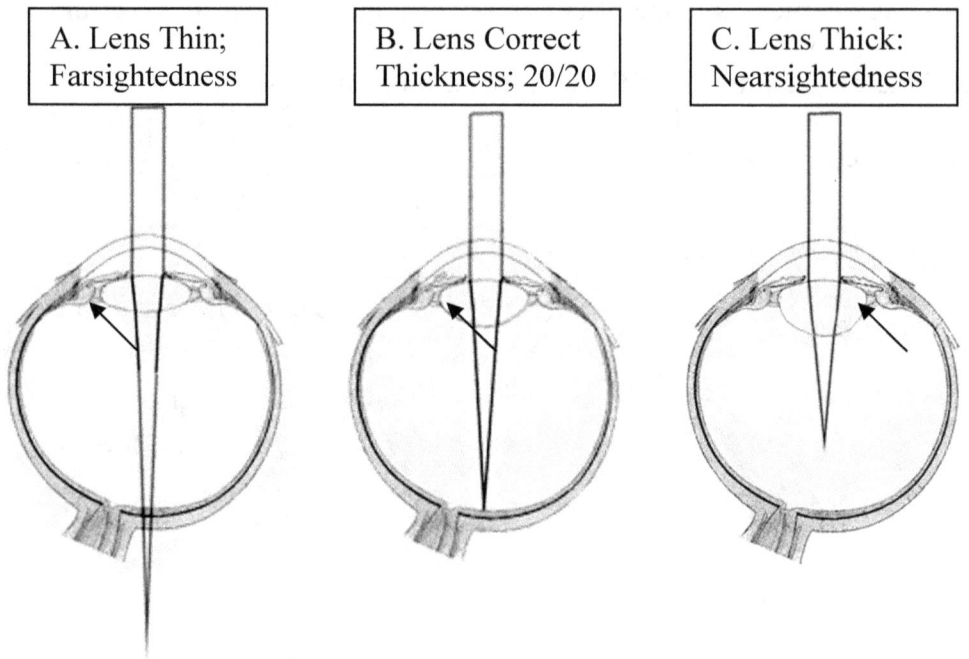

| A. Lens Thin; Farsightedness | B. Lens Correct Thickness; 20/20 | C. Lens Thick: Nearsightedness |

From Henry Gray, F.R.S, *Anatomy of the Human Body* 20th Edition, Lea & Febiger (Philadelphia & New York, © 1918) Page 1006[1] with modifications and additions by Rigney

The lens must change shape to change focus. The ciliary body muscle causes the lens to change shape in order to change focus from far to near and near to far. If the lens did not change shape, you could not see both near and far. You could only see near, but not far; or you could only see far, but not near. The eye is created to see both near and far - and *everything* in between. How can evolution explain the process of changing focus? It had to be planned and designed from the beginning. The ciliary muscle and the lens are interdependent, again we observe Interdependent Evidence of Creation, I.E.C.

11. Focusing Control

Again, this unique design **_defies_** evolution. The mechanics of the lens is an engineering marvel. The lens must *change shape* to focus from far to near or from near to far. The lens is encircled by a circular muscle which lines the inside of the eye called the ciliary muscle.[1] The ciliary muscle allows the lens to change shape causing it to change focus. The ciliary muscle allows the lens to become long and skinny to see far; then changes the shape and allows the lens to become short and fat to see near.

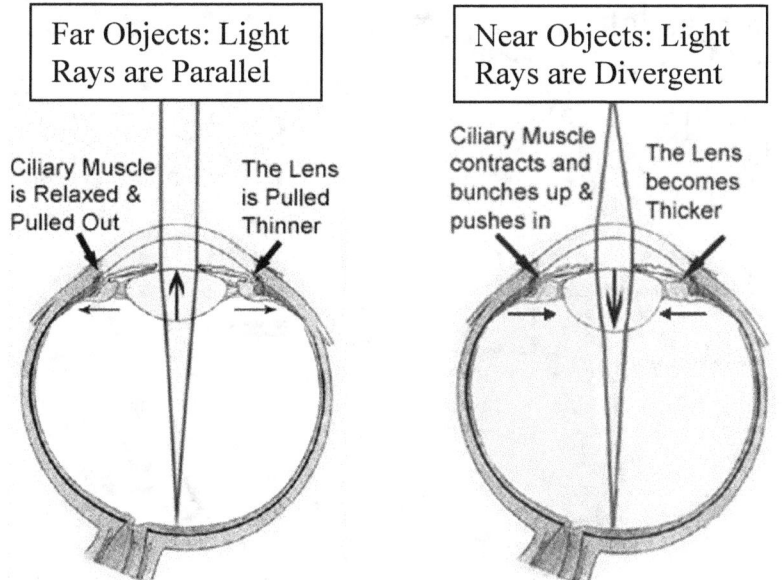

Far Objects: Light Rays are Parallel

Near Objects: Light Rays are Divergent

Ciliary Muscle is Relaxed & Pulled Out

The Lens is Pulled Thinner

Ciliary Muscle contracts and bunches up & pushes in

The Lens becomes Thicker

From Henry Gray, F.R.S, *Anatomy of the Human Body* 20th Edition, Lea & Febiger (Philadelphia & New York, © 1918) Page 1006[1] with modifications and additions by Rigney.

The Ciliary Muscle bunches up and moves inward (see arrows) allowing the Lens to become thicker, focusing the eye in for near objects. The thicker the Lens becomes, the more closely the eye can see. To refocus for far vision the Ciliary Muscle relaxes allowing the Lens to become thinner, changing the eye from near focus back to far focus. The Ciliary Muscle allows the Lens to become thinner for far vision (left), and thicker for near vision (right). The lens in its' natural state is the thicker focused-for-near state. So, when the ciliary muscles bunches up, it releases the tension on the lens and allows it to return to its' natural thick, focused-for-near state. When the ciliary muscle relaxes, it stretches the lens into the thinner focused-for-far state.

The lens is made of living tissue and living cells which are clear, like a window. If the lens were cloudy, you could not see. If the lens did not get its nutrients from the aqueous, it could not live; it would die, and you could not see. Without the *ciliary muscle* the lens cannot change shape to change focus. Without the sclera, the lens would have no base or platform on which to function and change shape. Each must be present- *all* together at the same time. There is no way for them to *all evolve* at the same time. Simultaneous complete *function* is *creation - Interdependent Evidence of Creation.*

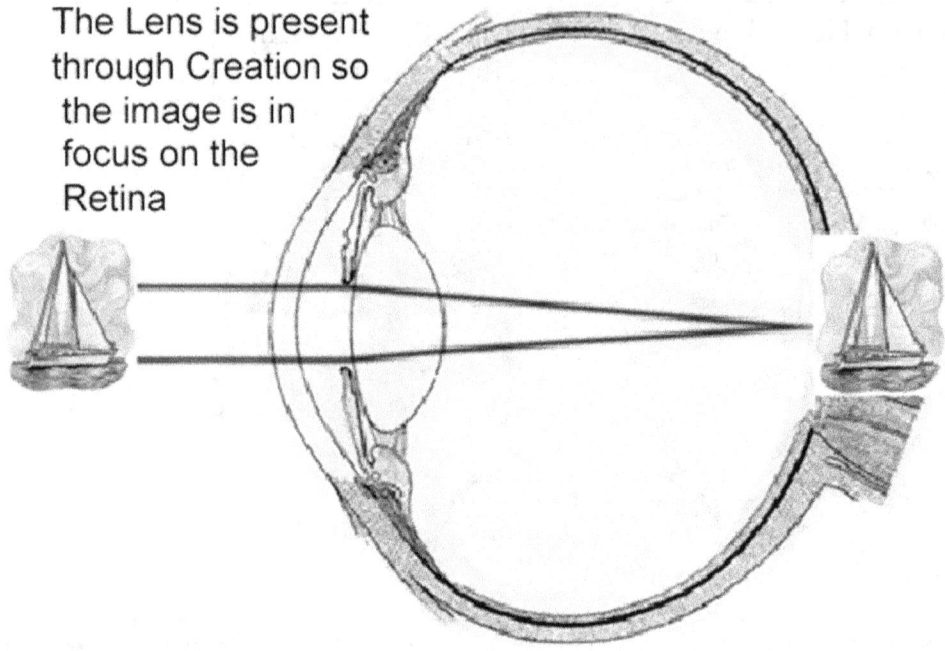

The Lens is present through Creation so the image is in focus on the Retina

In this example the lens has been "Created" and the eye sees clearly. The image on the retina is inverted: therefore, this drawing is not realistically illustrated. Isn't that amazing; even though the image on the retina is actually inverted, the brain compensates and re-inverts the image right side up.

From Henry Gray, F.R.S, *Anatomy of the Human Body* 20th Edition, Lea & Febiger (Philadelphia & New York, © 1918) Page 1006[1] with modifications and additions by Rigney

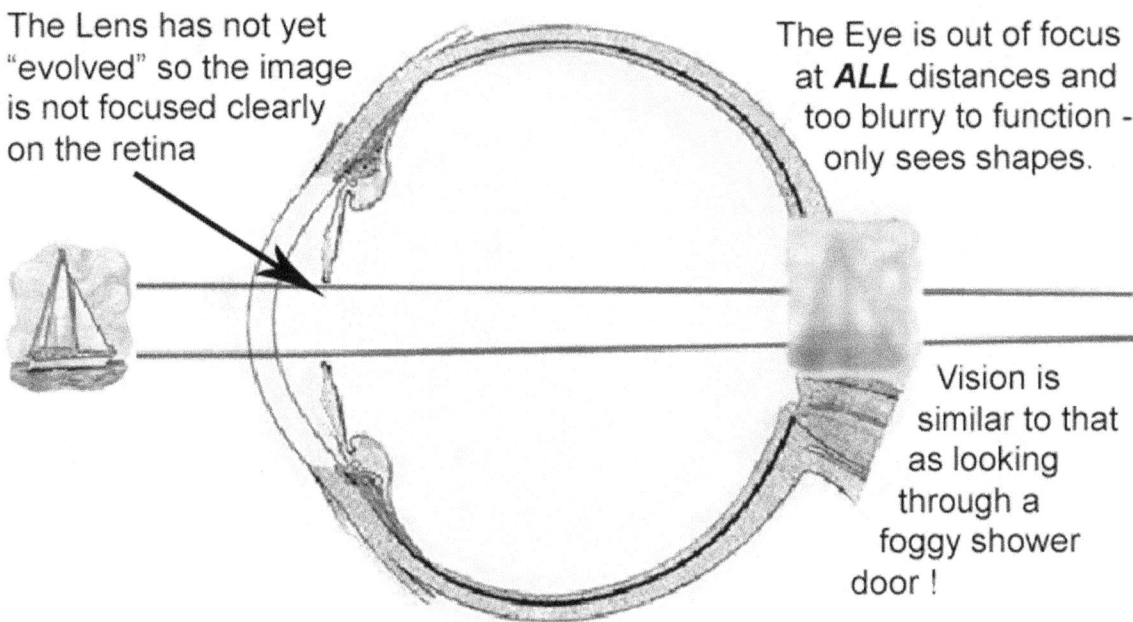

The Lens has not yet "evolved" so the image is not focused clearly on the retina

The Eye is out of focus at **ALL** distances and too blurry to function - only sees shapes.

Vision is similar to that as looking through a foggy shower door !

In this illustration, the lens has not "evolved" so the eye cannot see clearly -neither far nor near, and in this instance is considered "legally blind". Legally blind is where the eye can detect light, but vision is too blurry to see detail. Without the lens the eye could not see details - shapes would be a blur. The eye could not see clearly. A being could not survive while waiting on the lens to evolve.

(From Henry Gray, F.R.S, *Anatomy of the Human Body* 20th Edition, Lea & Febiger (Philadelphia & New York, © 1918) Page 1006[1] with modifications and additions by Rigney) (The image would be inverted.)

All 25 parts, systems and functions of the eye are totally interdependent. If one part is missing the eye cannot see or function and in some cases if a part is missing the eye will die. The focusing system requires nine different interdependent systems; 1. Lens, 2. Ciliary Body, 3. Cranial nerve #3 Oculomotor Nerve. 4.Sclera, 5. Zonules, 6. Aqueous, 7. Cornea, 8. Iris, 9. Retina. Totally interdependent parts and systems all with totally interdependent function can only happen through or by creation. We observe the simultaneous presence and their total interdependence is evidence of creation; **I**nterdependent **E**vidence of *Creation or I.E.C.*

12. Bones of the Eye

There are seven different bones[1] that fit together to form a hollow, circular opening in the head which holds the eye and the eye muscles. These seven bones form the eye socket, oftentimes referred to as the orbit. Each eye has seven bones so there are fourteen bones total which fit perfectly together to form two eye sockets.

The bones are designed to protect the eye from injury and to provide the eye muscles a platform on which to move the eye in all directions.

SEVEN BONES OF THE EYE SOCKET (ORBIT)

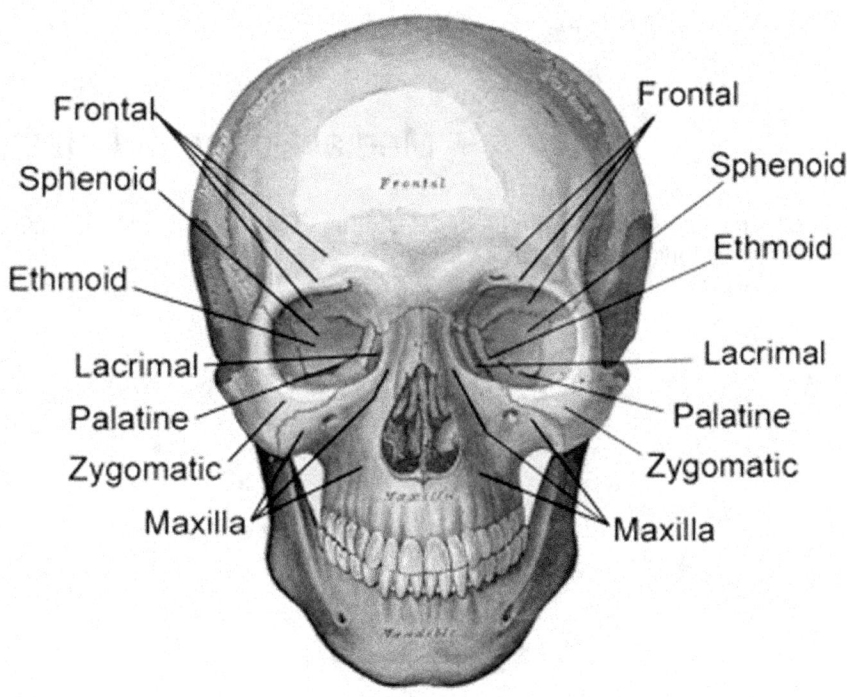

From Henry Gray, F.R.S, *Anatomy of the Human Body* 20th Edition, Lea & Febiger (Philadelphia & New York, © 1918) Page 186[1] with modifications and additions by Rigney

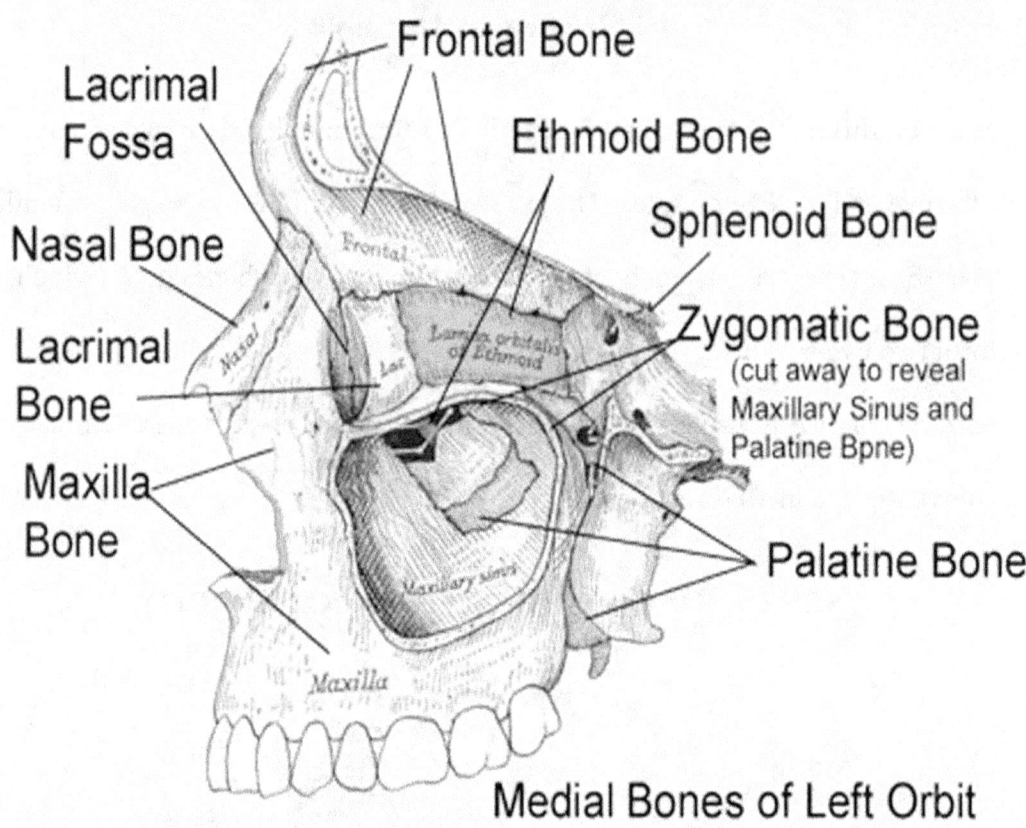

Lacrimal Fossa

Nasal Bone

Lacrimal Bone

Maxilla Bone

Frontal Bone

Ethmoid Bone

Sphenoid Bone

Zygomatic Bone
(cut away to reveal
Maxillary Sinus and
Palatine Bpne)

Palatine Bone

Medial Bones of Left Orbit

From Henry Gray, F.R.S, *Anatomy of the Human Body* 20th Edition, Lea & Febiger (Philadelphia, New York, ©
1918) Page 160[2] with modifications and additions by Rigney

The Floor of the Orbit

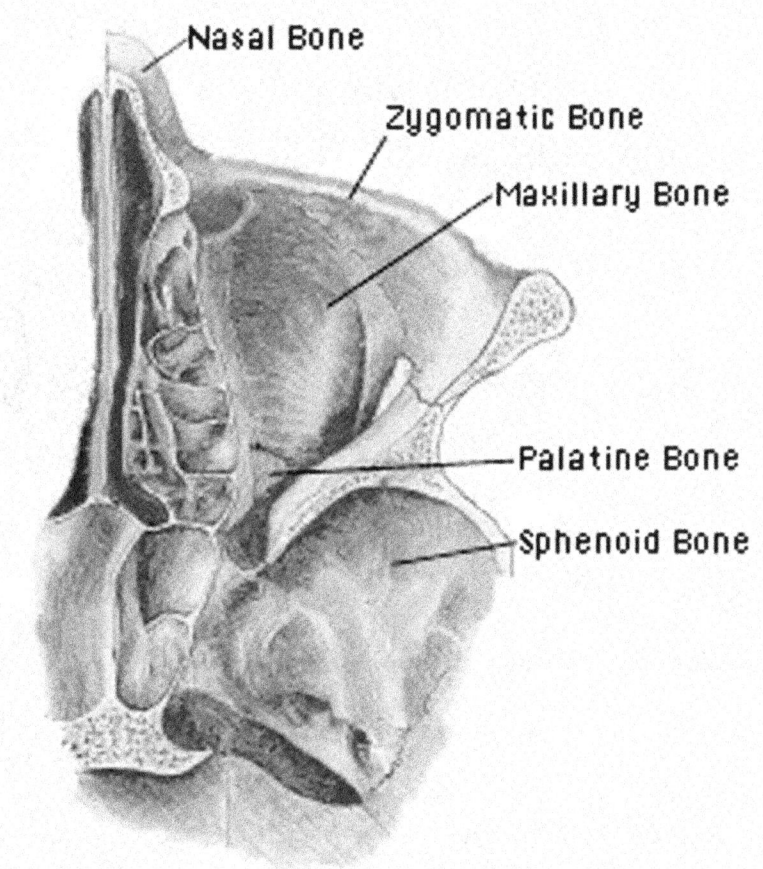

Nasal Bone

Zygomatic Bone

Maxillary Bone

Palatine Bone

Sphenoid Bone

From Henry Gray, F.R.S, *Anatomy of the Human Body* 20th Edition, Lea & Febiger (Philadelphia & New York, © 1918) Page 187[3] modifications and additions by Rigney

The bones are hard. The eye is soft. Hard bone tissue cannot evolve and turn into soft eye tissue. Soft eye tissue cannot evolve *through* hard bone tissue. Evolution would have you believe it all happened by chance over millions of generations. How could it ever happen, even in a million years? Evolutionists have great faith in chance. How is it any different from having faith in God, in each case you have to believe in something not seen. If you choose *not* to put your faith in God, you are choosing to put your faith in chance. It is one way or the other there is no other option. Chance? Or, God?

Creation: All the muscles and ligaments are present simultaneously

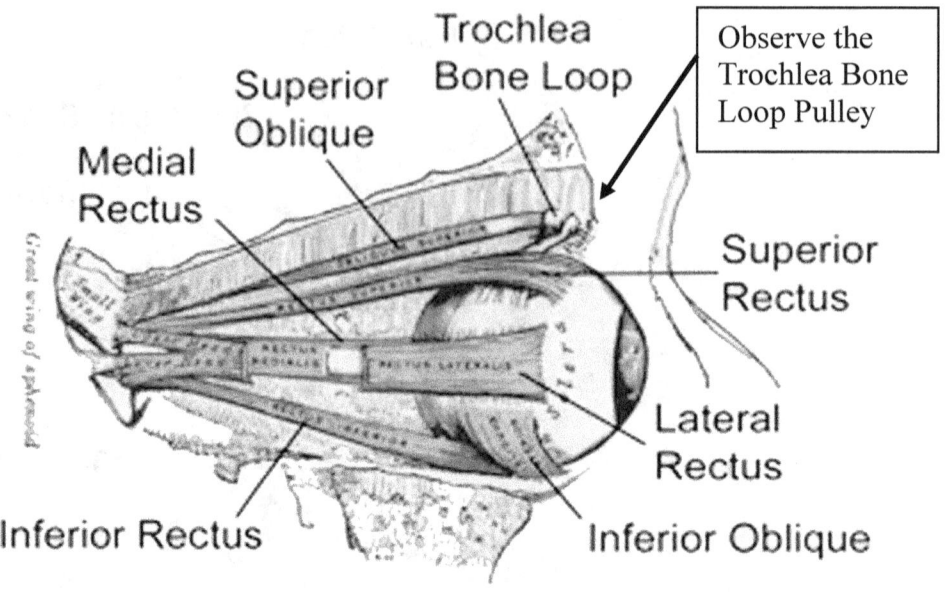

Evolution; Eye muscles and ligaments have not yet evolved, eyeball not connected!

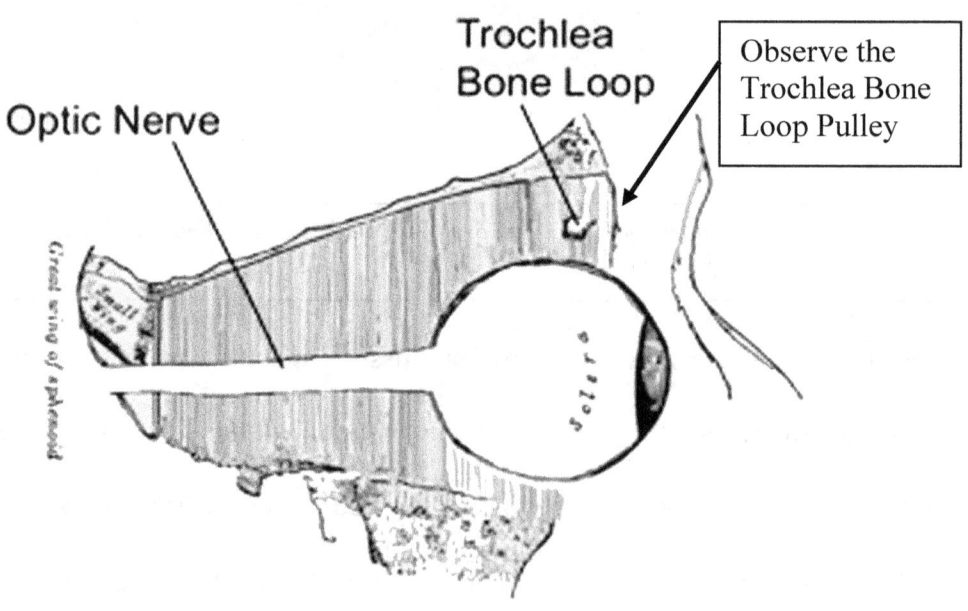

From Henry Gray, F.R.S, *Anatomy of the Human Body* 20th Edition, Lea & Febiger (Philadelphia & New York, © 1918) Page 1022[4] with modifications and additions by Rigney

Without bones there would be no way for the eyes to move. Without bones there would be nothing for the eye muscles and eyelids to connect to. Without the foramen (holes in the bones) the arteries and veins could not supply the necessary blood flow for the tissues to survive. Without the foramen, the nerves could not supply the necessary messages (impulses) for the structures to function. They all had to be present *simultaneously*; they could not have evolved because they *all* **must** be present. Their interdependence and simultaneous presence is evidence of creation; I.E.C.

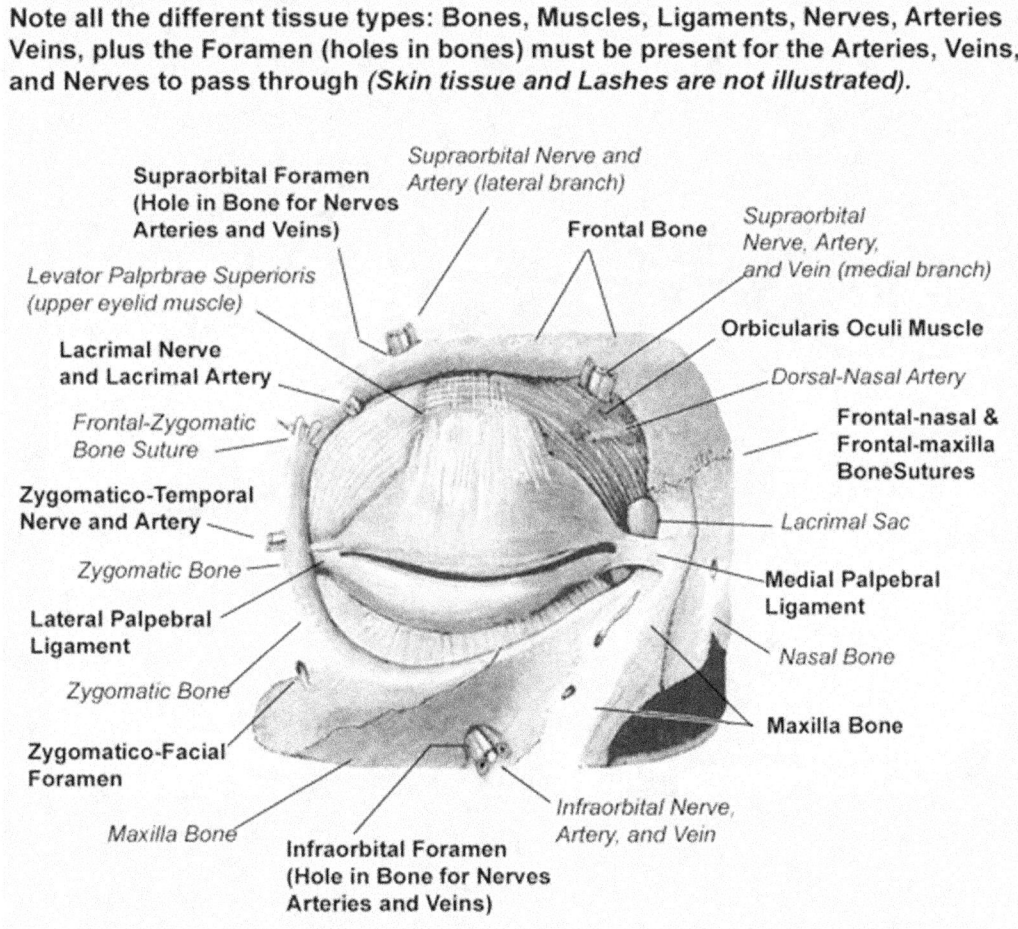

Note all the different tissue types: Bones, Muscles, Ligaments, Nerves, Arteries Veins, plus the Foramen (holes in bones) must be present for the Arteries, Veins, and Nerves to pass through *(Skin tissue and Lashes are not illustrated).*

From Henry Gray, F.R.S, *Anatomy of the Human Body* 20th Edition, Lea & Febiger (Philadelphia & New York, © 1918) page1027[5] with modifications and additions by Rigney

13. Foramen
(Pronounced for-A-men)

The brain is enclosed within the hard bones of the skull. The eyes are outside of the skull; they are not within the skull. The eyes are surrounded (except the front) by hard bones. Think about that. How could the eye evolve through hard bones? Or, how could hard bone tissue evolve to become soft eye tissue? Darwin said, "If it could be demonstrated that any complex organ existed, which could not possibly have been formed by numerous successive, slight modifications, my theory would absolutely break down."[1] Again, I must emphasize, he wrote this, so we don't have to guess or interpret what he meant, and thankfully he intentionally used the word ***ABSOLUTELY*** when he wrote this statement.

There are seven different bones which surround the eye to *Create* the eye socket. This intelligent design provides protection for the eye, yet enables the eye to function optimally. There must be holes (foramen) in the bones for arteries, veins, and nerves to pass from inside the skull to supply the eye, which is outside of the skull. The seven hard bones around each eye and the skull itself must have eighteen different holes for the soft arteries, veins, nerves, and tear ducts to pass through in order to provide the eye with the nutrients (from arteries and veins) and the nerves necessary for the eye to stay alive and function properly.

There are twelve Cranial Nerves. Cranial Nerves are nerves, which originate directly from the brain and then proceed to their specific end-organ, whereas Spinal Nerves supply their organs from the Spinal Cord. The nose, the tongue and vocal cords, the ear, the jaw muscles, the muscles of the face and throat, the lungs and digestive tract, the muscles of the neck, shoulder, and tongue, and the eye are examples of organs and structures supplied by the twelve Cranial Nerves. Of the twelve Cranial Nerves, six Cranial Nerves have direct connections to the eye. The foramen **_MUST_** be present for the nerves, arteries, and veins. The bones, eye, arteries, veins, tear ducts, nerves and foramen demonstrate Interdependent Evidence of Creation, I.E.C.

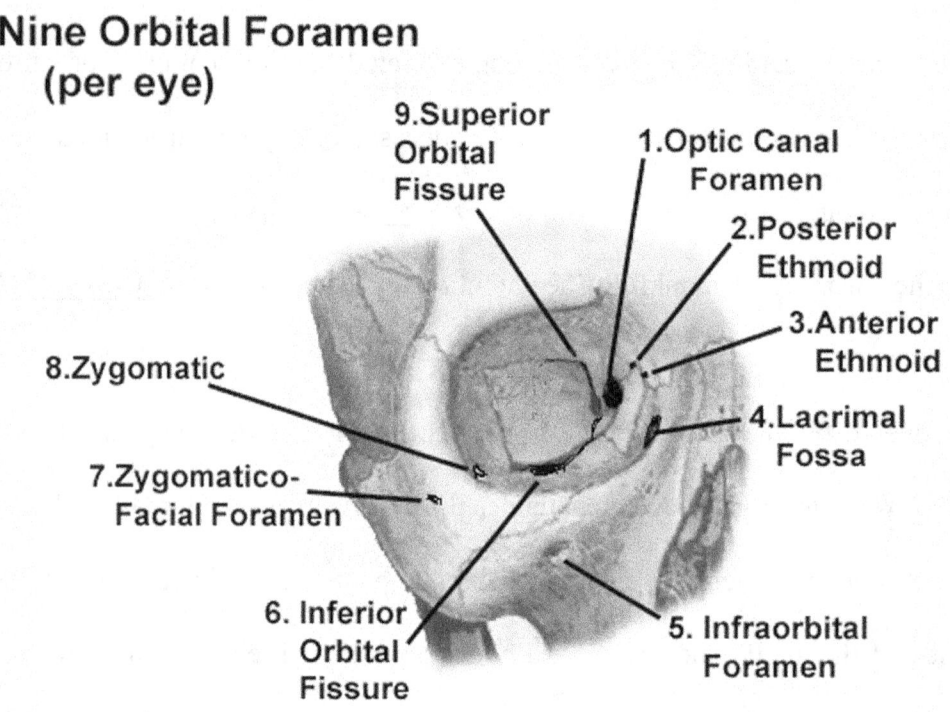

Nine Orbital Foramen (per eye)

9.Superior Orbital Fissure
1.Optic Canal Foramen
2.Posterior Ethmoid
3.Anterior Ethmoid
8.Zygomatic
4.Lacrimal Fossa
7.Zygomatico-Facial Foramen
6. Inferior Orbital Fissure
5. Infraorbital Foramen

From Henry Gray, F.R.S, *Anatomy of the Human Body* 20th Edition, Lea & Febiger (Philadelphia & New York © 1918) Page 186[2] with modifications and additions by Rigney

There are two eyes so we observe 14 bones, 12 nerves, and 35 foramen. ALL must form exactly, completely, totally, and simultaneously in order for the eye to live and function. They are all interdependent. Not one structure can function without any other structure. Each function must be fully completed for all of the components to work ***simultaneously.*** Not *one*, can wait on <u>ANY</u> *other* to evolve. We observe Interdependent Evidence of Creation! Darwin's theory "absolutely breaks down."[3]

The 12 Cranial Nerves

Of the 12 Cranial Nerves, 6 (2,3,4,5,6,& 7) connect directly to each eye *from each side.*
A total of 24 cranial nerves, 12 connect directly to the eye !

From Henry Gray, F.R.S, *Anatomy of the Human Body* 20th Edition, Lea & Febiger (Philadelphia & New York, © 1918) Page 817[4] with modifications and additions by Rigney

In addition to the nine Orbital Foramen *per eye,* there are nine ADDITIONAL FORAMEN OF THE SKULL *–PER EYE!* There are 18 holes (foreman) necessary, *per eye,* which ***must*** be in the bones of the entire skull for the eyes to function normally. All together there are *thirty-five* holes -

foramen[1,2,3] in the skull for both eyes. (The Foramen Magnum is one foramen which serves both eyes for the pupil dilator nerves; there is one foramen for each eye for the other seventeen foramen.)

The eighteen foramen are as follows:

1. Optic Foramen – penetration of Optic Nerve

2. Posterior Ethmoidal Foramen – Posterior Ethmoidal Nerve and Artery

3. Anterior Ethmoid Foramen – Anterior Ethmoidal Nerve and Artery

4. Lacrimal Fossa - Tear ducts to the inside of the nose

5. Infraorbital Foramen – Infraorbital Nerve, Zygomatic Nerve, Pterygopalatine Ganglion, Inferior Ophthalmic Vein

6. Infraorbital Fissure – Infraorbital Nerve, Zygomatic Nerve, Inferior Ophthalmic Vein

7. Zygomatico-Facial Foramen – Zygomatico-Facial Nerve and Artery

8. Zygomatic Foramen – Zygomatico-Facial Nerve and Artery (first penetration)

9. Superior Orbital Fissure- Oculomotor Nerve, Trochlear Nerve, Abduscens Nerve, and Ophthalmic division of Trigeminal Nerve

10. Foramen Rotundum – Maxillary division of Trigeminal Nerve (Indirectly connected to Eye via Trigeminal Nerve)

11. Foramen Ovale- Mandibular division of Trigeminal Nerve (Indirectly connected to Eye via Interconnections with the Trigeminal Nerve)

12. Foramen Lacerum – Internal Carotid Artery to Ophthalmic Artery for Blood Supply to the Eye, also Sympathetic Nerve Fibers to Pupil Dilator Muscle

13. Foramen Spinosum- Middle Meningeal Artery and Vein - blood supply to Optic Tract

14. Innominate canal - Lesser Superficial Petrosal Nerve input to lacrimal gland.

15. Internal Acoustic Meatus- Facial and Auditory Nerves serves Facial Muscles around Eye

16. Jugular Foramen – Glossopharyngeal Nerve, Vagus Nerve, Spinal Accessory Nerves and Transverse Sinus (and ophthalmic vein blood supply)

17. Hypoglossal Foramen –Hypoglossal Nerve (Indirectly connected to Eye via Interconnections with the Trigeminal Nerve)

18. Foramen Magnum – Sypmpathetic Nerve Fibers to the Pupil (One Foramen)

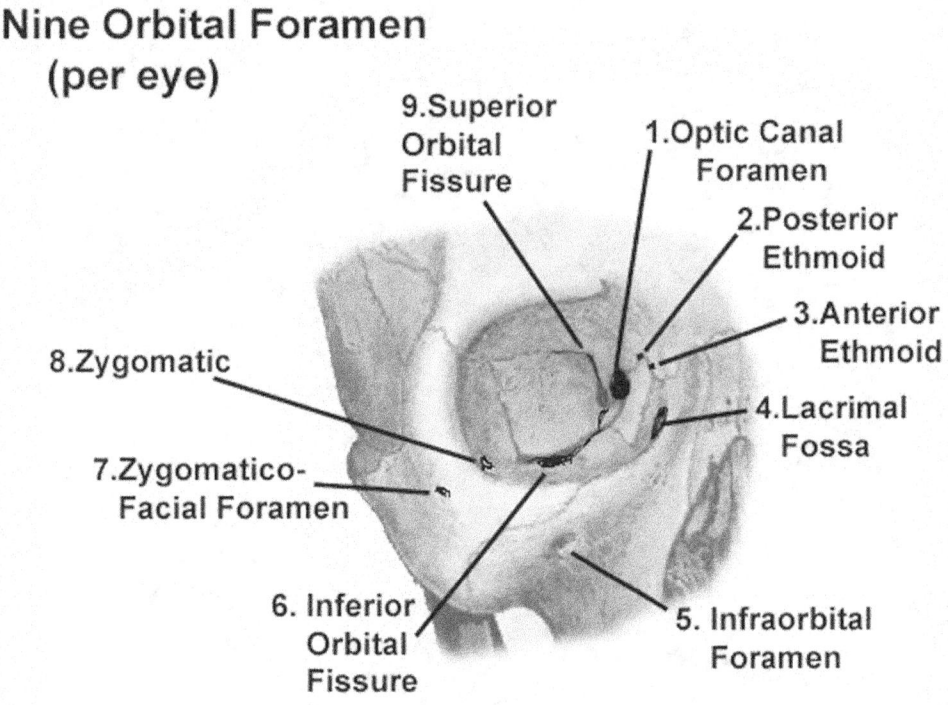

Nine Orbital Foramen (per eye)

9.Superior Orbital Fissure
1.Optic Canal Foramen
2.Posterior Ethmoid
3.Anterior Ethmoid
8.Zygomatic
4.Lacrimal Fossa
7.Zygomatico- Facial Foramen
6. Inferior Orbital Fissure
5. Infraorbital Foramen

From Henry Gray, F.R.S, *Anatomy of the Human Body* 20th Edition, Lea & Febiger (Philadelphia & New York, © 1918) Page 186[5] with modifications and additions by Rigney

9 foramen in the orbit of each eye !
18 Foramen must properly form for both eyes. If the foramen evolved, the eye cannot function until they have *ALL* evolved !

The Foramen prove Interdependent Evidence of Creation !

Base of inside of Skull as seen from above.

There are 9 additional Foramen of the Skull: #9-#18
(Foramen #1 is both an orbital foramen, and a foramen of the skull.)

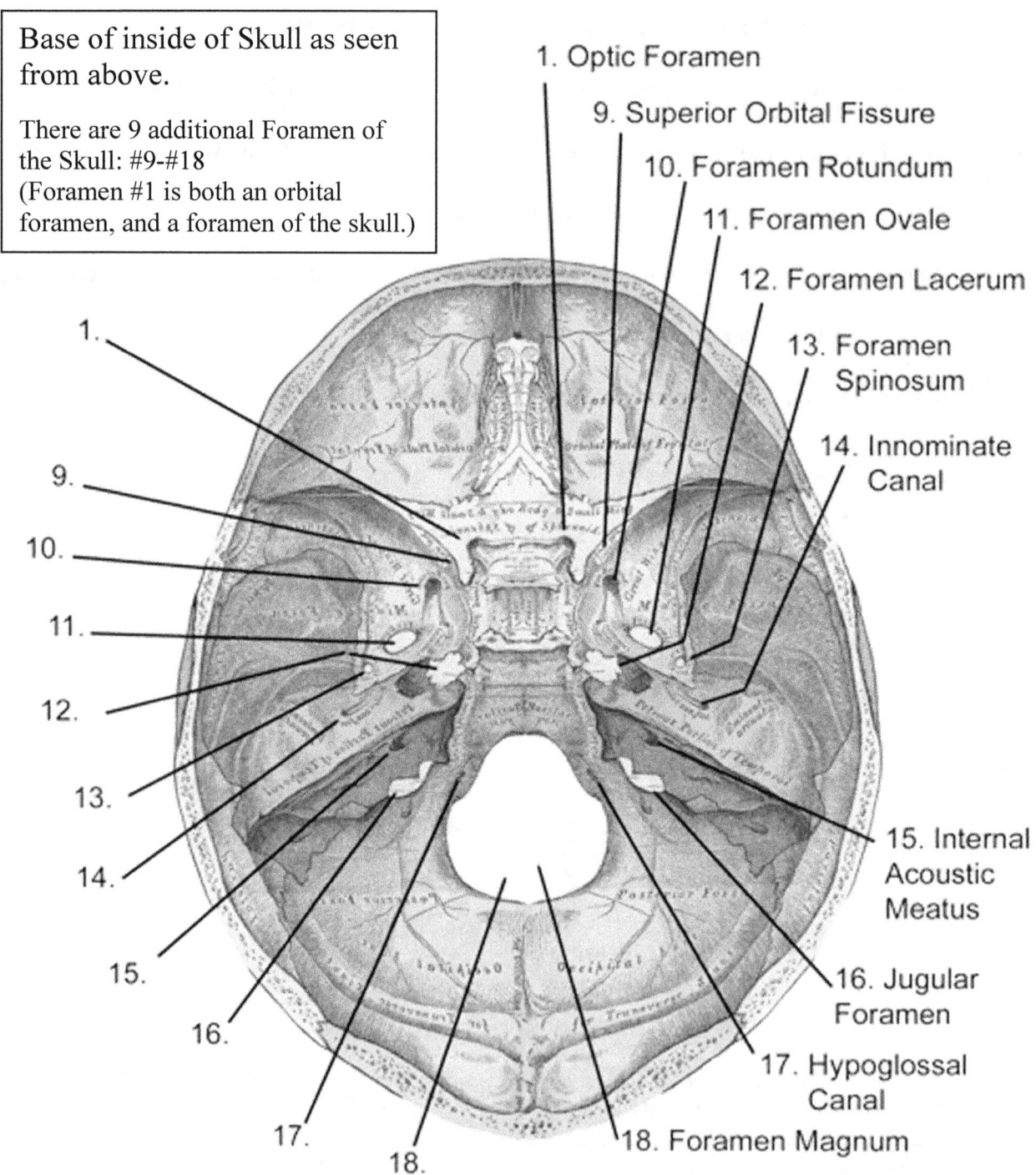

1. Optic Foramen

9. Superior Orbital Fissure

10. Foramen Rotundum

11. Foramen Ovale

12. Foramen Lacerum

13. Foramen Spinosum

14. Innominate Canal

15. Internal Acoustic Meatus

16. Jugular Foramen

17. Hypoglossal Canal

18. Foramen Magnum

From Henry Gray, F.R.S, *Anatomy of the Human Body* 20th Edition, Lea & Febiger (Philadelphia & New York, © 1918) Page 191[6] with modifications and additions by Rigney

Creation: The Bones, Foramen, Arteries, Veins, Nerves, Lacrimal Gland, and tear ducts are all interdependent and ***must*** form simultaneously. Not *one* can survive without *all* the others.

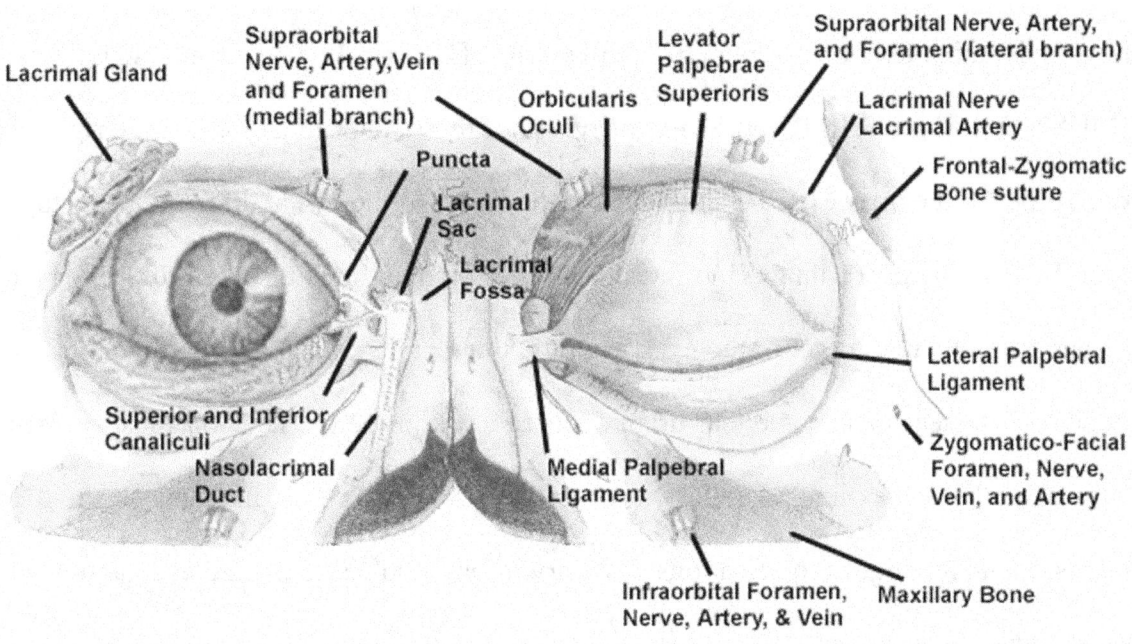

From Henry Gray, F.R.S, *Anatomy of the Human Body* 20th Edition, Lea & Febiger (Philadelphia & New York, © 1918) Page 1026, 1027, and 1029[7] combined with modifications and additions by Rigney

We observe ***Interdependent Evidence of Creation:*** The 1. muscles are dependent on the 2. ligaments, the ligaments are dependent on the 3. bones, the bones are dependent on the 4. arteries, the arteries are dependent on the thirty-five different foramen. Not *one* can exist without the other. They all must be present and they all must be present at the same time. The parts and structures are totally and completely interdependent on ALL the other structures. Each structure's function MUST be fully functional because their functions are totally interdependent, ***AND*** all the structures must be present and fully functional ***at the same time.*** This can ONLY be explained by Creation! For any creative process; all the parts must be made, all must be present, all must be assembled - *ALL* <u>at the same time.</u> When these conditions are met, the item has been created.

For example, think about the foramen. THERE ARE THIRTY-FIVE HOLES ALL TOGETHER THAT MUST FORM IN THE BONES. *All thirty-five* of these holes must form at the right place and at the right time for the arteries to be able to carry the blood to the eye and for the veins to carry the used blood back to the heart, lungs, and intestines. The foramen must also be present for the *twelve* different nerves to be able to get to the eyes from the brain. There must also be the holes (foramen) in the bones for the tear glands and tear drain ducts to work properly. They *all* must be present and fully formed at the same time—all thirty-five of them. They "could not have been formed by numerous, successive, slight modifications" as Darwin has said[8] because the eye cannot live while they are evolving.

The nerves, arteries, veins and tear ducts are soft tissue. Yet they pass through the hard bones. Soft tissue cannot *evolve* through hard tissue! The foramen must all be present, all at the same time! Again, Darwin's theory "absolutely breaks down."[8]

If the thirty-five different openings in the fourteen different bones were not present - or if the arteries, veins, nerves, and tear ducts did not form correctly to pass through the openings - and if they were not *all* present – all at the same time, the eye could not function and it would die. We observe *Interdependent Evidence of Creation,*

Evolution: In this example, the eye did not have adequate blood supply due to the arteries and veins not evolving when they should. Improper blood supply has caused abnormal development, tissue distortion and a sick lifeless eye. (Note from the Author: I left the trochlea bone-loop-pulley, and superior oblique muscle loop intact because they are evidence of creation. They could not have evolved)

Superior Oblique Muscle   Trochlea Bone Loop

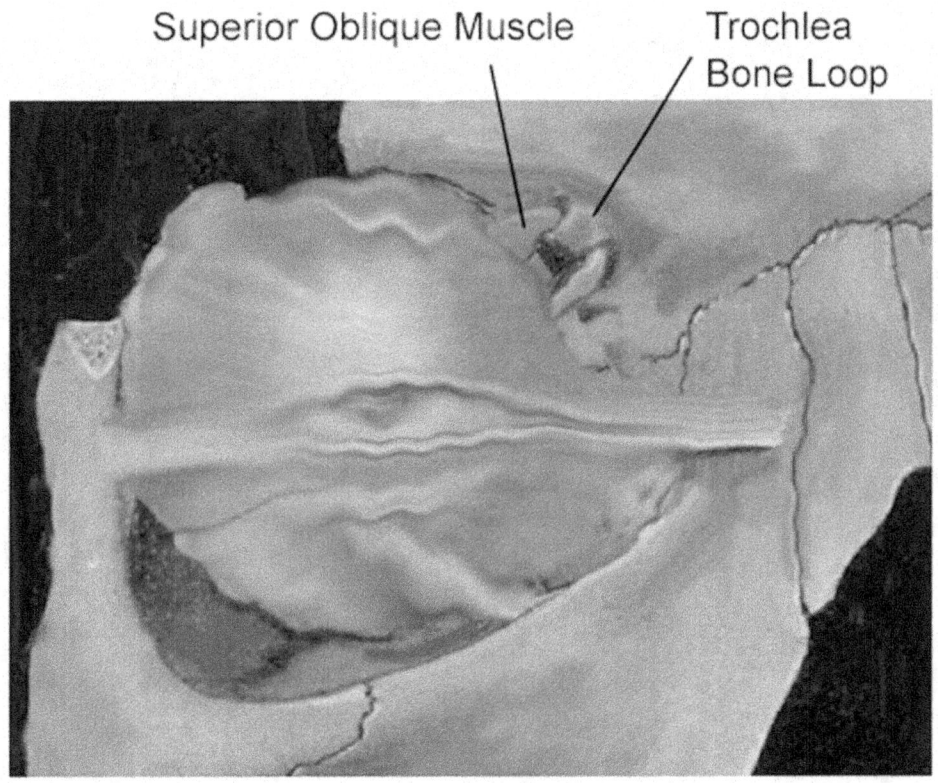

From Henry Gray, F.R.S, *Anatomy of the Human Body* 20th Edition, Lea & Febiger (Philadelphia & New York, © 1918) Page 186, p.1026, p. 1027 & p.1029[9] with modifications and additions by Rigney

14. Vitreous

The vitreous is the fluid-like gel which fills the back three-fourths of the eye. The vitreous helps keep the very delicate retina supported on the inside wall of the eye.[1] (This is important because the makeup of the retina is very delicate. The retina's makeup is about the same likeness as wet toilet paper). The *way* the vitreous is created—that is, the makeup of the vitreous, its viscosity, (how thin or thick of wateriness), and the arrangement of the tiny strands of fibers within the vitreous provides support to the retina during head and eye trauma. If the vitreous was thinner and more watery, it would not provide enough support during injuries. If it were thicker, the eye would not have enough *give* during direct injury. If the vitreous were too thin, a direct blow to the eye would be transmitted inward in a shockwave to the delicate retina, damaging the retina. The vitreous is not too thick and not too thin. How can the tissue of the eye *evolve* into a hollow void filled with fluid? Why is it not filled with blood, or bone, or fat, or air? Why is the density of the vitreous just right? Is it *Evolution or Creative intelligent design? You <u>choose</u> to <u>believe</u> one or the other.* We observe Interdependence of the Vitreous, Sclera, and Retina; 3 I.E.C.

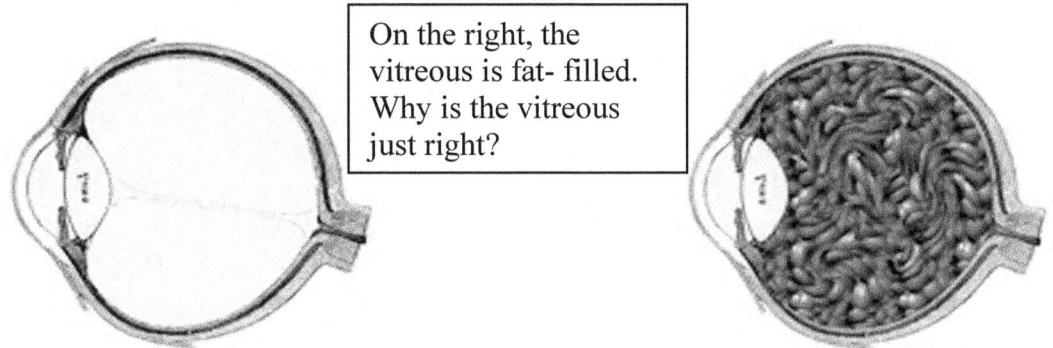

On the right, the vitreous is fat- filled. Why is the vitreous just right?

From Henry Gray, F.R.S, *Anatomy of the Human Body* 20th Edition, Lea & Febiger (Philadelphia & New York, © 1918) Page 1006[2], modifications and additions by Rigney

15. Sclera

The sclera is the white wall of the eye which makes up the back three-fourths of the eyeball. The *sclera* supports the *blood supply* the *choriocappilarris*, and the *vena voticosa* (artery-network, and vein-network respectively) to the *retina.*[1] The sclera also supports the six *muscles* of the eye which are attached to it. The sclera supports the *ciliary muscle* which focues the lens and holds the *lens* in place. Again, they *all* must be present, fully formed, and fully functional at the same time. The cornea, the retina, the six muscles, the lens, the ciliary muscle, the ciliary body, the choriocappillaris are all interdependent upon the sclera. They are interdependent on each other for the eye to function. We observe interdependent evidence of *Creation!*

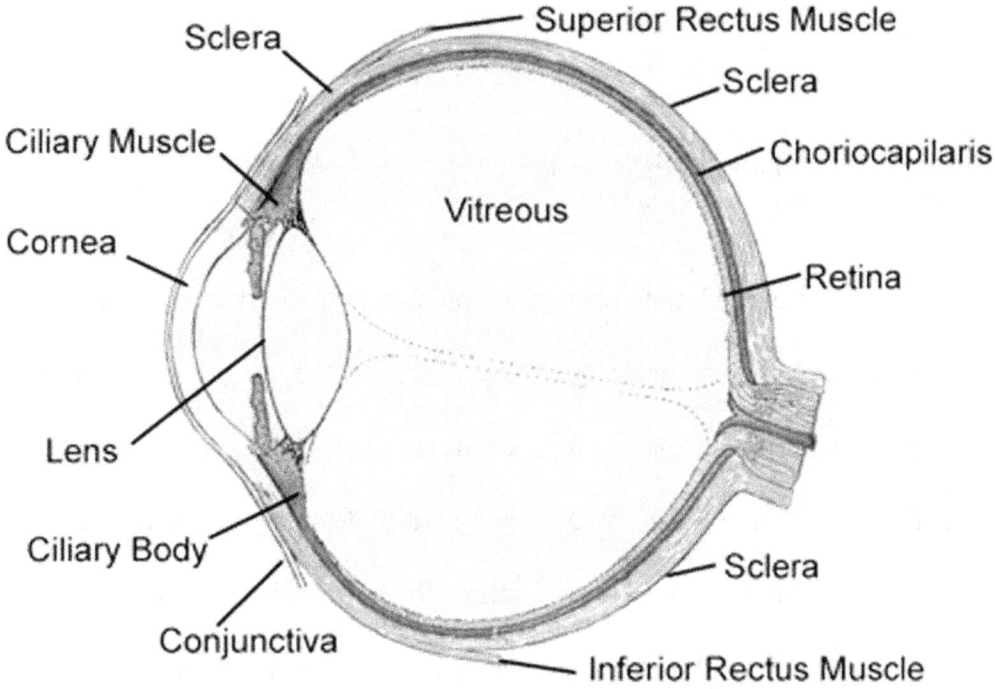

From Henry Gray, F.R.S, *Anatomy of the Human Body* 20th Edition, Lea & Febiger (Philadelphia & New York, © 1966) Page 1006[2] with modifications and additions by Rigney

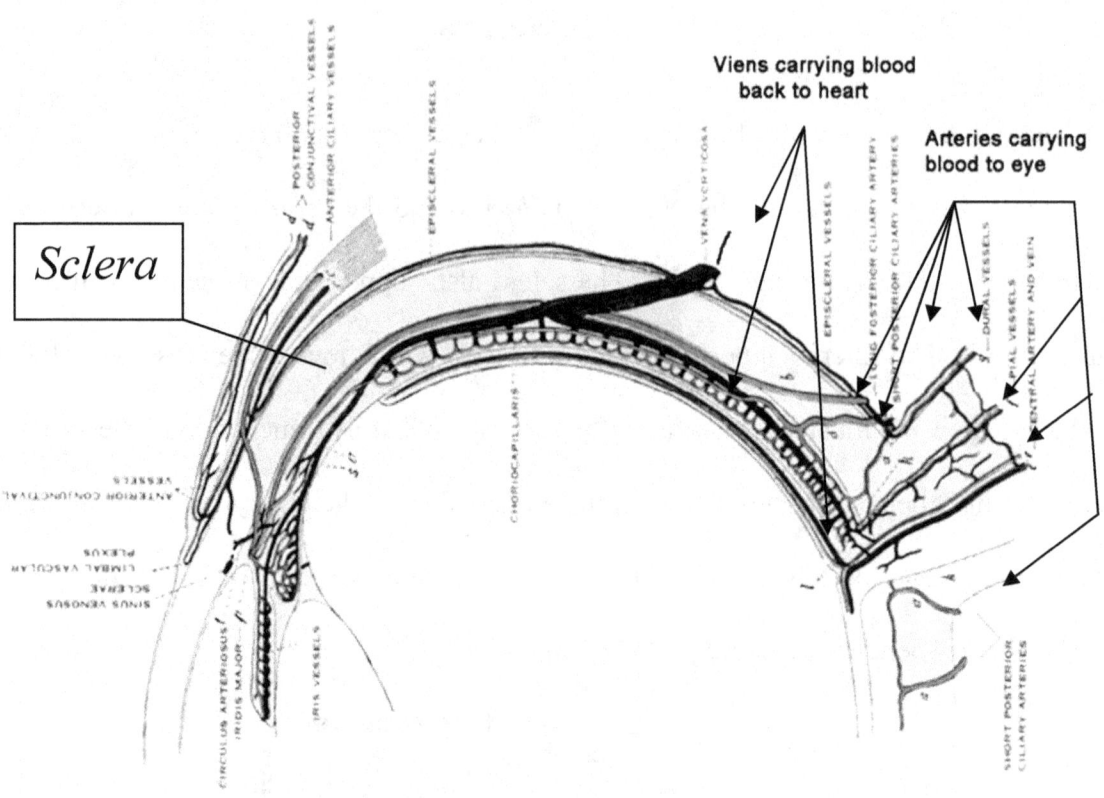

Sclera

From Henry Gray, F.R.S, *Anatomy of the Human Body* 20th Edition, Lea & Febiger (Philadelphia & New York, © 1918) Page 1012[3] with modifications and additions by Rigney

The retina must create a picture, then change the picture, and continue creating a new picture and update as the scene changes, every moment of every second the eye is open. For example, while driving 70 miles per hour in a car, the scene is changing every fraction of a second. The retina is creating and updating and changing the scene every fraction of a second. For the cells of the retina to create the pictures as they happen, they must continually be working. For cells to work they need nutrients and oxygen. As they work they create waste products. The retina must have a tremendous network of blood vessels (the Choriocapillaris) to supply the retina with the nutrients and oxygen it needs. The sclera provides the structure necessary for the blood vessels and the retina. The sclera, retina, chroriocapillaris, and vena vorticosa are all absolutely and totally interdependent. One cannot wait on the other to evolve. They must be simultaneously present and their function must be totally completed. Therefore, we again observe Interdependent Evidence of Creation, I.E.C.

The same applies to the six muscles which attach to the eye and the bones of the eye socket. The muscles are dependent on the sclera – which is dependent on the arteries and veins – which are dependent on the foramen – which are dependent upon the bones. The eye is dependent upon the bones for support and protection. The eye is dependent on the foramen in the bones because it must have blood flow. The eye is dependent on the lids, dependent on the bones and dependent on foramen. They are dependent upon each other. No one part can wait on the other part to evolve. They all must be present and their function must be complete and ALL simultaneously. EVOLUTION CANNOT EXPLAIN THEIR REQUIRED TOTALLY COMPLETED TOTALLY INTERDEPENDENT FUNCTION. It is Interdependent Evidence of Creation! I.E.C.

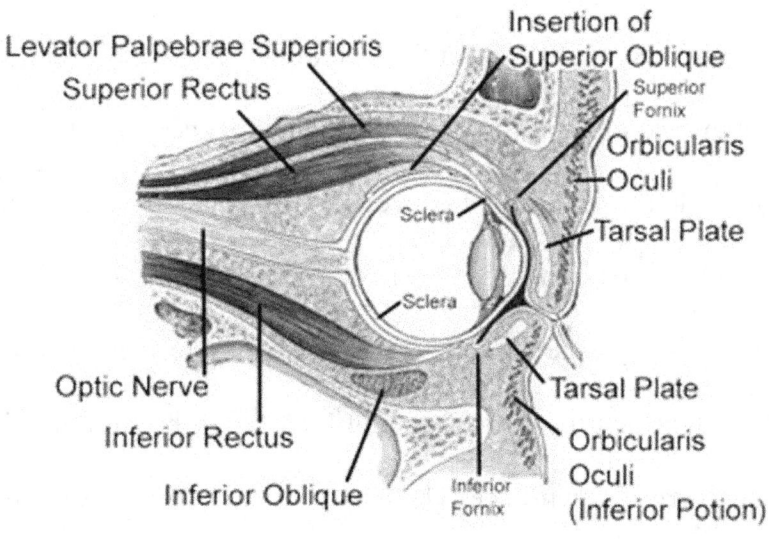

From Henry Gray, F.R.S, *Anatomy of the Human Body* 20th Edition, Lea & Febiger (Philadelphia & New York, © 1918) p.1021[4]

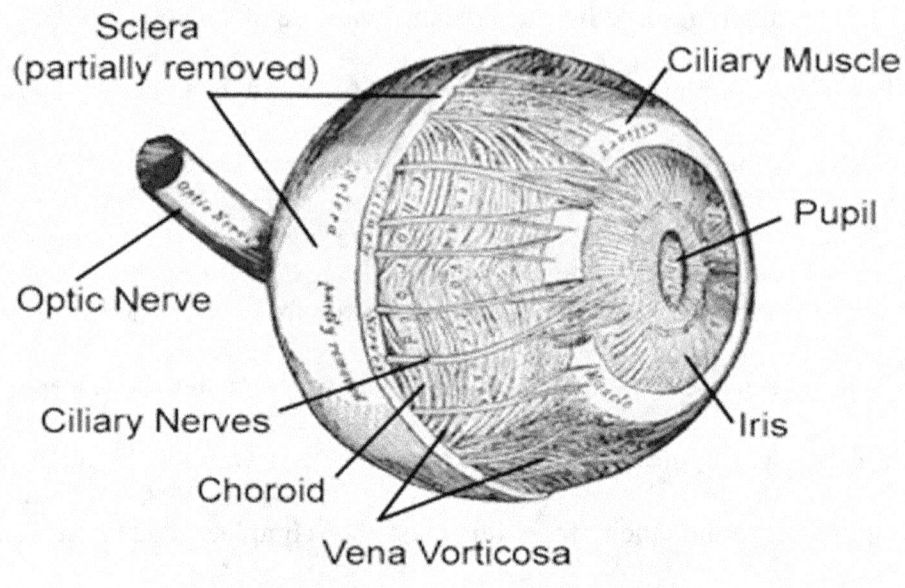

All the inner structures and the Cornea (not shown)
are absolutely dependent upon the Sclera.

From Henry Gray, F.R.S, *Anatomy of the Human Body* 20th Edition, Lea & Febiger (Philadelphia, Pennsylvania, © 1918) Page 1009[5] with modifications and additions by Rigney

Even more evidence of creation is the fact the sclera must be a very precise and exacting length. The precision is to the level of one millimeter or about the thickness of a credit card. If the globe of the eye provided by the sclera is too long, the eye is out of focus for far vision. If the globe of the eye provided by the sclera is too short the eye is out of focus for close vision. The length of the sclera must be *just right!* Only when the sclera is the exact correct length is the eye in *focus*! "Numerous successive slight modifications"[6] cannot be at work because a being cannot survive to propagate the next generation if it can't see. The length of the sclera must be exact and is critical to the thickness of a credit card. Again, Darwin's theory *"absolutely breaks down."* [6]

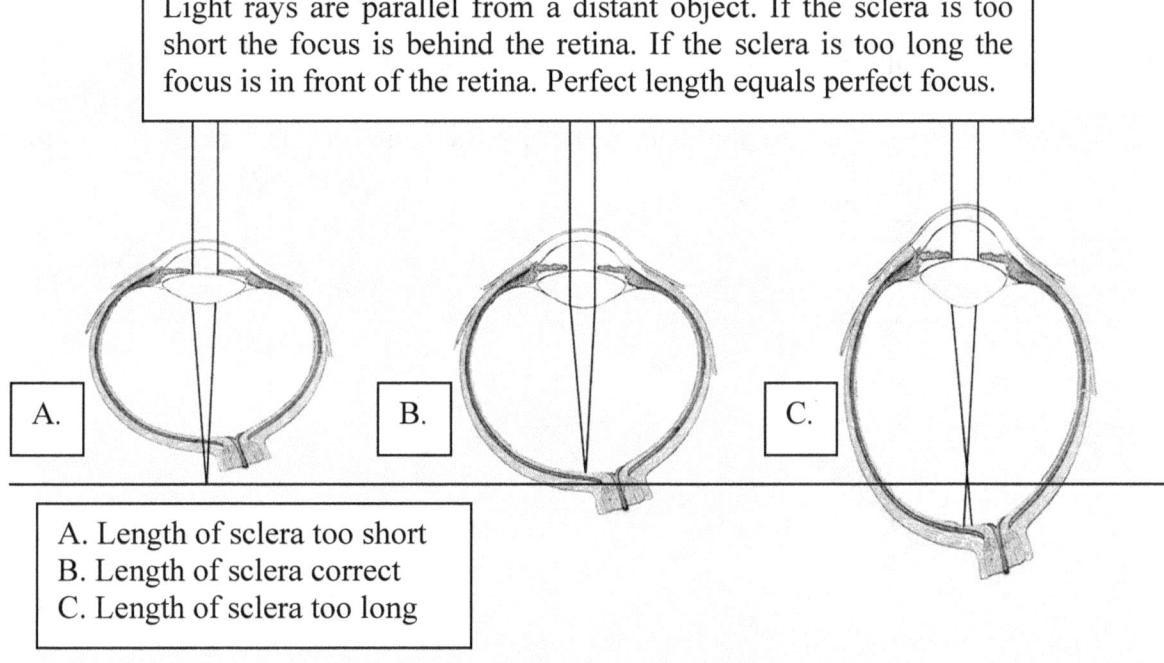

Light rays are parallel from a distant object. If the sclera is too short the focus is behind the retina. If the sclera is too long the focus is in front of the retina. Perfect length equals perfect focus.

A.   B.   C.

A. Length of sclera too short
B. Length of sclera correct
C. Length of sclera too long

From Henry Gray, F.R.S, *Anatomy of the Human Body* 20th Edition, Lea & Febiger (Philadelphia & New York, © 1918) Page 1006[7] with modifications and additions by Rigney

One millimeter too long or too short in the length of the sclera (or axial length of the eye) causes the eye to be about thee diopters out of focus and results in visual acuity of about 20/100 and a visual efficiency of about 48.9 percent.[8] Another way to say this is, (if the focus of the lens is relaxed and the cornea, aqueous, lens, vitreous, and retina are normal, that is. if there is no other eye problem) when the length of the sclera is about **1 millimeter** too long or too short, the vision is about 20/100 and about 50% blurred.[8] When the sclera is about *2 millimeters* too long or too short, the eye is about six diopters out of focus, and would have 20/400 vision, and be 96.7% blurred![8] (20/400 vision is considered "legally blind" which means the person can see light and dark, and gross shapes - but no details can be seen.). So, if the sclera was 2 millimeters or about the thickness of a dime too long or too short the person is legally blind.

Could the eye evolve to become the exact length? How could generations of beings function, while the eye was evolving? This is *additional evidence* of Creation!

16. Retina

The retina is made up of ten different layers of tissues and cells put together and arranged in a way to receive or *catch* light.[1] The retina contains the Photoreceptors—the *Rods* and *Cones* which receive the light rays and convert them into electrical impulses and send those impulses to the brain.[2]

Rods are present for night vision; Cones are present for day vision. Again, Rods and Cones *catch* and convert light rays or light particles into electrical impulses. When a Rod or Cone *catches* a light ray or light particle, the light causes the Rod or Cone's outer tail portion—*but only the outer tail portion*—to disintegrate or break apart.[2] This outer tail portion is called the Outer Segment. As the outer tail portion breaks apart, it causes a *spark of electricity,* (an impulse) to be created. Then the Rod or Cone sends this *spark of electricity* (impulse) to the Optic Nerve, which carries the *spark of electricity* (impulse) to the Brain.

As light continually enters the eye, every moment of every second, the Rods and Cones are continually disintegrating their outer tail portions (Outer Segments) with new inner portions being produced. The new inner portions are continually sliding down in such a way in which the new inner portions become new outer portions. (Somewhat like pushing toothpaste down and out of the tube.) As the outer portions continue to disintegrate, they continue sending impulses, and sparks of electricity, to the Brain.

The Rods and Cones are designed to disintegrate *at the tails* of the Outer Segments of the rods and cones; the body or the inner segments remain to rebuild new tails and new Outer Segments. This design keeps the rods and cones continually working without wearing out. Again, this *design* requires *foresight* and *preventive planning.* How is it *only* the outer tail portions disintegrate? How could the inner tail portions (inner segments) evolve to become self-replenishing? They had to be that way from the very beginning otherwise after only a few seconds they would wear out and the being could no longer see. If they *evolved* to become functioning – until they evolved the being cannot under ANY circumstances see. The being would be blind! ALL the components must be present for vision to occur. Leave any **<u>one</u>** component out and the being cannot see. Evolving to see just cannot happen by "numerous, successive, slight modifications[3]" - especially when we observe the retina. And; it is evident - ***<u>obviously evident,</u>*** Darwin did not know or understand this process when he states, "we ought in imagination to take a thick layer of transparent tissue, <u>with a nerve sensitive to light beneath.</u>"[4] This statement is in absolute error. The retina is "sensitive to light" not the nerve. The nerve transmits the impulses which the retina has converted from light.

Taking steps to stop something from possibly happening is *preventive planning.* Preventive planning demonstrates intelligent design, and is itself evidence of creation!

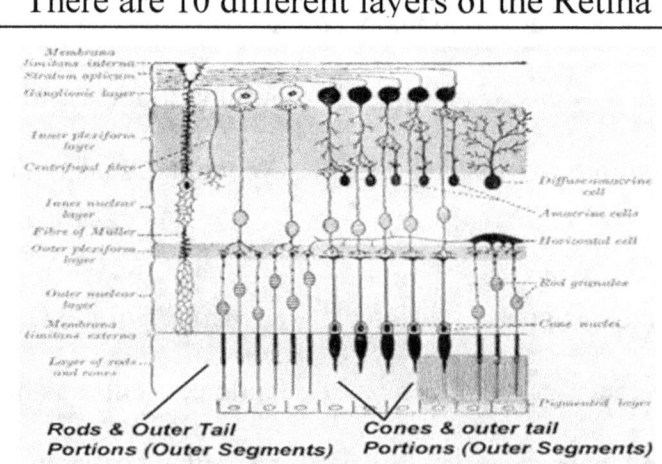

There are 10 different layers of the Retina

Rods & Outer Tail Portions (Outer Segments) **Cones & outer tail Portions (Outer Segments)**

From Henry Gray, F.R.S, *Anatomy of the Human Body* 20th Edition, Lea & Febiger (Philadelphia & New York, © 1918) Page 1016[5]

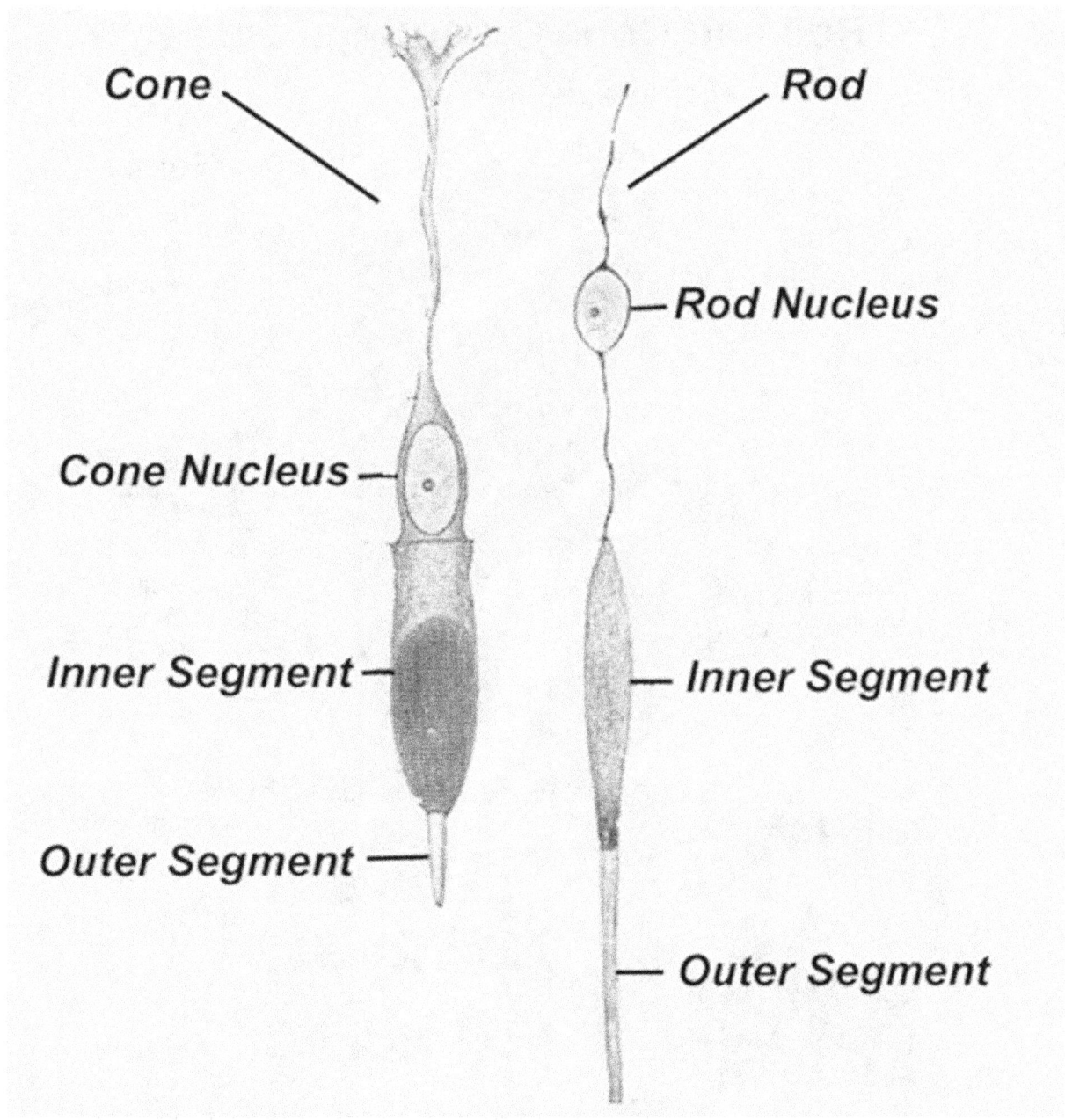

From R Greeff (1900) Handbuch der gesamten Augenheilkunde, 2nd ed, vol.1 Graefe and Saemisch, Leipzig.[6]

Rod and Cone Outer Segments

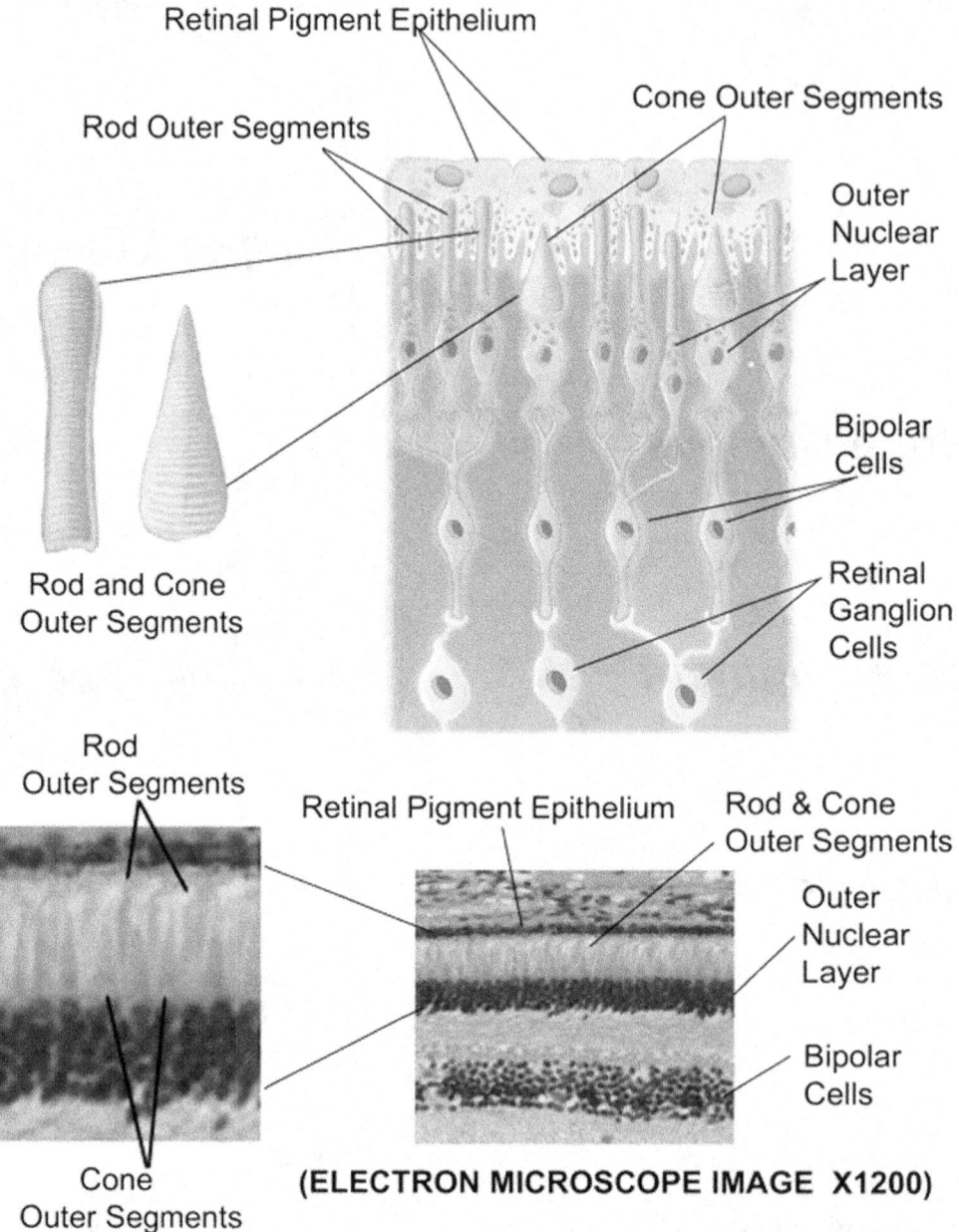

Retinal Pigment Epithelium

Cone Outer Segments

Rod Outer Segments

Outer Nuclear Layer

Rod and Cone Outer Segments

Bipolar Cells

Retinal Ganglion Cells

Rod Outer Segments

Retinal Pigment Epithelium

Rod & Cone Outer Segments

Outer Nuclear Layer

Bipolar Cells

Cone Outer Segments

(ELECTRON MICROSCOPE IMAGE X1200)

The images on this page were derived from: https://upload.wikimedia.org/wikipedia/commons/e/e8/1414_Rods_and_Cones.jpg
By OpenStax College [CC BY 3.0 (http://creativecommons.org/licenses/by/3.0)], via Wikimedi Commons.File:1414 Rods and Cones.jpg (File:1414 Rods and Cones - ru.svg)[7]
(They do not endorse me, or my use of their work)

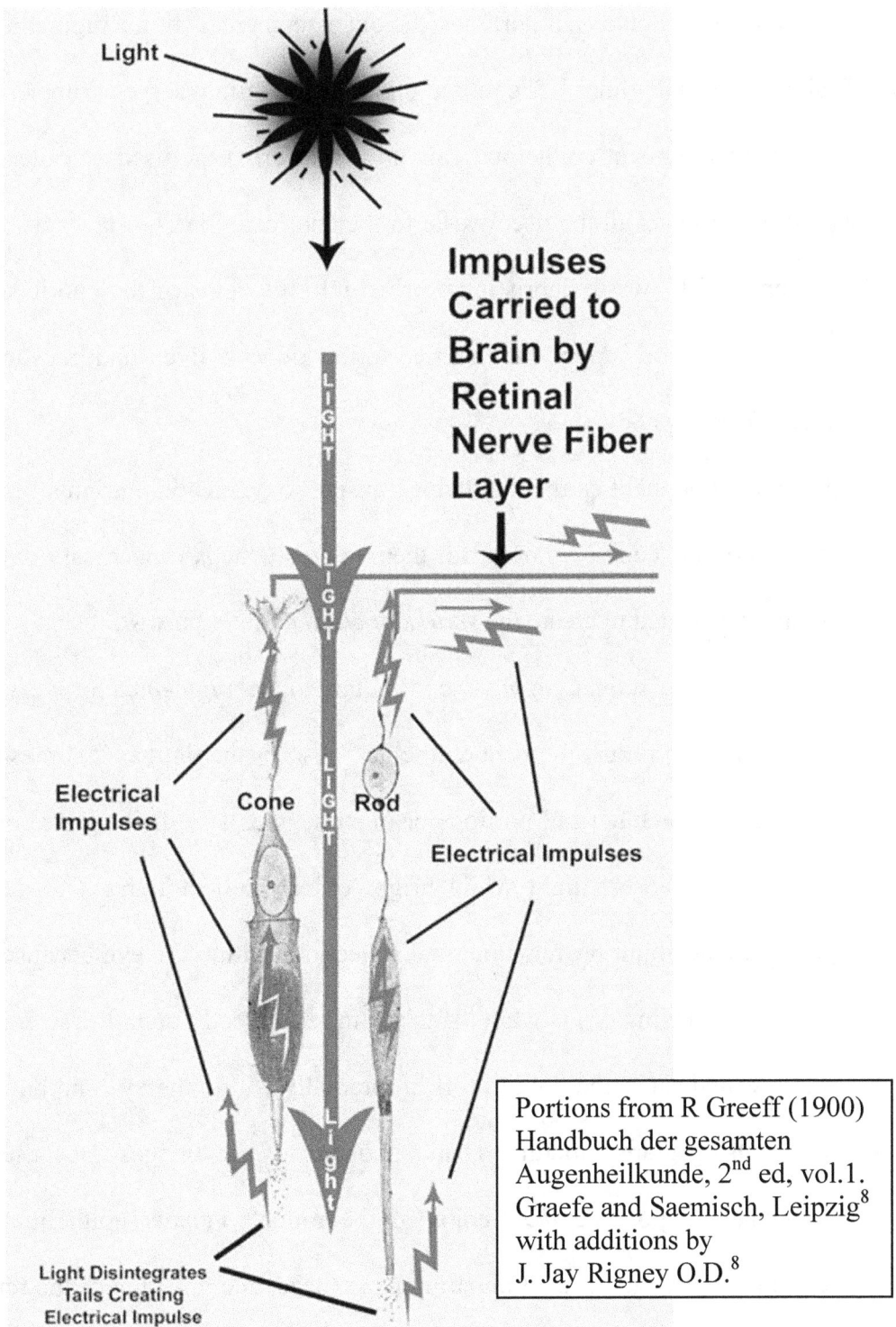

Light

Impulses
Carried to
Brain by
Retinal
Nerve Fiber
Layer

Electrical
Impulses

Cone

Rod

Electrical Impulses

Light Disintegrates
Tails Creating
Electrical Impulse

Portions from R Greeff (1900)
Handbuch der gesamten
Augenheilkunde, 2nd ed, vol.1.
Graefe and Saemisch, Leipzig[8]
with additions by
J. Jay Rigney O.D.[8]

The rod and cone's outer-tail portions (the outer segments) sit in a pigmented layer of cells called the retinal pigment epithelium.[9] The retinal pigment epithelium serves primarily three functions:

1. The retinal pigment epithelium eats up (or absorbs) the used-up outer segments[10] (left-over waste) and sends all the used waste to the choriocapillaris (a layer of blood which the retina sits upon). The waste debris is absorbed into the veins of the choriocapillaris and then into the venous blood supply, then carried to the kidneys, liver, and intestines for removal of the waste from the body.

2. The retinal pigment epithelium helps transport oxygen and nutrients from the choriocapillaris to the rods and cones[11] in order for them to create the *new* necessary outer-tail portions (outer segments) needed to create the *sparks of electricity (impulses)*.

3. God made the retinal *pigment* epithelium to be *pigmented* or very dark in color. This darkness of the retinal pigment epithelium absorbs the light as it comes into the eye (after the light causes the outer-tail portions or outer segments to disintegrate) by absorbing the light. Dark objects absorb light while bright objects reflect light. The darkness of the retinal pigment epithelium prevents internal reflections within the eye because the sclera (the tissue the retina is laying on) is white. If the retina sat directly upon the sclera, the whiteness of the sclera would cause the light to reflect internally within the eye and bounce around inside the eye, causing glare, cloudiness, and ghosting of the images. Because the retina is sitting immediately upon the *dark* color of the retinal *pigment* epithelium it prevents internal reflection within the eye, eliminating glare, haze, and ghosting of the retinal images resulting in clear vision without unwanted glare.[12] Observe the placement of the outer segments within

the retinal pigment epithelium. If the tails weren't within the retinal pigment epithelium the rods and cones could not receive the nutrients necessary to function and stay alive. Without the retinal pigment epithelium, waste products would accumulate and would have a toxic affect on the rods and cones causing them to degenerate. Do you think it just happened by chance? Evolution cannot support any of this. If the Retinal Pigment Epithelium were not present the rods and cones could not function and the eye could not see. Again, we see two tissues which are interdependent. We observe interdependent evidence of creation. I.E.C.

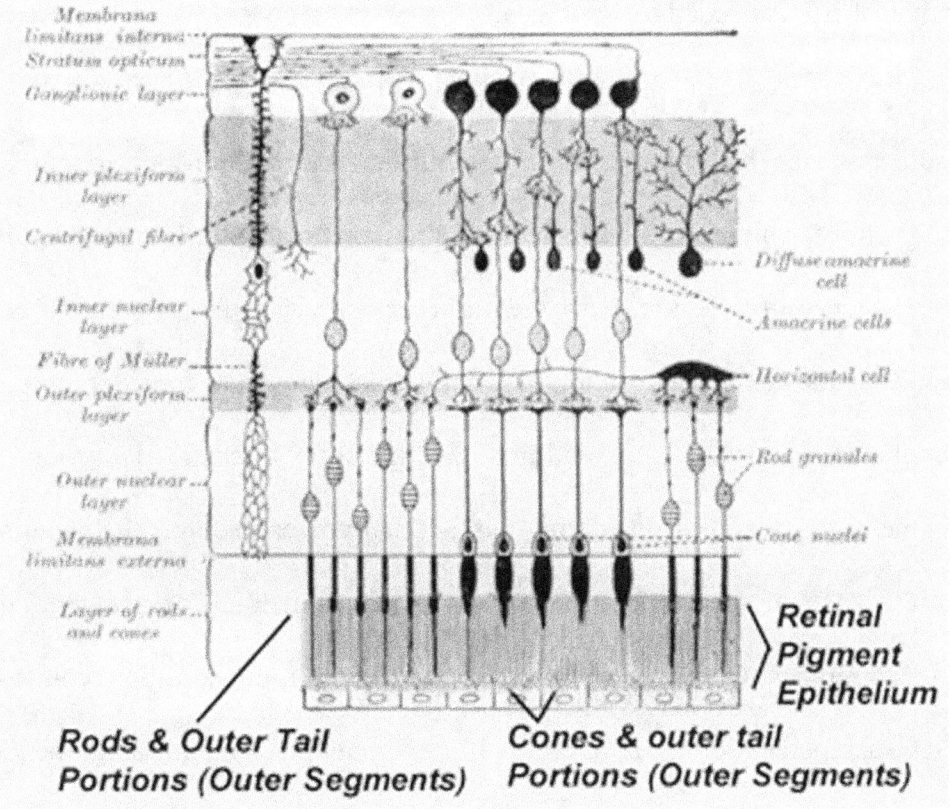

From Henry Gray, F.R.S, *Anatomy of the Human Body* 20th Edition, Lea & Febiger (Philadelphia & New York, © 1918) Page 1016[13] with modifications and additions by Rigney

The retina is made up of about 120 million rods and 6.5 million cones. That is 126.5 million rods and cones total per eye. That's 253 million rods and cones when we consider both eyes.[14] WOW! Rods are for night vision, and cones are for day and color vision. There are three different kinds of cones, which are designed for us to see the many different colors needed for color vision. There are three kinds of cones and one kind of rod, which totals to four[15].different types of photoreceptors. The retina and brain *together* mix the signals or *impulses* created by the 253 million rods and cones in order to create the *millions* upon millions of different colors, combinations of colors, shades of colors, and depths of colors needed for us to see.

The rods and cones work together with the brain so that a *spark of electricity (an impulse)* is created and somehow the brain turns all those millions and millions of impulses or *sparks of electricity* into pictures which are in the *brain* as they are in the *real* world. Again, we observe Interdependent Evidence of Creation. How the brain creates vision from sparks of electricity is not known,[16] only God knows. Darwin, and the evolutionist does not consider or contemplate the miracle of how the brain creates vision, and in so doing refuses to acknowledge God as the Creator and exchanges the truth for a lie. Considering the fact that you can see yet the experts can't tell you how, is itself evidence of God. Most people say when something is observed that cannot be explained it is a miracle. Converting light into electrical impulses, then converting electrical impulses into visual perception is a miracle. If only one piece is missing the being is blind. "Numerous, successive, slight modifications"[17] will not work when it comes to vision and perception. The eye sees, the brain perceives. Again we observe I.E.C.

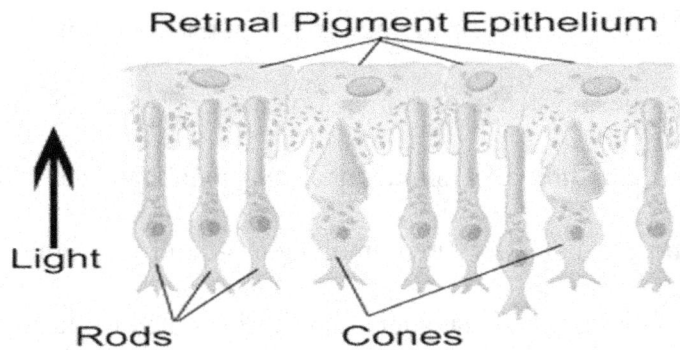

Retinal Pigment Epithelium

Light

Rods   Cones

From:https://upload.wikimedia.org/wikipedia/commons/e/e8/1414_Rods_and_Cones.jpg
By OpenStax College
[CC BY 3.0[18]
(http://creativecommons.org/licenses/by/3.0)], via Wikimedi Commons.
File:1414 Rods and Cones.jpg
(File:1414 Rods and Cones - ru.svg)[18]

An even *greater,* even more miraculous wonder is in addition to the 126.5 million photoreceptors *in each eye – 253 million rods and cones-* there are millions upon millions of additional *"Functional Retinal Cells"[19]* in each eye. These *"Functional Retinal Cells"* make the eye *function; and function automatically,[19]* whereas, the Photoreceptors make the eye *see.*[19] There are at this time (with today's level of technological investigation) at least fifteen different kinds [20,21] of *"Functional Retinal Cells."* They number around 506 million![22] So, all together there are approximately 759 million different visual processing cells in the retina of EACH EYE! That is .759 ***BILLION*** per eye for a total of 1.518 ***BILLION*** different visual processing cells <u>in just the retinas</u> of BOTH EYES.! ***1.52 BILLION!*** This number does not include the additional ***BILLIONS*** of pre-processing cells within the lateral geniculate body, or the additional ***BILLIONS*** of processing neurons within the visual cortex of the brain. Now, couple this with the fact they are all ***interdependent*** with each other, and all are working together to accomplish a specific task - VISUAL PERCEPTION! Again, something which is beyond man's comprehension![23] We observe millions, dependent upon *billions* dependent upon ***billions,*** *dependent upon* ***BILLIONS!*** The reason I am emphasizing the numbers here again, is because the cosmologists and statisticians say considering the billions of stars in the universe, the statistics are probable that there is life somewhere other than Earth. Yet those same cosmologists and statisticians when informed of the numbers of ***interdependent*** cells, tissues, systems, and organs of the eye and body; *and*, they know they are ***ALL*** not only interdependent; not only simultaneously- interdependent; not only simultaneous-totally-completed-interdependent, but they are simultaneous-totally completed-fully-functional-interdpendent. Those same cosmologists and statisticians when informed of these millions upon billions upon billions of numbers of simultaneously-totally-completed-fully-functional-interdependent cells, tissues, systems, and organs REQUIRED (not only REQUIRED, but ***ABSOLUTELY REQUIRED***) for the eye and body to function and live - don't consider the probabilities of this happening by chance. If you ignore all these facts and numbers, you CHOOSE to ignore the statistics and believe it all just happened by chance with no outside influence! You must CHOOSE TO IGNORE ALL RATIONAL THOUGHT! How is it they can look at statistics and say the odds of other life existing is *probable*, yet, they can look at all the statistical interdependent evidence of Creation, and not see that the probability of it happening by chance is NOT *PROBABLE!* In fact, somehow even with the facts and statistical evidence of Creation they can proclaim, not *theoretically* it happened by chance, but they proclaim *factually* it happened by chance as if it factually did happen by chance -even when the statistics show otherwise. However, we really should forget the statistical evidence and look at the physical evidence. The statistical evidence concludes the odds of it all happening by chance is extremely unlikely. Yet, that *implies* it could happen by chance – even though it is extremely, extremely, extremely, extremely, extremely unlikely. BUT, the evidence "demonstrates" Creation. The evidence observed is not; "a chance of statistical insignificance - yet a chance"; what is observed is ***Interdependent Evidence of Creation*** and no chance of evolution!

The 15 additional *Functional Retinal Cells*[24] of the retina are as follows:

1. *Midget Retinal Ganglion Cells* or the *singular interconnection* Retinal Ganglion Cells, Then, there are four different types of regular *retinal ganglion cells* or retinal ganglion cells with *multiple interconnections*. The regular retinal ganglion cells are of four types:

2. *Unistratified Retinal Ganglion Cells*

3. *Diffuse Retinal Ganglion Cells*

4. *Large Diffuse Retinal Ganglion Cells*

5. *Stratified Diffuse Retinal Ganglion Cells*

These four types of Retinal Ganglion Cells all have *multiple interconnections* whereas, the midget retinal ganglion cells have *singular interconnections*. Additionally, there are also the

6. *Amacrine Cells*

7. *Diffuse Amacrine Cells*

Then there are three different types of *Bipolar Cells*; see below):

8. *Rod Bipolar Cells*,

9. *Midget cone Bipolar Cells*,

10. *Flat Cone Bipolar Cells*

11. *Type A Horizontal Cells*

12. *Type B Horizontal Cells*

13. *Mueller Cells*

14. *Neuroglial Cells.*

15. Centifugal Fibers

Much of the function of these fourteen different types of *"Functional Retinal Cells"*[24] are not *exactly* known,[25] but electron microscopy does show they are present and each have distinct definite characteristics.

In summary and for simplification purposes, it is best to think of the purpose of these additional *Functional Retinal Cells* as to provide interconnections, relays, preprocessing, and processing of visual information within the retina itself[26] to serve several different *automatic functions* such as reflex activity, coordinated conjugant eye movements, and pupil reflexes.[27] Again, much of the function of these fifteen different types of additional *"functional retinal cells"* are not *exactly* known. It may be they help to preprocess the electrical sparks (impulses) before they are sent to the brain to aid in automatic functions. Wolff does note, "the retina, being derived from the optic cup is a part of the brain arising as a hollow outgrowth from the forebrain."[28] He later also says "the optic nerve is not a true nerve, but a fiber-tract connecting one part of the brain with another."[28] So, there definitely is some preprocessing and processing which occurs in the retina by these additional *"functional retinal cells."* Furthermore, when talking about the visual pathway and vision, Wolff notes that a small number of the optic nerve fibers "establish mesencephalic connections which indicate reflex activity, such as eye movements and pupillary changes."[29] Such functions I (Rigney) think,[30] possibly _may_ include the following[30] (but have not been verified[30]):

- They may help the brain interpret depth i.e. depth perception.

- They may help the brain to see in 3D.

- They may help prevent the images from fading.

- They may help the brain to *scrutinize* or study and screen wanted objects from unwanted objects (like watching a ball game through a chain link fence).

- They may help keep the eyes tracking and fixated or *locked onto* moving targets.

- They may enable the eyes to immediately jump from one object the eye is looking at then change immediately to an object seen in the periphery without having to search—changing gaze from one object directly to another object without searching or missing.

- They may help to see in different depths of intensity (or *shades of gray* and *shades of colors.*)

- They may help the eyes *scan* over objects while looking for specifically needed objects.

- They may help to return the eyes, and position the eyes, when looking from left to right and from far to near.

- They may keep the eye fixated on a target while the body is walking, running, or moving.

It may be these *functional retinal cells* guide the way our eyes move, track, search, *lock onto*, and change looking from one target to another, They may adjust the light intensity and adjust and help to perceive differing depths and shades and saturation of colors.[30] Again, much of the function of these additional retinal cells is unknown[31] so I am speculating as I am making these statements. Additionally, because they are not exactly known, I further speculate these millions upon millions of *functional retinal cells* provide the *automatic functions* we take for granted. These *automatic functions* occur without our even thinking about it. These millions upon millions of *functional retinal cells* are needed for the eyes to *function* whereas the photoreceptors (rods and cones) are needed to *see*. These *functional retinal cells* may function like internal relay switches within the retina to provide the necessary interconnecting cells for your eyes to work properly without one having to even think about it, because we do know it happens and it happens *automatically*.[31]

It is evident these functions of the additional retinal cells are planned, and preprogrammed *functions* designed into the retina.[32] Their many functions defy evolution and require foresight and planned *purposed function.*[32] The *functions* of the eye cannot wait for them to evolve; they must be present—all of them fully functional, all at once and all at the same time. If any *one* were missing while the eye was waiting for it to *evolve* the eye would not function properly and the individual could not see properly - if at all.[32]

The reason these *functions* are so very important is that if something is attacking me or thrown at me and I am not paying attention; without having to even think about it, my eye sees it and knows something is coming at me. Not only do I know it is coming at me, I know how fast it is coming at me, and I know how exactly where it is going to hit me. I know it is going to hit me on my leg, on my body, on my arm, on my foot, on my head, from the left, from the right, from the bottom, from the top, or any combination of the above. And because of these billions and billions of interconnections in the retina, and thankfully, because within the brain itself there are billions and billions of *more* interconnections from the eyes to the other neurological centers within the brain which interconnect with the neck muscles, leg muscles, arm muscles, ears, and on and on. Without thinking, I can avoid and protect myself from injuries. *Protection* and *reflexes* are words which portray *planning and prevention* which requires *foresight and knowledge.* Design and engineering are *required* for protection and planned prevention. *We know the eye and brain perform all these functions automaticaly. We observe that the eye does indeed function in this manner.* Evolution cannot explain all these internal interconnections relays and reflexes. The conclusion based upon the evidence observed is Interdependent Evidence of Creation, not evolution. In fact, of the twelve *cranial nerves,* (cranial nerves are nerves that come directly from the brain rather than from the spinal cord; cranial nerves are considered central nerves; whereas, nerves coming from the spinal cord are considered peripheral nerves) six of the twelve cranial nerves are directly connected to the eye.[33] The other six cranial nerves have either direct or *indirect* interconnections with the eye through either; anatomic relations, cranial nerve nuclei, or central connections through central nerve nuclei such as the Edinger-Westphal nucleus, and the mesencephalic nucleus. Other peripheral and intermediate connections are; the nucleus of the tractus solitarius, the nucleus of the spinal tract of the trigeminal nerve, and other cranial nerve ganglion such as the sphenopalatine ganglion and the superior cervical ganglion and others.[34,35] Therefore, there are direct and indirect interconnections within my head muscles, my neck muscles, my arm muscles, my leg muscles, my heart and my eyes.[34,35] I am aware of things directed at me which may hurt me, and I can duck or dodge the object - I don't even have to think about it! Again, these are preplanned *preventive* mechanisms *built into* and taken into consideration *ahead of time* with foresight and knowledge. There are billions upon billions of them! These *functions* enable the being to function and provide for itself when looking for food or hunting prey. They prevent danger or damage from happening to the individual, and they help the individual survive when attacked. It is *survival by design* not *survival of the fittest.*[36] If these necessary functions were not present in the beginning, *all* together, *all* fully functional, and *all* at the same time, the being (or person) could not function, survive or provide for itself. The eye would not function properly if any one component were missing. For all of them to be fully developed and fully functional at the same time there could be no possibility of evolution. Fully developed, fully

functional, interdependent function that is totally completed - all of it present, all at the same time *is*

creation; Interdependent Evidence of Creation, I.E.C.

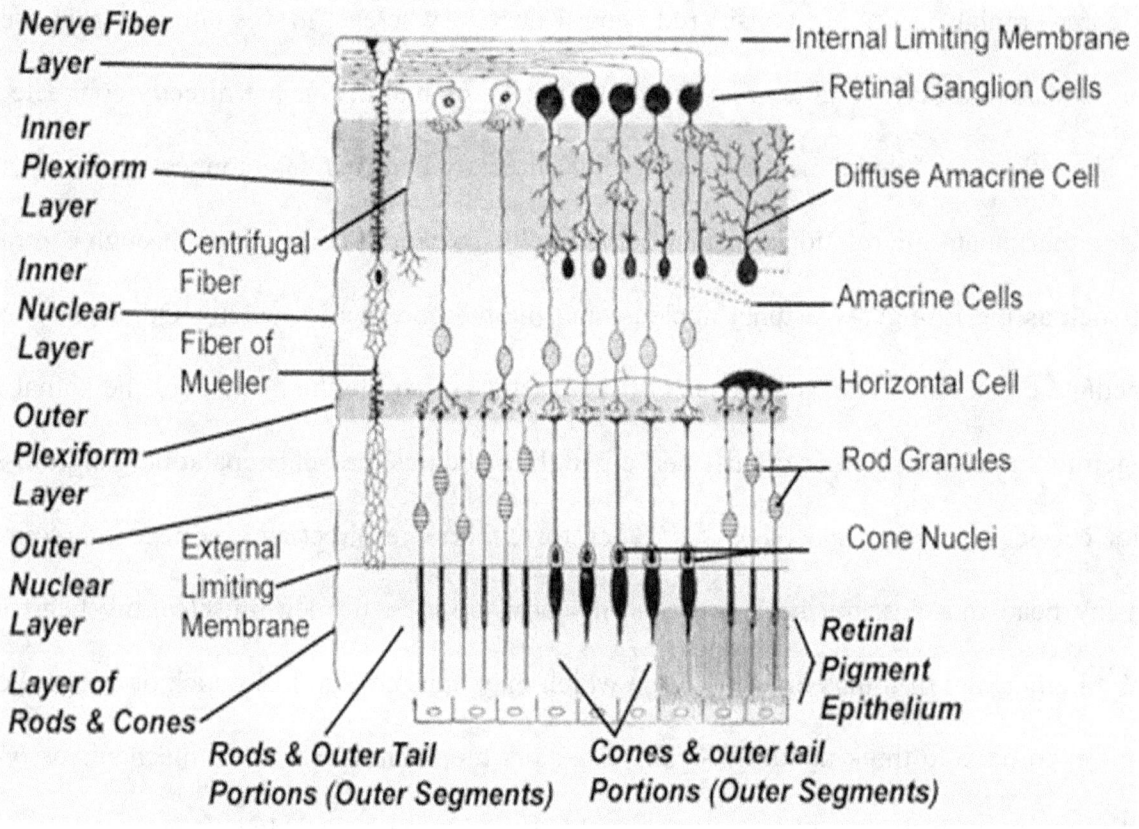

From Henry Gray, F.R.S, *Anatomy of the Human Body* 20th Edition, Lea & Febiger (Philadelphia & New York,
© 1918) Page 1016[37] with additions and modificationsby Rigney

The *Functional Retinal Cells* provide pre-processing of visual information to provide

instantaneous reflex responses to changes in the visual field. The Cranial Nerves are interconnected

within and through the nerve nuclei to work together automatically, and reflexively automatically.

The evidence demonstrates intelligent ***design, knowledge, and foresight. Plus, the numerous interconnections are additional Interdependent Evidences of Creation.***

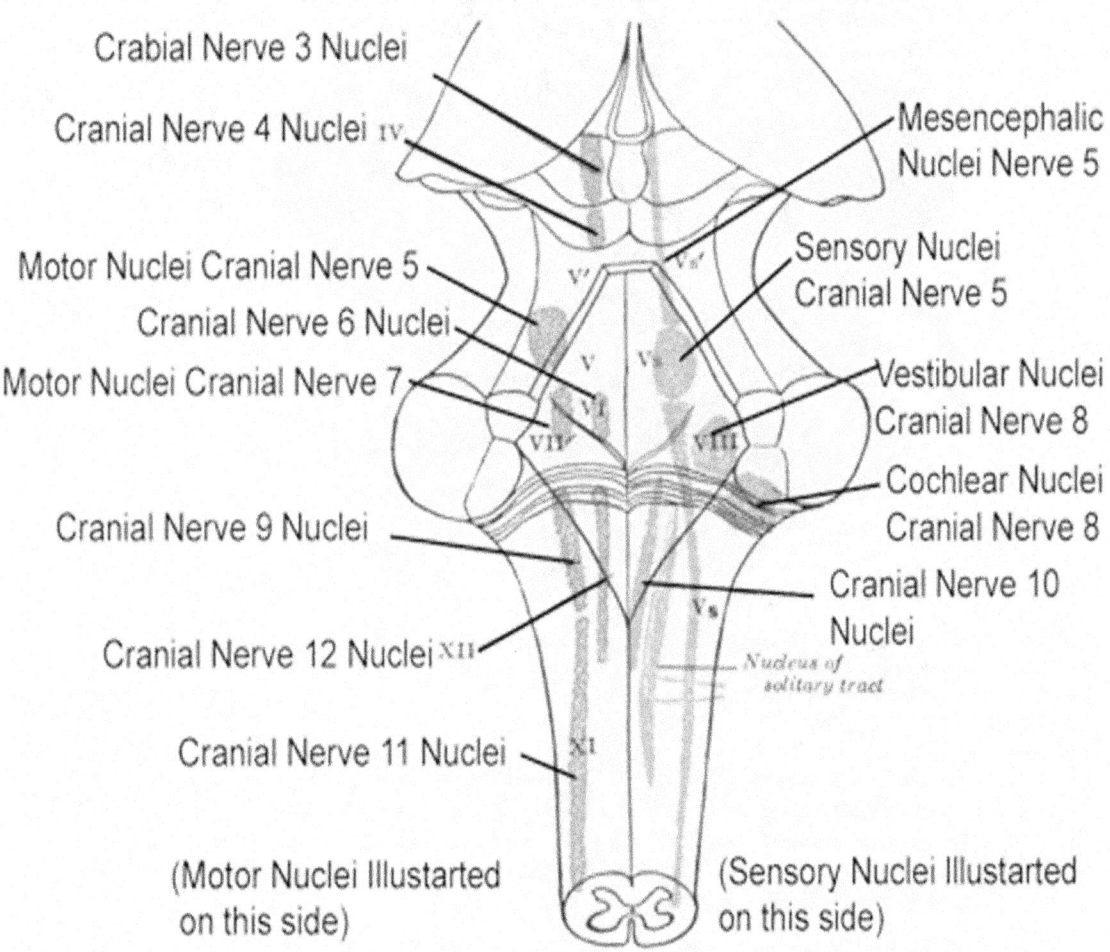

CRANIAL NERVE NUCLEI IN BRAIN STEM

(Motor and Sensory Nuclei Are actually on both sides, they have been separated here for ease in observation)

Crabial Nerve 3 Nuclei

Cranial Nerve 4 Nuclei IV

Motor Nuclei Cranial Nerve 5

Cranial Nerve 6 Nuclei

Motor Nuclei Cranial Nerve 7

Cranial Nerve 9 Nuclei

Cranial Nerve 12 Nuclei XII

Cranial Nerve 11 Nuclei

Mesencephalic Nuclei Nerve 5

Sensory Nuclei Cranial Nerve 5

Vestibular Nuclei Cranial Nerve 8

Cochlear Nuclei Cranial Nerve 8

Cranial Nerve 10 Nuclei

Nucleus of solitary tract

(Motor Nuclei Illustarted on this side)

(Sensory Nuclei Illustarted on this side)

From Henry Gray, F.R.S, *Anatomy of the Human Body* 20th Edition, Lea & Febiger (Philadelphia & New york, © 1918) Page 781[38] with modifications by Rigney

Interconnection of Cranial Nerve Nuclei enables the eyes to function automatically and also provides automatic reflex functions necessary for survival.

Picture Source: Art Explosion 525,000, Purchased By J. Jay Rigney O.D.[39]

Picture Source: Art Explosion 525,000, Purchased By J. Jay Rigney O.D.[40]

These automatic preprogrammed *functions* of the additional retinal cells, and the interdependent interconnections within the brain, the nerve nuclei, other interconnections within the midbrain, and interconnections within the sympathetic chain enable the being to function and provide for itself when looking for food or hunting prey. They prevent danger or damage from happening to the individual, and help the individual to survive when attacked. These interdependent functions are additional evidence of creation. We observe Interdependent Evidence of Creation.

Because they work so efficiently and effectively we take these automatic functions for granted. However, when one of them ceases to function you realize how important they are. Consider, as Darwin "imagines"[41] what your life would be like if you had to hunt for food to survive. "Imagine"[41] walking through the woods, navigating your body over rocks, through tress and bushes, stepping over sticks, logs, holes, and ditches, having to think about your every move as you roam through the woods hunting for something to eat. "Imagine"[41] you are looking for nuts or berries, or a small animal you need to eat to survive. Now "Imagine"[41] being able to find your food if you were blind - waiting on your vision to evolve. "Imagine"[41] walking through the woods while blind. This alone could not be done. After bumping into trees, tripping over rocks, getting tangled up in vines and thorns you would stop; becoming prey for other organic beings. "Imagine"[41] trying to find water if you were blind and in the woods. How long do you think you could survive? This scenario would be the same for "all the organic beings which have ever lived on this earth,"[42] because as Darwin says, they; "have descended from some one primordial form into which life was first breathed."[42] You may say, "No! Rigney is taking what Darwin has said out of context and bent or twisted Darwin's words out of context to fit Rigney's interpretation of what Darwin has said, to make Darwin's statements appear foolish. Unfortunately, this is NOT the case. Within *the same paragraph* on page 380 of *The Origin of Species,* Darwin makes these statements (I will quote the entire paragraph word for word so there is no question of taking Darwin's *WRITTEN WORDS* out of context.)

Darwin writes on page 380, "Analogy would lead me one step further, namely, <u>to the belief that all animals and plants have descended from some one prototype</u>. But analogy may be a deceitful guide. Nevertheless, all living things have much in common, in their chemical composition, their germinal vesicles, their cellular structure, and their laws of growth and reproduction. We see this even in so trifling a circumstance as that the same poison often similarly affects plants and animals; or that the poison secreted by the gall-fly produces monstrous growths on the wild rose or oak-tree. Therefore, I should infer from analogy that <u>probably all the organic beings which have ever lived on this earth have descended from some one primordial form, into which life was first breathed</u>."[43] (Underlining by Rigney. Now, just for clarification, re-read these underlined statements above. They are contained within the same paragraph.)

Thus, as Darwin states, all organic beings, *plants and animals* evolved from the same "one primordial form."[43] If animals evolved from the same "one primordial form" as plants; animals would HAVE TO EVOLVE FEET, LEGS, HANDS, ARMS, HEADS, AND ORGANS SUCH AS KIDNEYS, LIVER, LUNGS, STOMACH, INTESTINES, and BRAINS.

Then again you may say, NO! Rigney *AGAIN* is taking what Darwin has said out of context and bent or twisted Darwin's words out of context to fit Rigney's interpretation of what Darwin has said, to make Darwin's statements appear foolish. Again, unfortunately, this is NOT the case. Darwin specifically addresses this on pages 156-172 of his book *The Origin of Species* under the heading

"Organs of extreme perfection and complication."[44] While these six pages are too numerous too directly quote as I did Darwin's paragraph on page 380; I will say, it is here Darwin discusses the evolution of major complex organs such as the lungs, the digestive system, the stomach, the auditory apparatus, organs of flight, muscle, bones, tails, feet and of course the eye. It is within this section, pages 156-172 "Organs of extreme perfection and complication"[44] Darwin states, (page 159) "If it could be demonstrated that any complex organ existed, which could not possibly have been formed by numerous, successive, slight modifications, my theory would absolutely break down."[45]

Therefore, when Darwin says, "all animals and plants have descended from some one prototype"[46] and, "all the organic beings which have ever lived on this earth have descended from some one primordial form, into which life was first breathed."[46] Coupled with the fact he devotes sixteen pages explaining how "Organs of extreme perfection and complication"[47] have evolved. Then, when he says, "If it could be demonstrated that any complex organ existed, which could not possibly have been formed by numerous, successive, slight modifications, my theory would absolutely break down."[48] I certainly am **_NOT_** taking **_anything_** he has said out of context for his statements to appear foolish. He *"believes"* - HIS THEORY TEACHES; "all animals and plants have descended from some one prototype"[46] and, "all the organic beings which have **_ever_** lived on this earth have descended from some one primordial form, into which life was first breathed"[46] – (quoted word for word!) "**_Any_** complex organ" –which means; all organs, stomach, bladder, arms, legs, brain, hands, liver, feet –even the eye (all the while knowing how intricate and detailed it is) "formed by numerous, successive, slight modifications." [48] And "**_all_** organic life which have ever lived on this earth[46]" ALL SPECIES; bacteria, fleas, redwoods, hummingbirds, whales, dinosaurs, bats, bears, cacti (cactus), giraffes, spiders, clams, octopuses, penguins - **_EVERYTHING_** "descended from some one primordial form, into which life was first breathed,"[46] according to Darwin's "THEORY". (He knew plants, insects, and dinosaurs are organic life.)

So, when I said, *"imagine"[49] what life would be like if you had to hunt for food to survive. "Imagine"[49] walking through the woods, navigating your body over rocks, through tress and bushes, stepping over sticks, logs, holes, and ditches, having to think about your every move as you roam through the woods hunting for something to eat. "Imagine"[49] you are looking for nuts or berries, or a small animal you may need to eat to survive. Now, "Imagine"[49] being able to find your food if you were blind and waiting on your vision to evolve. "Imagine"[49] just walking through the woods while blind. That alone could not be done. After bumping into trees, tripping over rocks, getting tangled up in vines and thorns eventually you would stop. becoming prey for other organic beings. "Imagine"[49] trying to find water if you were in the woods and blind. How long do you think you could survive? This scenario would be the same for "all the organic beings which have ever lived on this earth."[50]*

So, when I said, "imagine"[51] trying to survive while waiting on your vision to evolve. I was being VERY generous. Because Darwin's theory would not only apply to the eye, but to **_ALL_** the organs within the body, *and **_ALL_*** physical structures of the muscles, bones, arms, and legs also. He said all organs evolved.[52] He said muscles, bones, and organs of flight all evolved.[52] And he said they **_all_** evolved from "some one primordial form into which life was first breathed."[53] "Imagine"[54] evolving arms legs, hands and feet! If "all animals and plants have descended from some one prototype" [55] and "all the organic beings which have ever lived on this earth have descended from some one primordial form, into which life was first breathed" [55] at some point evolution from crawling to walking, and evolution of limbs had to occur. "Imagine"[56] the process of evolving legs and feet, and arms and hands. Note, more importantly, just because you can "imagine"[56] it, it does NOT mean it "*actually*" occurred. And even more importantly, where is the *evidence* of it ever having occurred? "Imagine"[56] the process of the evolution of walking or the "evolution of flight". According to Darwin it would have had to occur, just as according to Darwin on page 159 of *The Origin of Species,* the eye evolved by "numerous, successive, slight modifications."[57] "Imagine"[58] evolving from non-seeing to seeing. When we observe the eye, the conclusion reached is; Interdependent Evidence of Creation.

The retina is dependent on: 1.The Cornea which has thirty-six I.E.C.,, 2.The Aqueous which has thirty-two I.E.C., 3.The Tabecular Meshwork to keep the Eye pressurized which has six I,E.C., 4.The Lens which has six I.E.C., 5. The Vitreous which has three I.E.C., 6. The Sclera, (the same I.E.C. as the Vitreous), 7. The Rods, 8. The three different Cone types, 9. The Retinal Pigment Epithelium, 10. The Choriocappilaris, 11. The Vena Voticosa, 12. The Optic Nerve, and 13. The visual cortex. Plus, there are 15 additional Functional Retinal Cells. Therefore, all together the Retina has 111 directly interdependent systems and structures. We observe 111 Interdependent Evidences of Creation, I.E.C.

17. Optic Nerve

The optic nerve is the main nerve of the eye. You could think of it as working somewhat like an electrical cord. The Optic Nerve carries the "messages" of the 253 million photoreceptors and the "messages" from the 1.53 billion *additional Functional Retinal Cells* of the Retina to the Brain. These "messages" are electrical impulses, or sparks of electricity.

There are two optic nerves, one for each eye. Each optic nerve is made up of about 1.2 million individual nerve fibers[1, 2] (like 1.2 million strands of wires in an electrical cord) which carry the messages of the *sparks of electricity* (impulses) created by the 253 million rods and cones and the 1.53 billion *additional functional retinal cells* in the retina to the brain.

Somehow (only God knows) the optic nerve carries all these millions of different *sparks of electricity (impulses)* to the brain, and somehow the brain and the eye combines all those *sparks (impulses)* to create a picture in your *mind* as it *is* in the *real world*. Some preprocessing of vision takes place in the retina.[3] In fact, the retina and the optic nerves are actually an extension of the brain.[3] Again, according to Wolff, "the optic nerve and retina are an outgrowth of the brain. The structure of the Optic Nerve is identical to the neural structure of the brain. The fibers of the Optic Nerve have no neurolemmal cells, though other satellite cells occur; it is surrounded by meninges, and the primary and secondary sensory neurons of the neural pathway, both in the retina, The retinal ganglion cells corresponding similarly to those in the gracile or cuneate nuclei in the medulla oblongata."[3]

Darwin was not even close; he had *no remote idea* as to what was happening inside the retina, optic nerve and visual cortex, when he said, "several facts make me suspect that any sensitive nerve may be rendered sensitive to light."[4]

I doubt if he knew today what we *now* know if he would stand behind that statement. Furthermore, I doubt <u>anyone</u>; knowing what we know today about the retina, the optic nerve and the visual cortex, would agree with that statement. And if "any sensitive nerve"[4] cannot be "rendered sensitive to light"[4] again, his "theory absolutely breaks down!"[5]

Each optic nerve derives its fibers from the retina. The retina has its nerve fibers split into four quadrants.[6] The nerve fibers are split left from right, exactly down the middle, and top from bottom exactly across the middle. The nerve fibers proceed from each eye's retina, backward to the visual cortex located in the lower back portion of your brain (the visual cortex is the part of the brain where vision is processed). As the optic nerve fibers proceed back to the visual cortex, the inside nasal nerve fibers from each eye (exactly half of them) split off exactly down the middle and then cross over and join up in the exact same location in the visual world with the outside temporal optic nerve fibers. In this manner, the temporal nerve fibers of one eye join up with nasal nerve fibers of the other eye and continue *together* backward to the visual cortex. This crossing over happens in what is called the *optic chiasm.*[6] The *optic chiasm* is located about one-third of the way back from the front of the head. The optic chiasm defies evolution. This miraculous intentional design causes what is seen in the right-hand field of vision to be detected by the left side of *each eye's retina* - because light travels in a straight line.

The way it works is as follows: When something is in the right-hand field of vision, light travels in a straight line and is therefore detected by the outside retina of the left eye and the inside retina of the right eye. The inside retina fibers of the right eye then split off and cross over in the optic chiasm and then join up with the outside retina fibers of the left eye and proceed *together* backward to the visual cortex.[6] This design is so specific whatever is seen in the visual cortex in the right eye by one of the 126.5 million photoreceptors (a rod or one of the three cones) is seen by the exact same one of the 126.5 million photoreceptors of the left eye. Each of the "exact" 126.5 million photoreceptors of the right eye is located in the exact same spot in the right eye as it is in the left eye! *Point for point,*[7] photoreceptor for photoreceptor, nerve fiber for nerve fiber, in the same spot in the right eye as it is in the left eye. This specific precise design causes the 126.5 million photoreceptors and their respective 1.2 million nerve fibers in the right eye to be located and in the same exact corresponding spot in the left eye. Top of the right eye equals top of the left eye—point for point. Bottom of the right eye equals bottom of the left eye—point for point. Left side of the right eye equals left side of the left eye—point for point. Right side of the right eye equals right side of the left eye—point for point![7] This miraculous *intentional* design causes what is seen in the right eye to exactly be seen *the same* in the left eye so the picture of the right eye overlaps and is joined *together* with the picture of the left eye so only one picture is seen. If your eyes were not intentionally *wired* in this manner, they would not work together properly and you would see double.

(The illustrations to follow on pages 157 and 158 were derived from:
Les Voies Optiques:
http://www.campusdanatomie.org/sites/default/files/users/admin/voiesoptiques_sans_legende.pdf
https://creativecommons.org/licenses/by-sa/3.0[8]
(They do not endorse me, or my use of their work)

The entire visual system is divided exactly down the middle!

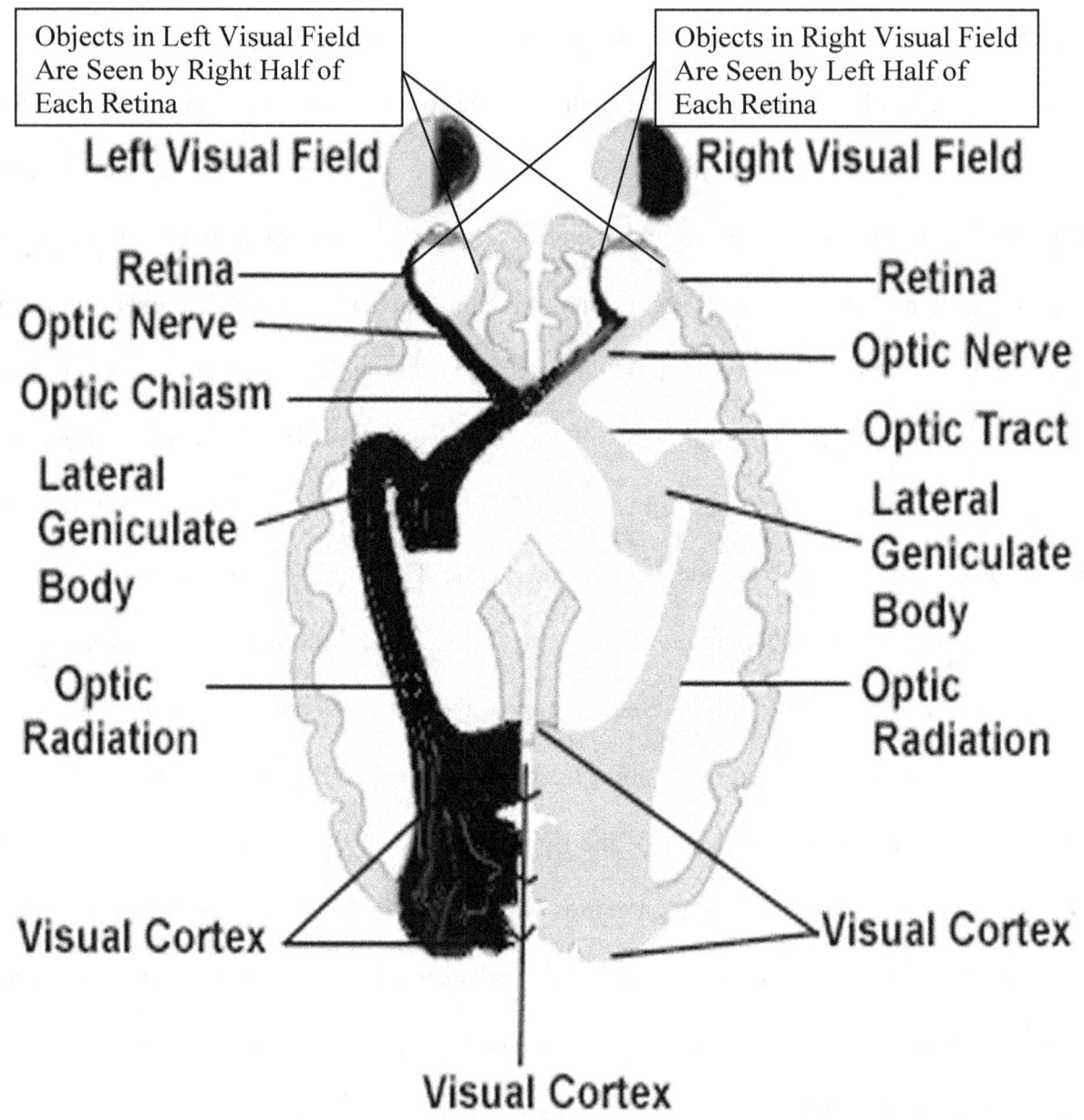

Objects in Left Visual Field Are Seen by Right Half of Each Retina

Objects in Right Visual Field Are Seen by Left Half of Each Retina

Left HALF of Field of Vision (Visual Field) is seen by right half of Retina, right half of Optic Nerve, right half of Optic Tract, right half of Lateral Geniculate Body, right half of Optic Radiation, and right half of Visual Cortex. (In this illustration; Gray part of Field of Vision is seen by Gray part of Visual System.)

Right HALF of Field of Vision (Visual Field) is seen by left half of Retina, left half of Optic Nerve, left half of Optic Tract, left half of Lateral Geniculate Body, left half of Optic Radiation, and left half of Visual Cortex. (In this illustration; Black part of Field of Vision is seen by Black part of Visual System.)

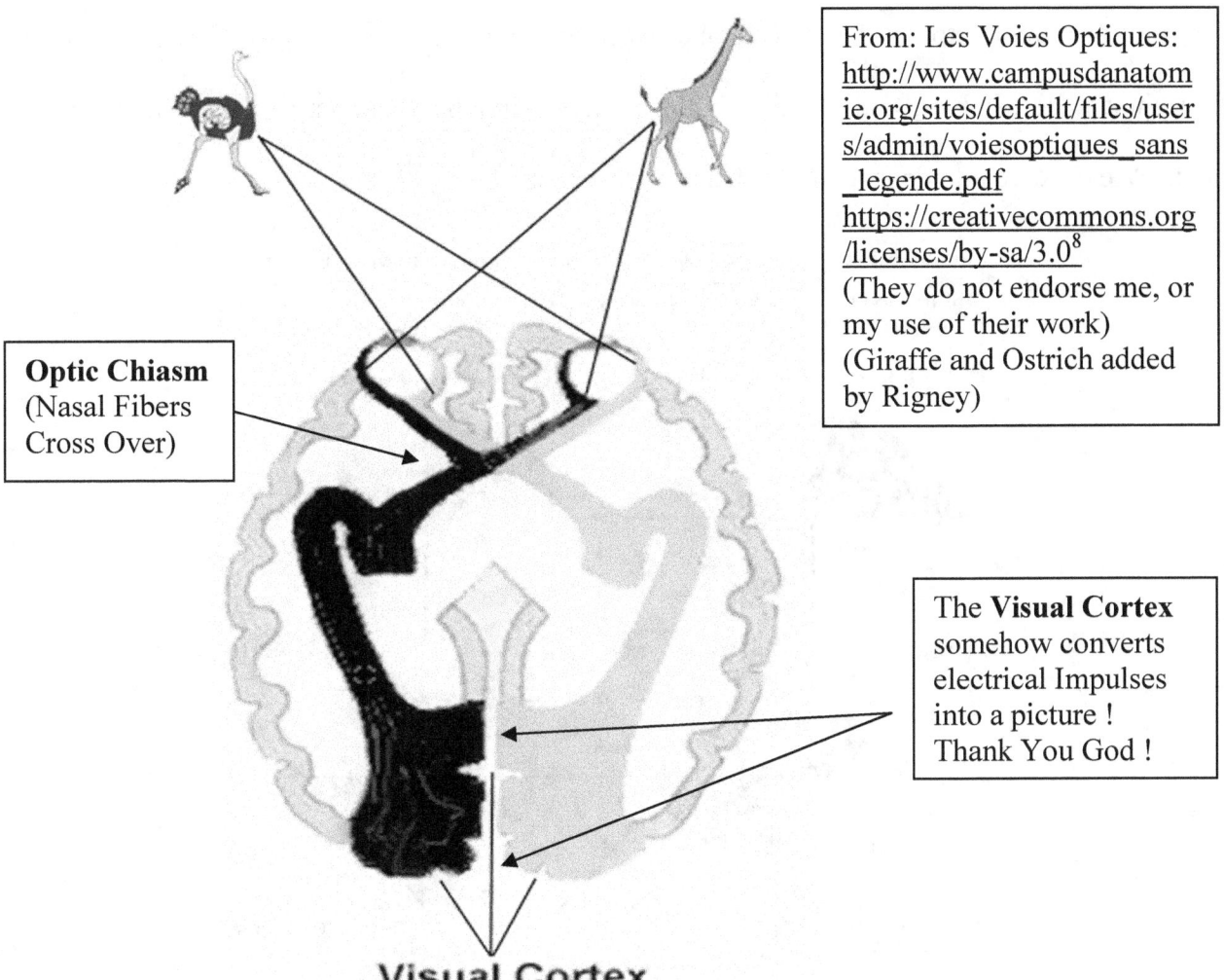

Optic Chiasm
(Nasal Fibers
Cross Over)

The **Visual Cortex**
somehow converts
electrical Impulses
into a picture !
Thank You God !

Visual Cortex

The exact left half Field of Vision (the ostrich) is seen by the same exact right (gray) side of each eye. The exact right half Field of Vision (giraffe) is seen by the exact left (black) side of each eye. The outside retina of the right eye (gray) sees the left field of vision. The inside retina of the right eye (black) sees the right side field of vision *and* the outside retina of the left eye (black) sees the right field of vision. The inside retina of the left eye (gray) sees the left side field of vision. The inside fibers of each eye cross over in the optic chiasm and join the outside fibers of the other eye in the visual cortex of the brain.

Evolution cannot explain this precise and exact splitting of nerve fibers - exactly and precisely into right and left halves, and top and bottom halves. Intelligent intentional design is "clearly" evident. We observe interdependent Evidence of Creation, I.E.C.

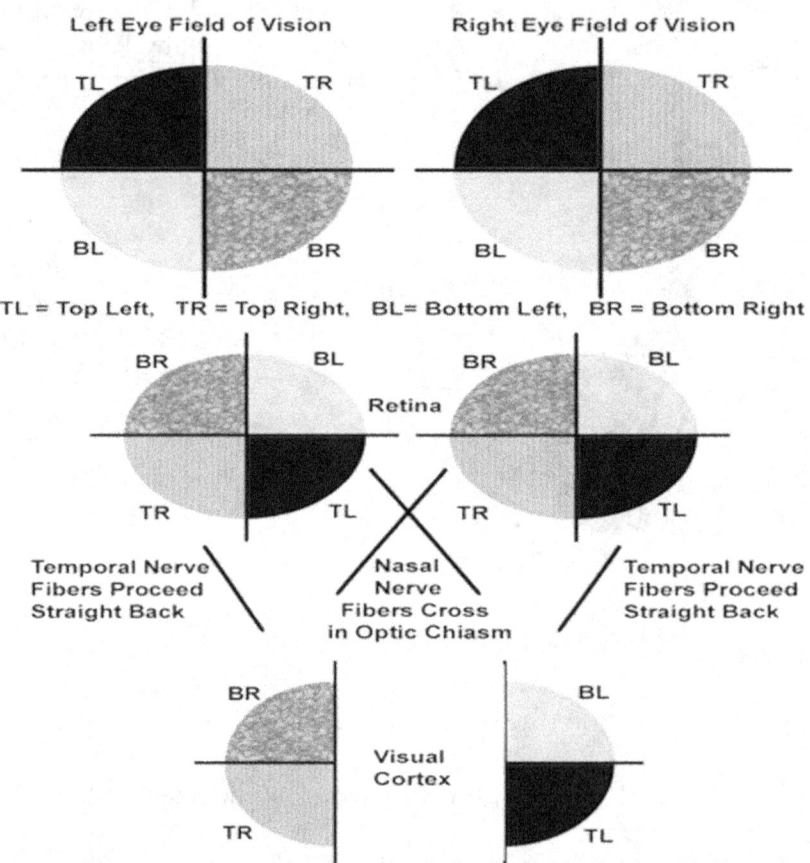

Each Visual Field is split into exactly left & right half and top & bottom half in each eye! Exactly the inside & top half and inside & bottom half of each eye split in the Optic Chiasm and then join up with the exact coresponding nerve fibers of the opposite eye in the Visual Cortex - point for point !

Left Eye Field of Vision   **Right Eye Field of Vision**

TL = Top Left,   TR = Top Right,   BL= Bottom Left,   BR = Bottom Right

Retina

Temporal Nerve Fibers Proceed Straight Back

Nasal Nerve Fibers Cross in Optic Chiasm

Temporal Nerve Fibers Proceed Straight Back

Visual Cortex

Left half of Vision is seen by exactly the right side half of each Retina.
Right half of Vision is seen by exactly the left side half of each Retina.
Exactly half of the inside nerve fibers cross over in the Optic Chiasm !
The left Visual Cortex Processes the right half of the Field of Vision.
The right Visual Cortex Processes the left half of the Field of Vision.

Note: The retina and the nerve fiber pathway of the entire visual system are divided into quadrants; left and right halves and top and bottom halves – exactly down the middle left and right, and exactly down the middle top and bottom, on both the right eye and the left eye, "exactly" the same in each eye, at exactly the same time in both eyes! This design is preserved throughout the entire nerve fiber pathway such that the Visual Field is represented locally consistently throughout the anatomy of the visual pathway. From the Retina, to the Optic Nerve, to the Optic Chiasm, to the Optic Tract, to the Lateral Geniculate Body, to the Optic Radiation, to the vision processing centers of the Visual Cortex. The top left visual field is seen by the lower right portion of the retina. The top left visual field is processed neurologically by the botttom right portion of the visual system - all the **solid dark black portions.** And at the same time, in each eye, the top right half of your vision is processed by the bottom left half of each eye – **all the gray shades with vertical lines.** As a result, the left half of your brain processes the right half of your field of vision. The only way to accomplish this is for the nasal half of the visual fibers ***of each eye*** to split off and then join up with the temporal fibers of the other eye. They split off exactly half way in the middle of each eye- then come together with the temporal fibers of the other eye so the retinas of the left eye and the right eye are represented point for point in the visual cortex! This splitting of exactly half of the nerve fibers exactly down the center of the retina occurs in the Optic Chiasm. Each point in the right eye is in the exact same spot in the left eye. Evolution cannot explain this exact nerve fiber splitting design. It is obvious; "numerous successive slight modifications"[9] are not at work because until the evolution process has completed the "numerous successive slight modifications"[9] the eyes cannot function together and the being would see double. What is observed is Interdependent Evidence of Creation; exactly half, <u>exactly top and bottom,</u> *<u>exactly down the middle,</u>* *IN EACH EYE,* **AT EXACTLY THE SAME TIME!**

18. Visual Cortex

The part of the brain which creates vision is called the *visual cortex*. It is located in the lower back portion of your head and brain. This location is the best location possible to protect and preserve the vitally important function of vision.

The eyes are in the front of the head, yet the vision-processing center (the visual cortex) of the brain is in the lower back portion of your head. How could the brain evolve the vision processing part in the back - while the eyes are in the front? If they had evolved, you would think they would be located immediately adjacent to each other. How can the 126.5 million photoreceptors, the 506 million additional functional retinal cells, and 1.2 million nerve fibers[1] for each eye (253 million total photoreceptors, 1.12 billion total additional functional retinal cells, and 2.4 million total nerve fibers) get from the front, and evolve all the way to the back? How could they happen to evolve to split into quadrants? How could these soft nerve tissues evolve through the hard bones of the eye socket? "Numerous, successive, slight modifications"[2] as Darwin's *theory of evolution* would have you believe, can't explain this intentional, front-to-back. protective design.

Each retina contains about 126.5 million rods and cones and is actually an extension of the brain[3] and the retina actually pre-processes the signals, as there are 126.5 million photoreceptors per eye providing impulses for 1.2 million optic nerve fibers per eye.[4, 5,6] Each optic nerve contains about 1.2 million nerve fibers all bundled together like a fiber optic cable and is about the diameter and thickness of a regular writing pencil, yet these 1.2 million nerve fibers which are transmitting electrical impulses, somehow don't short each other out, and don't mix the electrical impulses all together and become a confused, mixed-up, trashy signal. Light->Impulses->Perception=Vision; A miracle!

Light enters the eye. The rods and cones convert the light into electrical impulses. After these *sparks of electricity* created by the rods and cones are sent to the visual cortex from the retina by the optic nerve, the visual cortex converts these millions of millions of *sparks of electricity* (electrical impulses) into a picture in the brain exactly as it is in the real world! How this is done is not exactly known. While there is much written regarding the structure and arrangement of the neurons within the visual cortex, Eugene Wolff, one of the leading authors regarding structure and function of the human eye ultimately says, "the visual pathway becomes a most intricate array of interacting conductors, at each level of which a further grade of complex "processing" occurs. How far each level in this neuronal hierarchy can be said to contribute directly to the subjective phenomena which we designate as "seeing" is currently undetermined."[7] He is more or less saying, we do not know *how* the visual cortex creates a picture in the mind as it is in the real world.

I wanted to see if there was anything more current, possibly on the Internet, explaining how the brain creates a picture. After a rather exhaustive search, as of the date of this writing, from what I found; I think no one really knows. I think at this time only God knows.

Darwin did not contemplate the complexity of the eye and the process of the brain *creating* vision. He compared the eye to a telescope and then went on to say, "we know this instrument (the telescope) has been perfected by the long-continued efforts of the highest human intellect; and we naturally infer that the eye has been formed by a somewhat analogous process."[8] It is apparent he did not understand the complexity of the eye and the reasons why and how the eye works. And if the eye formed as he says, "by a somewhat analogous process" how is it Darwin can acknowledge "this instrument [the telescope] has been perfected by the long-continued efforts of the highest human intellect" but he does not acknowledge ***THE*** Highest Intellect that "perfected" the eye? "Perfected in this connotation *IS* CREATED. He then also says, (I am quoting him word for word) "If we must compare the eye to an optical instrument, we ought in <u>imagination</u> to take a *thick layer of transparent tissue*, with a *nerve sensitive to light* beneath, and then <u>suppose</u> every part of this layer to be continually changing slowly in density, so as to separate into layers of different densities and thicknesses, placed at different distances from each other, and with the surfaces of each layer slowly changing in form."[8] (Underlining by Rigney)

He ought to not "*imagine*" he ought to instead look at the evidence! Why would anyone "<u>naturally infer</u> that the eye has been formed by a somewhat analogous process"[8]? Is it because *he* says we should? And who is "*we*" in this statement? It is apparent he is trying to instill his thoughts into your mind and at the same time get you to agree with him. But, just because he says it, does NOT make it true. He does not consider or understand the requirements necessary for tissue to actually be "transparent"; that the thickness and arrangement of the layers of the cornea must be arranged precisely to the specificity of the wavelength of light, 1/100,000th of a millimeter, or 10^{-9} meters, (or about $1/100^{th}$ of a human hair) and just as the natives he refers to in his book *The Origin of Species,* p. 429[9] where he said, "When we no longer look at an organic being as a savage looks at a ship, as at something wholly beyond his comprehension"[9,10] Darwin did not understand the overwhelming evidence in front of him when he considered evolution of the eye. He even said he thought (*page 157, "The Origin of Species"*) "any sensitive nerve may be rendered *sensitive to light.*"[11]

It is clear he did not know or understand the process by which the eye converts light into electrical impulses and the "nerve" is not sensitive to light, it instead is the transporter of the messages the retina has produced from reacting to light. It also is apparent he does not understand or contemplate; in fact, he does not even consider how the rods and cones can convert light into impulses, and the genius of the outer segments degrading and being replenished by the inner segments. Darwin did not have the ability in the 1800s to know then, what we know now; and just as Darwin stated when he was speaking of classifying of species... "When we no longer look at an organic being as a savage looks at a ship, <u>as at something wholly beyond his comprehension</u>, When we regard every production of nature as one which has had a history; when we contemplate every complex structure and instinct as the summing up of many contrivances, each useful to the possessor, nearly in the same way as when we look at any great mechanical <u>invention</u> as the summing up of <u>labor</u>, the experience, the reason, and even the blunders of <u>numerous workmen</u>; when we thus view each organic being, how far more interesting, I speak from experience, will the study of natural history become"[12] (underlining by Rigney). He either ignores, does not know, or could not understand it - because it was "something wholly beyond *his* comprehension;"[12] the fact the retina is converting the impulses (sparks of electricity) into messages which are carried by the optic nerve, and the brain is creating those impulses into perceptual vision *(somehow)* in the brain. He does not even consider how the brain *creates* vision. Thereby because of *his* perceptual blindness[13] he does not acknowledge the Creator, just as the natives he wrote about did not acknowledge the early explorer's ships because they were beyond their comprehension.[13] Ironically, in the example referenced above, Darwin ***does*** acknowledge the "workmen"[14] and "labor"[14] responsible for an "invention"[14], yet he ***does not*** acknowledge THE INVENTOR of ***life***, which requires a much more complex and higher level of knowledge to "invent". In fact, he is describing, in this example, exactly what I am talking about when we observe ***Interdependent Evidence of Creation***. Darwin says, and I quote: "if we look at ***ANY*** great mechanical invention as the summing up of labor, the experience, the reason, and even the blunders of numerous workmen"[14] these underlined words of Darwin[14] "labor" [14], "experience" [14], "reason" [14], "blunders" [14] are ***evidences*** of THE FACT there was a creator. In FACT, he acknowledges there is a creator of the telescope or "any mechanical invention"[14] when he uses the words "numerous workmen"[14] - yet, he didn't observe them create the telescope or the invention! How did he ***know*** there were workmen that created the telescope and "any mechanical invention"? He observed Interdependent Evidence of their existence by observing their Creation – the telescope, or as Darwin states, "any mechanical invention"[14] demonstrates and *proves* the theory Interdependent Evidence of Creation!

Furthermore, when he says, "If we look at[14]" - this ***IS*** observation. Therefore, in this statement, Darwin affirms my theory ***Interdependent Evidence of Creation***. He observes ("looks at") a telescope or "any mechanical invention[14]" and acknowledges there is "workmen[14]" necessary to create the telescope or any mechanical invention. Yet, when he observes (looks at) life he doesn't acknowledge ***THE*** Creator – even though there is evidence of Creation! How is it he acknowledges the workmen of a simple telescope, or 'any mechanical invention" for that matter, but he does not acknowledge the Creator of complex life? IT IS THE EXACT SAME PRINCIPLE!!! We ***observe***, or as Darwin states, "if we **look at** any great mechanical invention"[14] - if we "look at" [14] life we observe EVIDENCE OF A CREATOR. In his example it is the evidence of "workmen."[14] My theory; *Interdependent Evidence of Creation* is therefore confirmed by Darwin's statement of looking at the telescope, or "any mechanical invention" ***KNOWING*** there were workmen necessary to "invent" **(CREATE)** the telescope.

If one ***does*** what Darwin ***says to do*** in observing and concluding the origin of "any mechanical invention" or the telescope[15]. That is, if one is to observe ("look at") "any mechanical invention" and conclude there was "labor" and "workmen" required to "invent" the "mechanical invention" or telescope; after ***observing*** life (which is the actual science behind Darwin's theory, namely ***observation***) one would conclude infinitely more information is conveyed OF THE CREATOR when observing life than of workmen when observing a telescope. How anyone can conclude labor, workmen and invention occurred in the making of a telescope, or "any mechanical invention"[14] - but then at the same time conclude life *just happened* - and not recognize the "labor"[14] "workman"[14] and "invention"[14] associated with the invention and creation of life is troublesome. Think about it, a telescope, "ANY mechanical invention"[14] conveys a "laborer,"[14]"workman,"[14] and an "inventor,"[14] but somehow life does not. When one *knows* and *understands* the basic principles necessary for life and concludes that it happened by "chance" it is *very* troublesome. To propagate and promote the ***theory*** as fact is alarmingly troublesome. The principle is *"exactly" the same.* If you can acknowledge the workmen and the inventor, you should also acknowledge the Creator –infinitely more so! Come on Darwin, BE CONSISTENT! It ***IS*** as you say, "an analogous process[11]"!

Light, converted to electrical impulses, converted into visual perception, according to one of the world's leaders in ocular anatomy and physiology, Eugene Wolff, is "beyond our comprehension."[15] Again, when something occurs which cannot be explained we call it a miracle. The miracle of visual perception occuring in the visual cortex demonstrates "labor," a "workman" and an "inventor" - the ultimate inventor of the miracle of vision and perception, **_THE_** Creator. We should acknowledge Him for Who He IS and bow in reverential AWE of Him. Instead some *choose* to exchange the truth for a lie - even in the presence of ALL THE EVIDENCE, just as Darwin has done. He can acknowledge the workman - but even after observing the evidence, Darwin does not acknowledge the Creator. You too are choosing to either believe or not believe in the Creator. You have learned much of the evidence of Creation by observing the eye - *Interdependent Evidence of Creation*, (there is still much more over-whelming evidence yet to come in the remaining chapters). Examine the evidence. God has shown himself to you plainly in the evidence of HIS CREATION. The evidence is all around you and *plainly* observed. In fact, it is so plain that God says, in His Word the Bible, you have no excuse for not believing He exists. Romans 1: 19-20 says *"For that which is known about God is evident to them and made plain in their inner consciousness, because God Himself has shown it to them. For ever since the Creation of the world His invisible nature and attributes, that is, His eternal power and divinity, have been made intelligible and clearly discernible in and through the things that have been made. So men are without excuse; altogether without any defense or justification."* [16] Thus we observe that even God's Word proclaims **_Interdependent Evidence of Creation_** – and so strong is the evidence God says you are without excuse. He has revealed himself to you! Observe the evidence; the Visual Cortex is dependent upon all 24 of the other interdependent systems, and at the same time, each one of the 24 systems exhibit multiple interdependent structures and functions. We observe exponential Interdependent Evidence of Creation. Observe the evidence. Look at the trochlea and the superior oblique eye muscle! Look at the design and function of the cornea! Look at the routes of nerve supply to the pupil! Look at the foramen! Look at the aqueous inflow/outflow system! Look at the tear in-flow/out-flow system! Look at the lens and ciliary body focusing control system! Look at the retina, optic nerve, and visual cortex vision perception system! You *know* He is THE CREATOR, He has revealed Himself through His Creation- ***PLAINLY!*** You have NO EXCUSE! **CHOOSE GOD!**

"For that which is known about God is evident to them and made plain in their inner consciousness, because God Himself has shown it to them. For ever since the Creation of the world His invisible nature and attributes, that is, His eternal power and divinity, have been made intelligible and clearly discernible in and through the things that have been made. So men are without excuse; altogether without any defense or justification." [16]

19. Nerves of the Eye

Six different nerves supply all the nerves of the eye. These six nerves are called Cranial Nerves because they originate directly from the brain stem rather than coming off of the spinal cord. There are twelve Cranial nerves. Cranial nerves are considered *central nerves,* whereas nerves that supply your leg or arm etc. are considered *peripheral nerves*. Because central nerves come directly from the brain they tend to have more automatic or subconscious functions; that is, you do not have to think about it for them to work. Of the twelve central cranial nerves six are *directly* connected to the eye. The other six cranial nerves have *indirect* connections to the eye.[1,2]

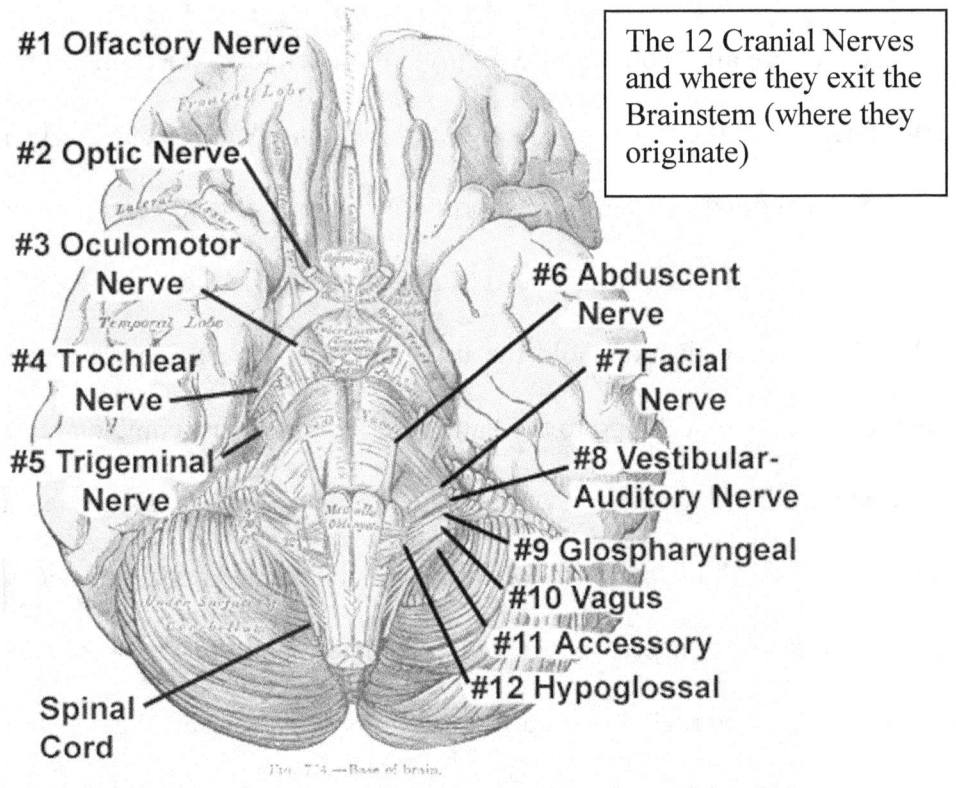

#1 Olfactory Nerve
#2 Optic Nerve
#3 Oculomotor Nerve
#4 Trochlear Nerve
#5 Trigeminal Nerve
Spinal Cord

The 12 Cranial Nerves and where they exit the Brainstem (where they originate)

#6 Abduscent Nerve
#7 Facial Nerve
#8 Vestibular-Auditory Nerve
#9 Glospharyngeal
#10 Vagus
#11 Accessory
#12 Hypoglossal

From Henry Gray, F.R.S, *Anatomy of the Human Body* 20th Edition, Lea & Febiger (Philadelphia & New York, © 1918) p.817[1] with additions by Rigney

Other automatic subconscious functions which are designed, automatic, pre-planned functions necessary for life we take for granted include; regulation of body temperature, heart rate, blood pressure, respiration (or breathing), digesting of food, water requirements (thirst), intestinal control, bladder control, hormone control, maintenance of the immune system in order to fight infection, sensing of tissue damage, wound repair and healing, regulation of fluid balances, electrolyte/mineral balance and ph of the body and the blood. Without these functions the individual could not survive. They are *all necessary* and are *all interdependent*. Remove any single **_one_** of them and the individual will die. Again, we observe Interdependent Evidence of Creation. They all must be present and fully functional **_all at the same time_**. No one system can function while waiting on another system to evolve because they are all absolutely and totally interdependent. We should thank God daily that He provided all these automatic functions. Just think how complicated your life would be if you had to monitor all these functions and make the appropriate adjustments to keep your body working. Think about how miserable your life would be if some of these functions were not present. We should thank God daily that our bladders hold the urine and that it just doesn't run out all the time. We should thank God every time we go to the bathroom it doesn't just come out at *any* time. We should thank God when we lay down and get up our heart and blood vessels sense the needed change in blood pressure and we don't pass out when we get up from laying down (which is what would happen if there were not the necessary sensors in the heart, and sensors in the arteries of the neck and the brain). We should thank God we are not too hot or too cold all the time and we don't have to spend hours warming ourselves in the sun, as do the reptiles. We should thank God when we injure ourselves we don't have to figure out how to fix it. We should thank God we are not sick all the time. These are just some of the *subconscious* functions. We should thank God too for the many conscious functions. We can walk, run, lift, bend, and move. Think about the many people confined to wheel-chairs etc. Right now thank God you can _____, and _____, and _____, ……. and on and on and on (you fill in the blanks). On top of this He has given us senses which make life more enjoyable and pleasurable; touch, feel, taste, smell, hear, and of course, sight. While these make life more pleasurable they are also truly *necessary*. To think all these many, functions *evolved* is absurd. Wait 'till one of them goes wrong for *you* then see what life would be like. When you lose a bodily function, you realize how precious these things we take for granted are. Think about what life would be like if you could not walk, if you could not lift, if you could not run, if you simply could not bend over. What if you could not chew? What if you had kidney stones all the time? What if you *constantly* had to go to the bathroom because your bladder did not evolve or the bodily wastes just came out at any or all times? What if you could not hear? Or what if you just heard static or ringing in your ears all the time. What if you were always cold, or always hot? What if you caught a cold but never got over it? What if you got a sore, it became infected, and kept getting worse? What if you could not see clearly? What if *everything* were always blurry? What if you could not see at all, if you were totally blind? We have *SO* much to be thankful for. Thank Him, (thank God -not Darwin). Thank Him right now! "Lean on, trust in, and be confident in the Lord with all your heart and mind and do not rely on your own insight or understanding. In all your ways *acknowledge Him*, and He will direct and make straight and plain your paths. Be not wise in your own eyes; reverently fear and worship the Lord and turn [entirely] away from evil." Proverbs 3:5-7[3] "Thank you God. You have provided everything so I may have and enjoy life, and have it in abundance (to the full, till it overflows)." John 10:10[4]. "Thank you God. You have provided all the nerves necessary for me to see, and you have provided everything necessary so I can continue to keep on seeing. Thank you that the functions of my eyes are automatic and that they work and keep on working without me having to ever think about it."

Of the twelve automatic, cranial nerves, six directly supply the eye, and the other six have indirect connections with the eye.[5,6]

Cranial nerve #1 is the Olfactory nerve, the nerve which helps you sense smell. The eye is indirectly interconnected with the Olfactory nerve, (the nose and sense of smell) by way of the Trigeminal Nerve, Cranial nerve #5, (the maxillary division) through interconnection in the pterygopalatine ganglion. These fibers have influence on the *lacrimal gland* and cause the eyes to water when irritated by allergens (allergies) and strong smells[7].

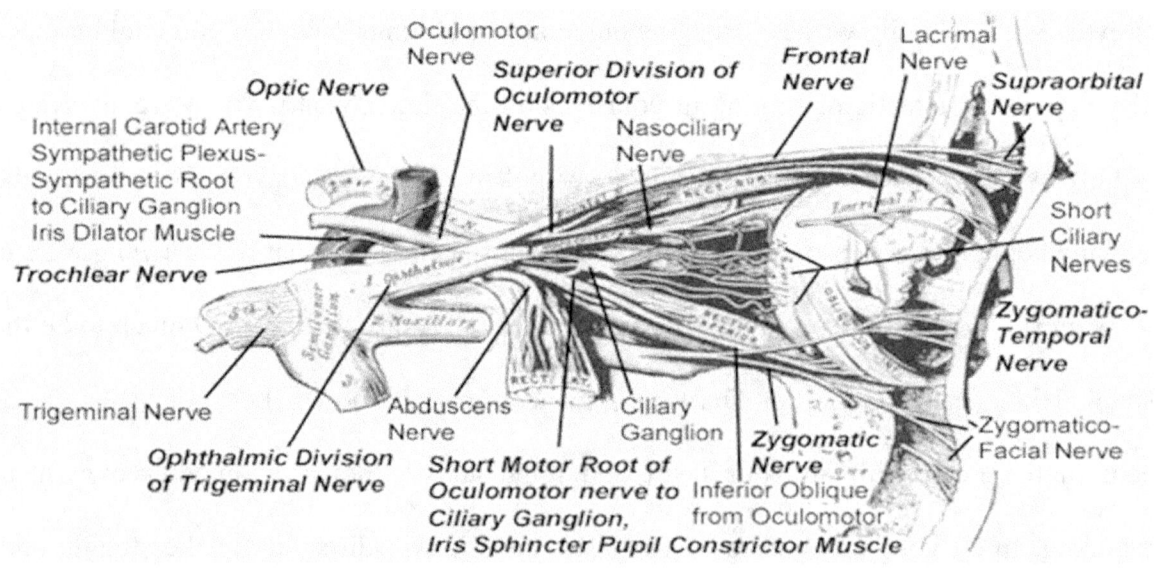

From Henry Gray, F.R.S, *Anatomy of the Human Body* 20th Edition, Lea & Febiger (Philadelphia & New York, © 1918) Page 887[8] with additions/modifications by Rigney

Cranial nerve #2 is the Optic Nerve. It carries the messages of vision. The Optic Nerve is directly connected to the eye.

Cranial Nerve # 3 is the Oculomotor Nerve. The medial rectus, inferior rectus, inferior oblique, and superior rectus eye muscles are all supplied by branches from the oculomotor nerve. The oculomotor nerve also supplies the main muscle of the eyelid which lifts the eyelid open, the levator palpebrea superioris. The oculomotor nerve is directly connected to the eye. The medial rectus pulls the eye in, the inferior rectus pulls the eye down, and the inferior oblique rotates the eye outward and also pulls the eye up while you are looking in and up. The oculomotor nerve is directly connected to the eye.

Cranial nerve #4 is the Trochlear nerve. The trochlear nerve supplies the superior oblique eye muscle. It rotates the eye inward and pulls the eye down when you look in. It is directly connected to the eye.

Cranial nerve # 5 is the Trigeminal nerve. It supplies the eye and eyelid with pain and touch *sensory* fibers, and also supplies and controls tearing. The trigeminal nerve also supplies pain and touch sensory fibers to the face and the mouth, and it has a motor division which supplies the jaw muscle for chewing. It is directly connected to the eye.

Cranial Nerve #6 is the Abduscens nerve, the nerve which supplies the lateral rectus muscle. The lateral rectus eye muscle pulls the eye out. It is directly connected to the eye.

Cranial Nerve #7 is the Facial Nerve. The facial nerve supplies the muscles of the face, some of the salivation glands, and some of the sensations of taste[9,10]. The *orbicularis oculi muscle,* the muscle which encircles the eyelids on both top and bottom and causes the eye to squeeze shut is also supplied by the facial nerve. Cranial Nerve #7 the facial nerve is directly connected to the eye.

Cranial Nerve # 8 is the Auditory nerve. The eye is indirectly interconnected with the Auditory nerve, as there is direct input from the inner ear semicircular canals.[11] (The semicircular canals in the ear provide information to the brain regarding balance and head position. When the tilt or turn, or rotation of our head changes there are messages sent to the *eye muscles* for the eye to maintain a fixed or *locked-on* position so when your head and/or body moves your eyes don't drift off position or off target.)

Cranial Nerve #9 is the Glosopharyngeal nerve, it supplies, the parotid gland, and some of the taste buds, and some of the muscles of the neck.[12, 13] The eye is indirectly interconnected with the Glosopharyngeal nerve,[14] (and so indirectly connected to the mouth and the pharynx and sense of taste) by way of the Trigeminal Nerve cranial nerve #5 (the maxillary division), and a division of the facial nerve (the Greater Petrosal Nerve) by way of both the sympathetic chain, and pterygopalatine ganglion.[15, 16] These fibers also have influence on the *lacrimal gland*[16] and also cause the eyes to water when irritated by strong taste.

Cranial Nerve # 10 is the Vagus Nerve. The Vagus nerve supplies the pharynx, the larynx, the trachea, the lungs, the heart, the liver, the kidneys, the esophagus, the small and large intestines.[17, 18] The eye is indirectly interconnected with the Vagus nerve by way of the superior cervical ganglion in the sympathetic tree.[18]

Cranial nerve #11 is the Accessory Nerve. The Accessory nerve supplies the intrinsic muscle of the larynx, some parasympathetic fibers for the Vagus nerve, the muscle of the side of the neck (sternocleidomastoid), and the muscles of the top of the shoulders (trapezius muscle).[19, 20] The eye is indirectly interconnected with the Accessory nerve by way of the Vagus nerve (the Nodose Ganglion) by way of the superior cervical ganglion in the sympathetic tree.[21]

Cranial Nerve #12 is the Hypoglossal nerve. The Hypoglossal nerve supplies the muscles of the tongue and some of the muscles of the neck.[22] The eye is indirectly interconnected with the Hypoglossal nerve by way of the Vagus nerve plexus (the Nodose Ganglion) and by way of the superior cervical ganglion in the sympathetic tree.[23]

In Summary, of the twelve cranial nerves, six directly supply the eye, and the other six indirectly have input into functions of the eye.

All six of the nerves which directly supply the eye come from different areas of the brain therefore they all have different paths to the eye. They pass through different foramen. All the nerves have their own arterial blood supply and venous supply. *All* must be present and fully functional - all at the same time! If any one were not present, the eye would not function properly. Each nerve had to be fully developed and fully functional all at the same time. Fully-developed, fully-functional simultaneous function is *creation!* Evolution cannot explain *all* these processes happening at the same time. We know they all had to occur simultaneously because they are all absolutely interdependent. They *all* MUST be present and they *all* MUST be fully functional *all* at the same time! Fully developed, fully functional, simultaneous function of interdependent parts can only happen by creation! Observe the illustrations on pages 178 -183. Look at the complexity; do you really think it could have evolved? We observe interdependent evidence of creation.

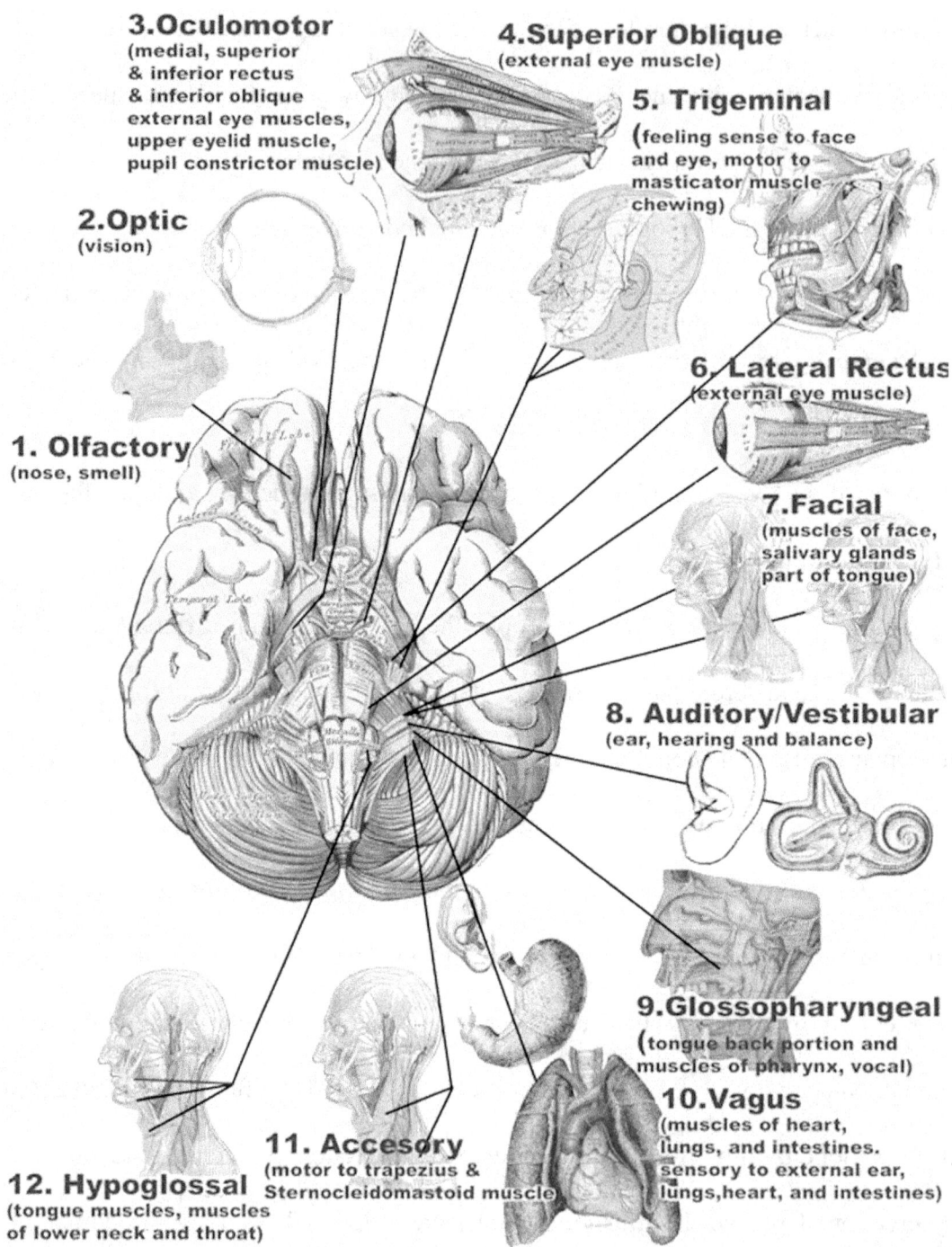

From Henry Gray, F.R.S, *Anatomy of the Human Body* 20th Edition, Lea & Febiger (Philadelphia & New York, © 1918) Page 817[24] with additional illustrations by Henry Gray from pages 881, 1006, 1022, 890, 898, 1022, 927, 379, 1034, 1048, 909, 527, 1165, 1034, and 379[26] combined and arranged by Rigney.

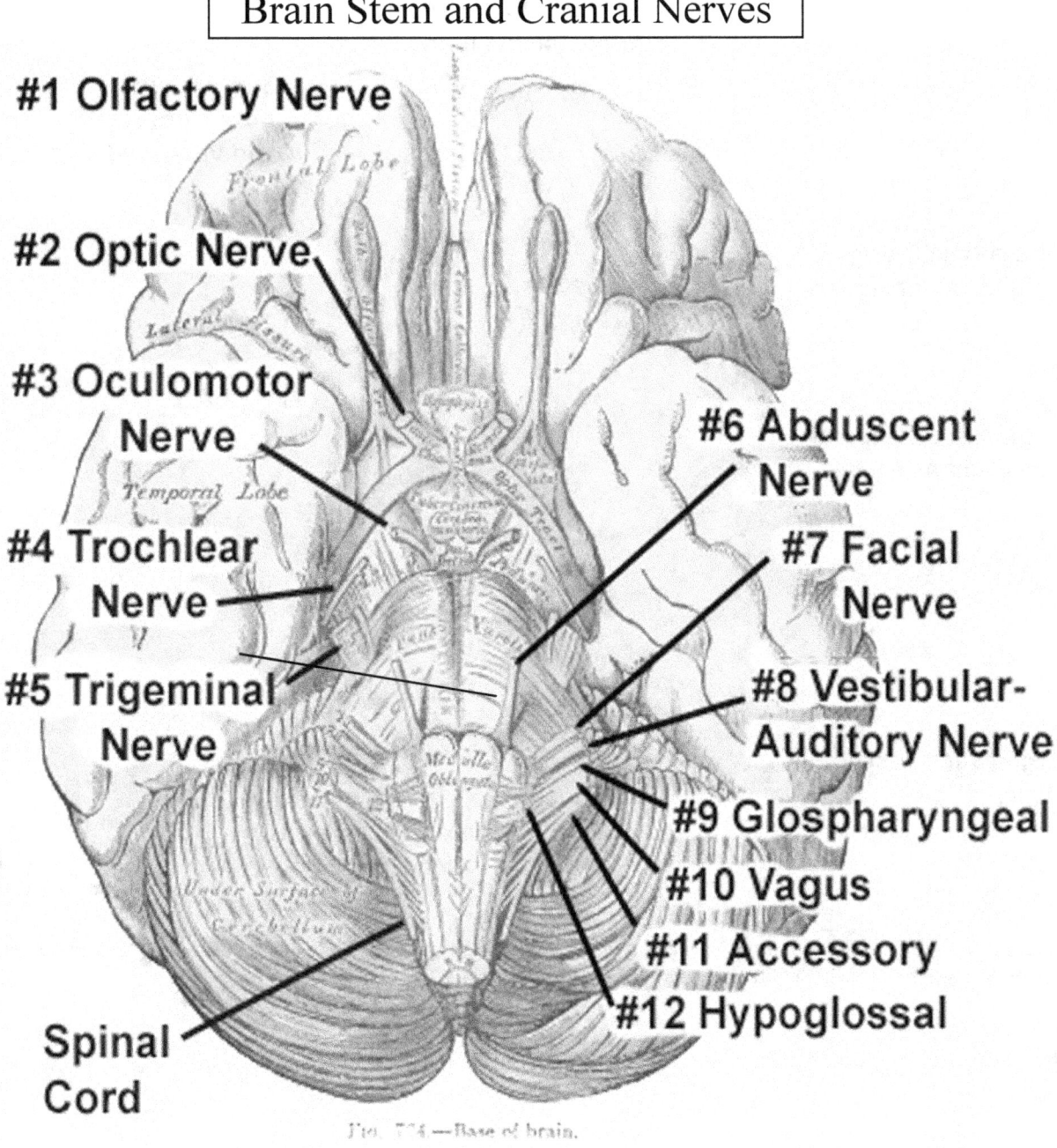

Fig. 714 —Base of brain.

From Henry Gray, F.R.S, *Anatomy of the Human Body* 20th Edition, Lea & Febiger (Philadelphia & New York, © 1918) Page 817[25] with additions and modifications by Rigney

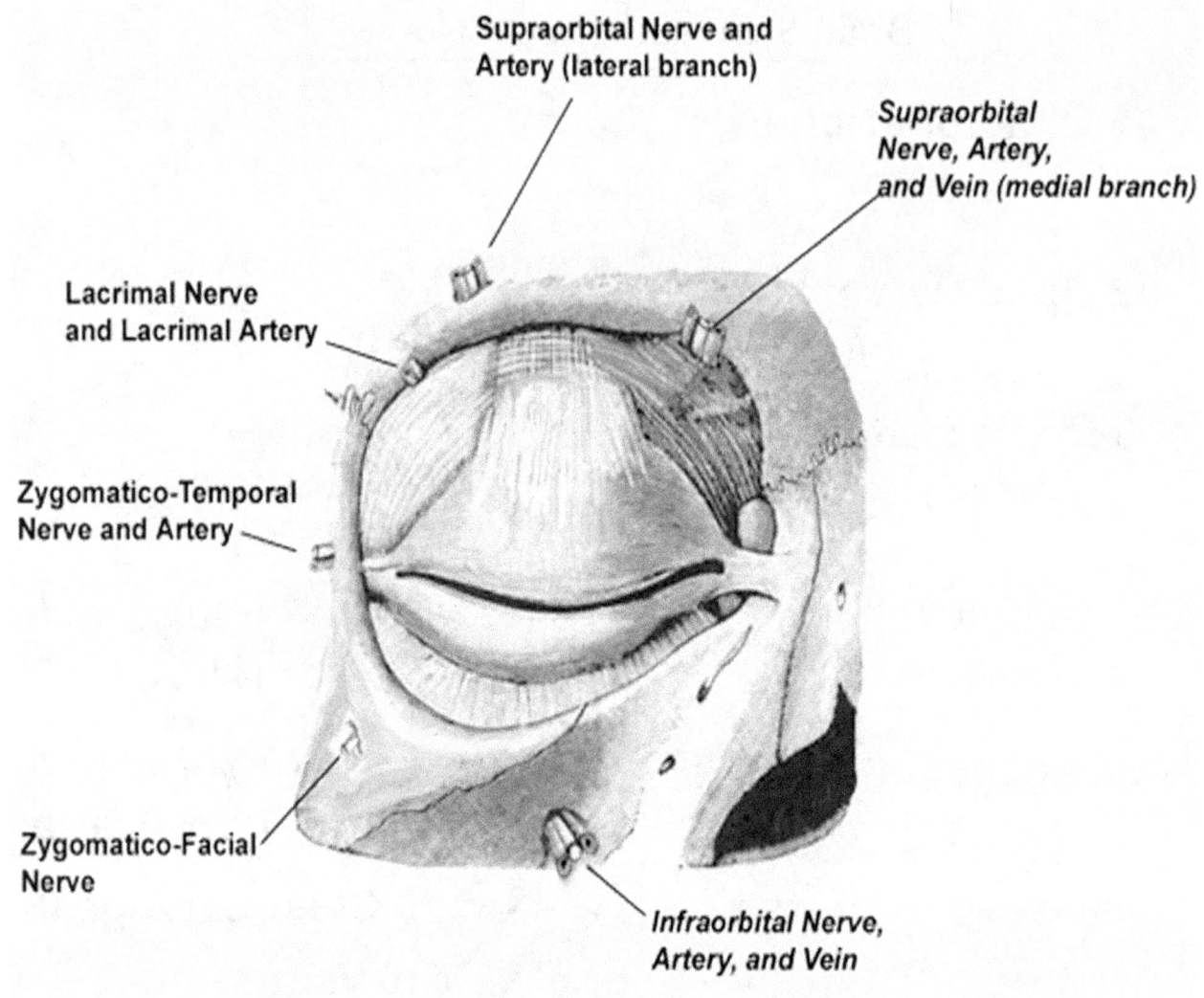

Supraorbital Nerve and Artery (lateral branch)

Supraorbital Nerve, Artery, and Vein (medial branch)

Lacrimal Nerve and Lacrimal Artery

Zygomatico-Temporal Nerve and Artery

Zygomatico-Facial Nerve

Infraorbital Nerve, Artery, and Vein

From Henry Gray, F.R.S, *Anatomy of the Human Body* 20th Edition, Lea & Febiger (Philadelphia & New York, © 1918) Page 1027[26] with additions and modifications by Rigney

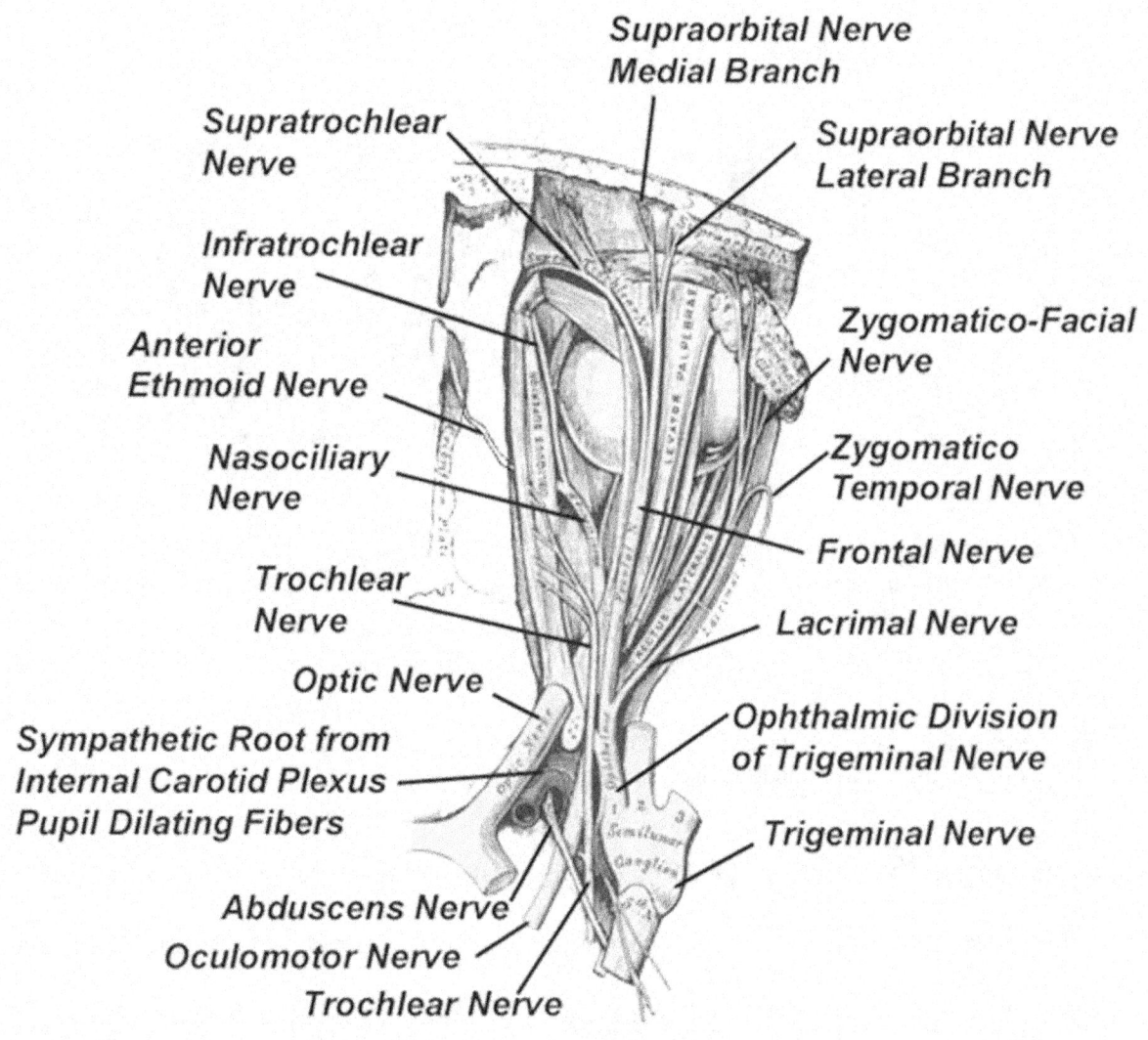

From Henry Gray, F.R.S, *Anatomy of the Human Body* 20th Edition, Lea & Febiger (Philadelphia & New York, © 1918) Page 885[27] with additions and modifications by Rigney

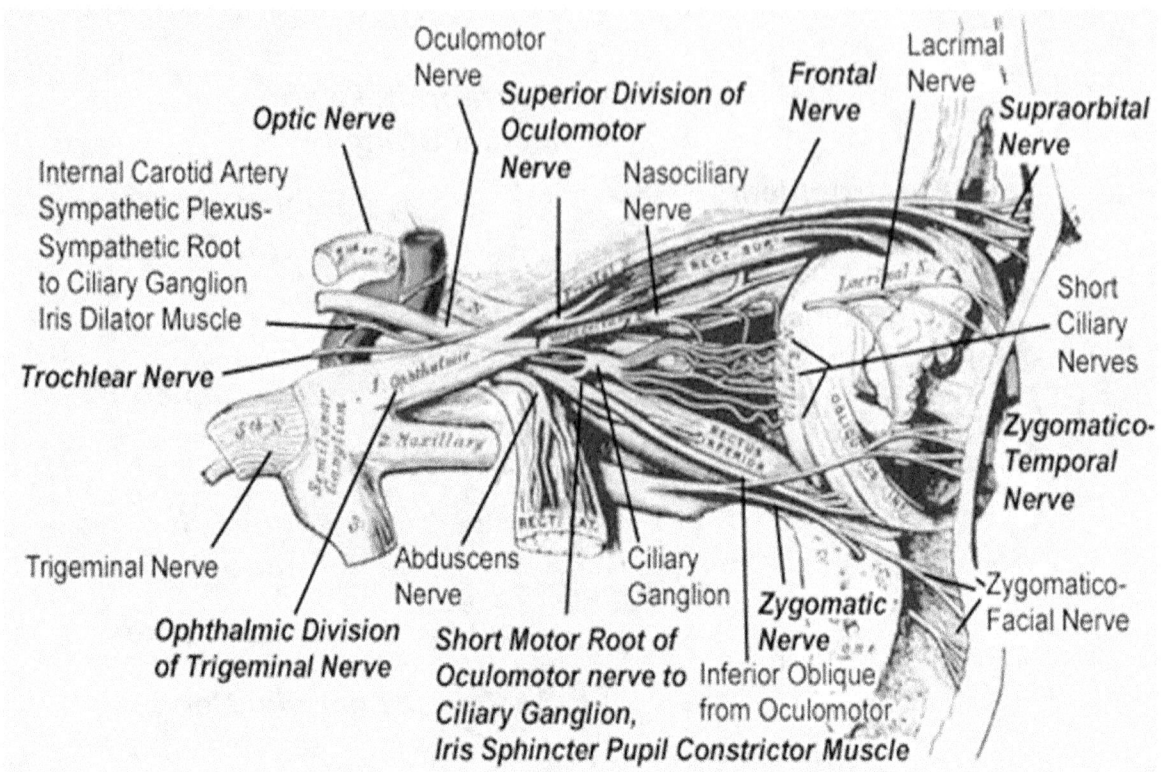

From Henry Gray, F.R.S, *Anatomy of the Human Body* 20th Edition, Lea & Febiger (Philadelphia & New York, © 1918) Page 887[28] with additions/modifications by Rigney

Look at the complexity of design! Observe the numerous interdependent structures! "Numerous, successive, slight modifications"[28] cannot explain all these different nerves and their specific ***SIMULTANEOUSLY REQUIRED FUNCTIONS!*** They are all interdependent, and they all must be fully functional, and fully developed or the eye cannot function. The eyes will not line up straight, they will not move together in unison and maintain single vision, the eye will dry out, and the eyes could not blink or tear - just to name a few of the required interdependent functions. Lose any ONE and it is fatal to the eye and ultimately the being. We observe Interdependent Evidence of Creation. Darwin's theory absolutely breaks down.[29]

Neural Contents of the Right Orbit

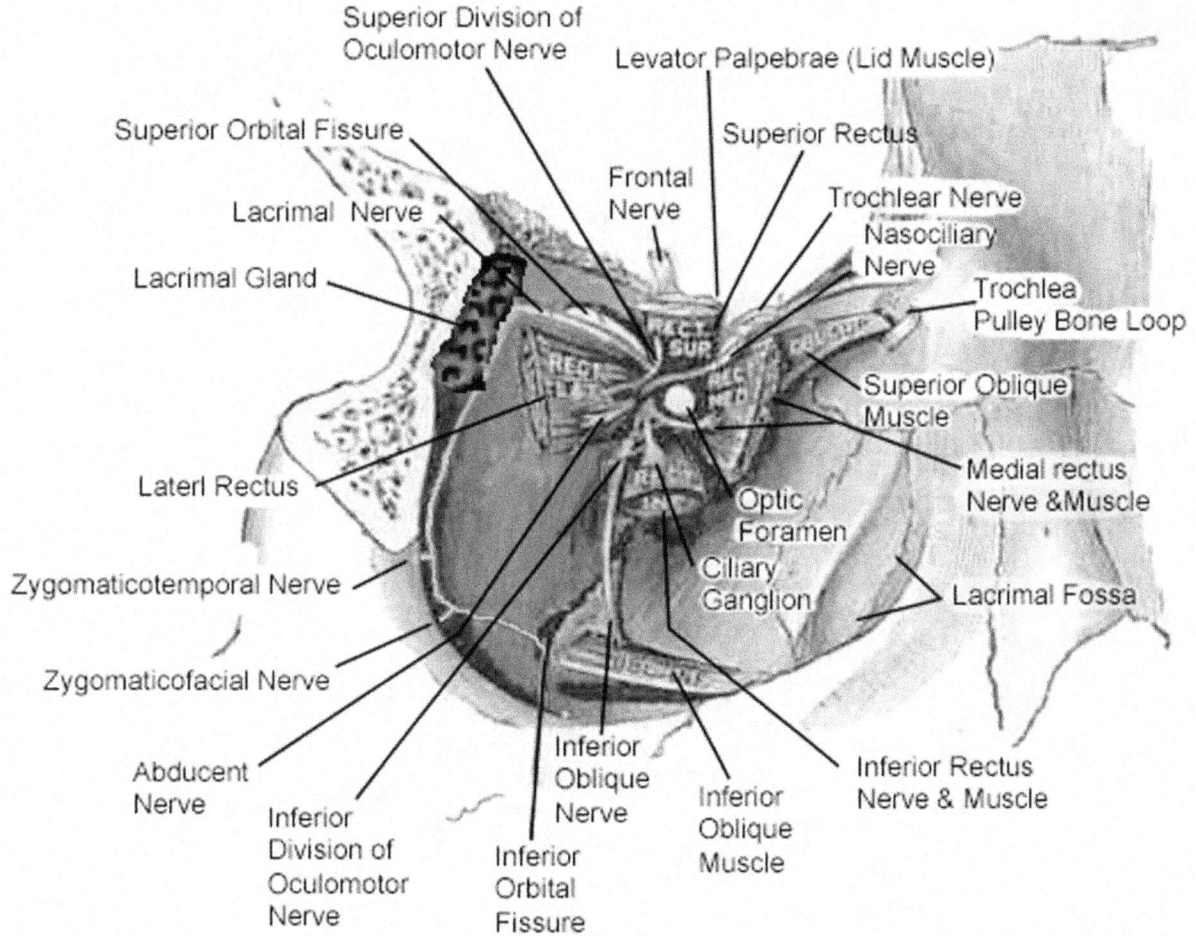

From Henry Gray, F.R.S, *Anatomy of the Human Body* 20th Edition, Lea & Febiger (Philadelphia & New York, © 1918) Page 900[30] with additions and modifications by Rigney

We observe: 7 Bones, 35 Foramen, 18 Nerves; ALL 60 ARE ABSOLUTELY INTERDEPENDENT!

Darwin's theory absolutely breaks down."[31]

20. Blood Supply to the Eye

Blood carries the nutrients and oxygen needed for the cells of the body to live and function. Without blood flow to the eye, the eye could not live. The arteries carry blood from the heart and lungs *to* the eye. Blood in the arteries has the oxygen from the lungs and the nutrients from the intestines needed for the eye to live and function. After the blood flows through the intestines to receive needed nutrients, it flows to the right side of the heart and then to the lungs for transfer of carbon dioxide (used air, expelled when you exhale) with oxygen (unused air we receive when we inhale) and then it goes back to the left side of the heart where the heart pumps the blood through arteries to the rest of the body and the intestines. Arteries therefore carry blood *with* oxygen and nutrients *to* the tissues. Veins collect the blood back from the tissues after they use the oxygen and nutrients, and the veins carry the blood back to the heart. Blood therefore *circulates* through the body. Blood travels through the body three times per minute[1] (or 4,320 times per day). In one day blood circulates through the body 12,000 miles! That's four times the distance across the US from coast to coast in just one day[1]! There is on average 100,000 miles of blood vessels in the average adult body[2]. Evolution cannot explain it. The individual would die while waiting on the blood vessels to evolve. They had to be fully developed and fully functional.

From the heart, the aorta is the main and first artery that carries blood with oxygen from the heart. All other arteries come from branches off the aorta artery. On the top of the aorta comes the innominate artery, which branches into the right subclavian artery, right vertebral artery, and right common carotid artery. A branch for the left common carotid artery comes off the aorta and a branch for the left subclavian artery comes off the aorta artery. From the right and left common carotids come branches for the right and left internal carotid arteries[3]. The ophthalmic artery branches off the internal carotid in about the middle of the brain and then goes to the eye. The ophthalmic artery branches into fourteen more branches in order to supply the eye, eye muscles, eyelids, forehead, and side of the nose[4] - that's twenty-eight branches for both eyes!

After the eye, eye muscles, the eyelids, the forehead, and the side of the nose use the oxygen in the blood, the veins carry the blood back to the intestines, lungs, and heart to dump off the waste products and the used oxygen (carbon dioxide) and to pick up new nutrients and a new supply of oxygen to repeat the process. All fourteen different branches of arteries from the Ophthalmic artery have to form completely and fully and must be directed by a specific route to the appropriate tissue of the eye. *And,* all the fourteen different veins of the eye must form properly and fully to carry the blood back to the intestines lungs and heart. Again, *all* must form fully and simultaneously because they are all wholly dependent and totally interdependent. If any one artery were missing or any one vein were missing -or if they were not all present at the same time, the tissue which lacked an artery or vein would die. One tissue could not survive while it is waiting for the other to evolve. They *all* must be present and *all* must be fully developed and *all* must be fully functional *all* at the same time. Fully-developed, fully-functional interdependent tissue systems which must develop simultaneously can *only* be explained by creation. In fact, this describes any act or process of creation.

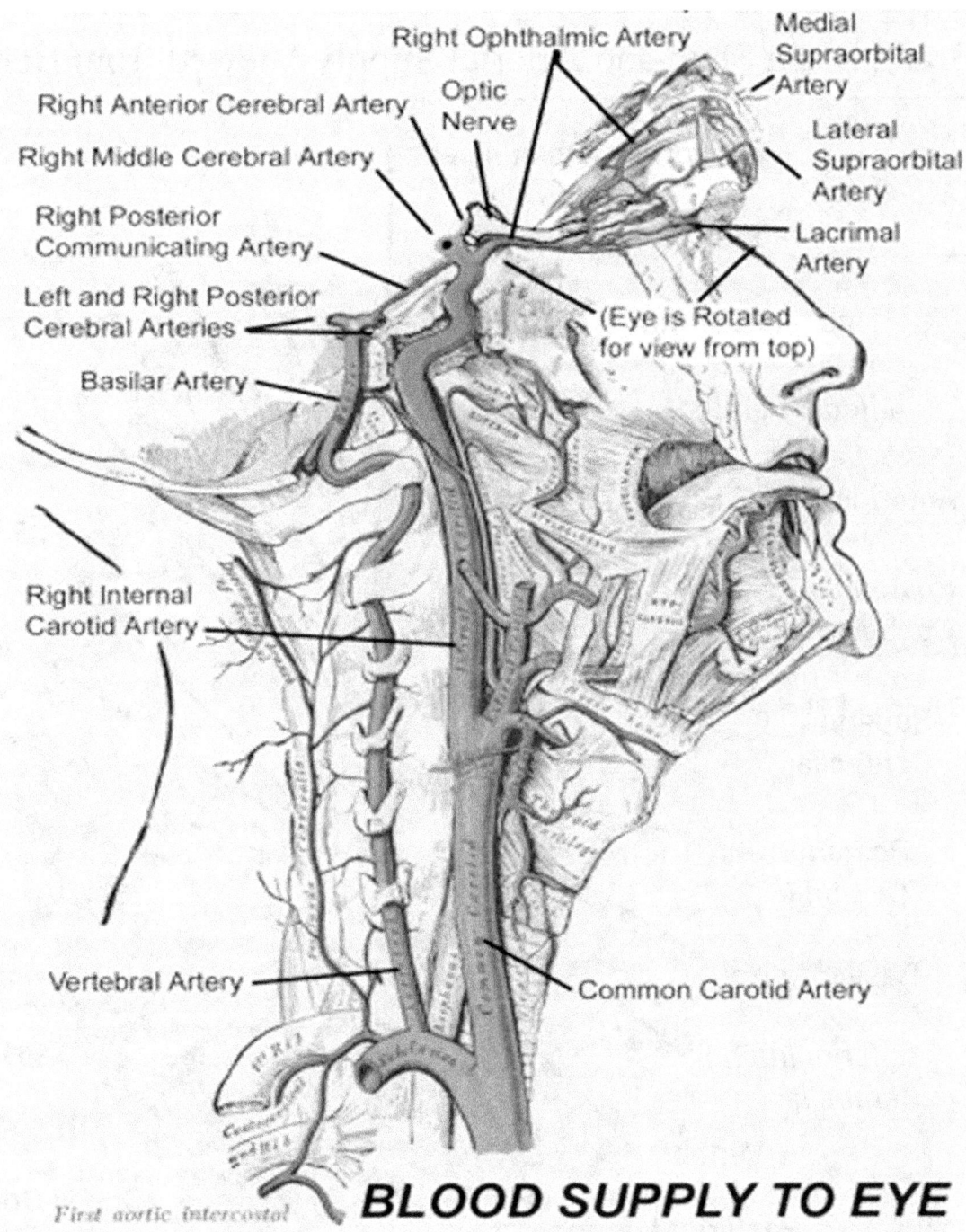

BLOOD SUPPLY TO EYE

From Henry Gray, F.R.S, *Anatomy of the Human Body* 20th Edition, Lea & Febiger (Philadelphia & New York, © 1918) Pages 566, 885[5] with additions and modifications by Rigney

J. Jay Rigney, O.D.

Brain, Brain Stem and Arterial Supply Viewed from Below

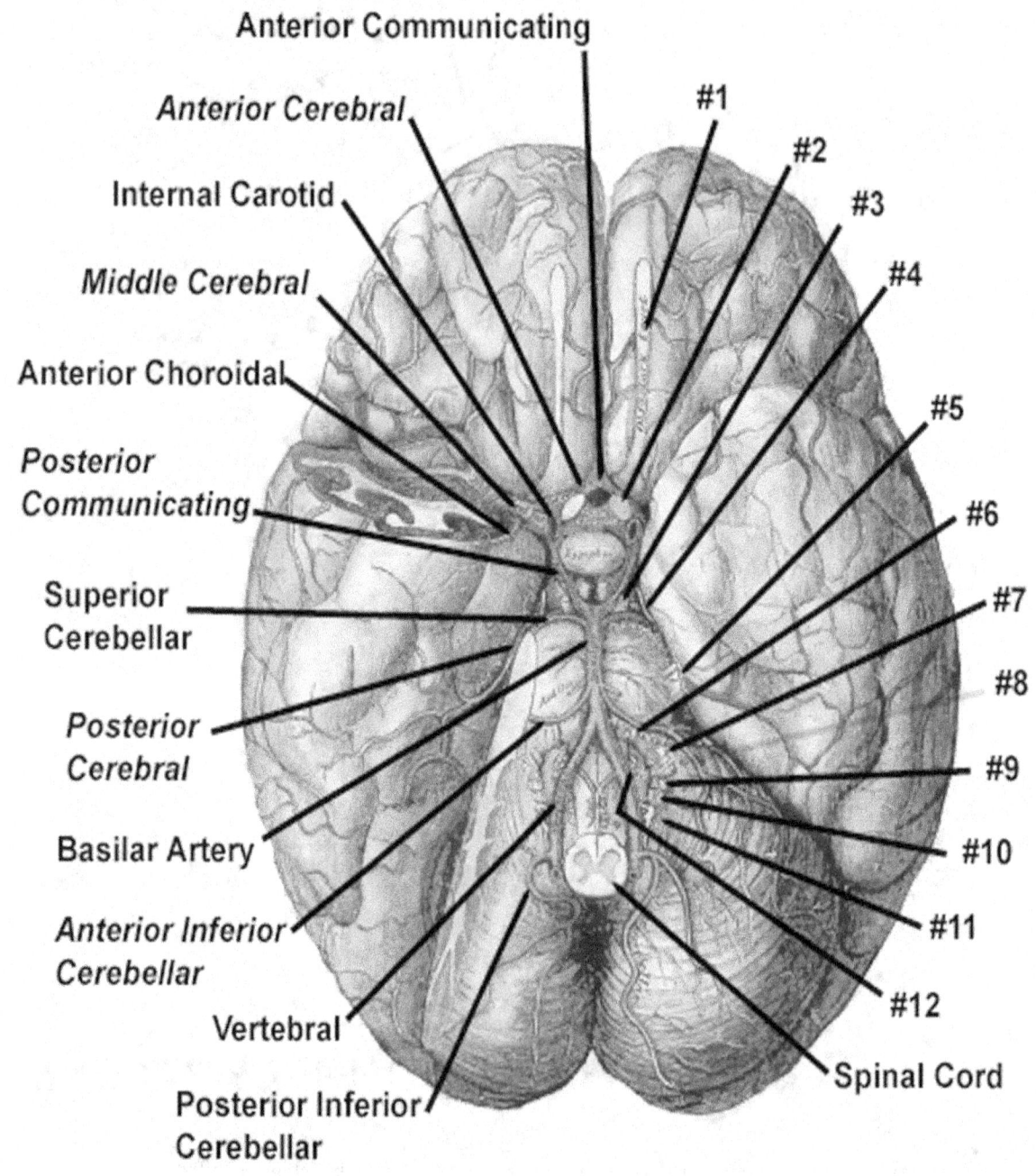

Anterior Communicating

Anterior Cerebral

Internal Carotid

Middle Cerebral

Anterior Choroidal

Posterior Communicating

Superior Cerebellar

Posterior Cerebral

Basilar Artery

Anterior Inferior Cerebellar

Vertebral

Posterior Inferior Cerebellar

#1
#2
#3
#4
#5
#6
#7
#8
#9
#10
#11
#12
Spinal Cord

From Henry Gray, F.R.S, *Anatomy of the Human Body* 20th Edition, Lea & Febiger (Philadelphia & New York © 1918) Pages 572, 574[6] with additions by Rigney

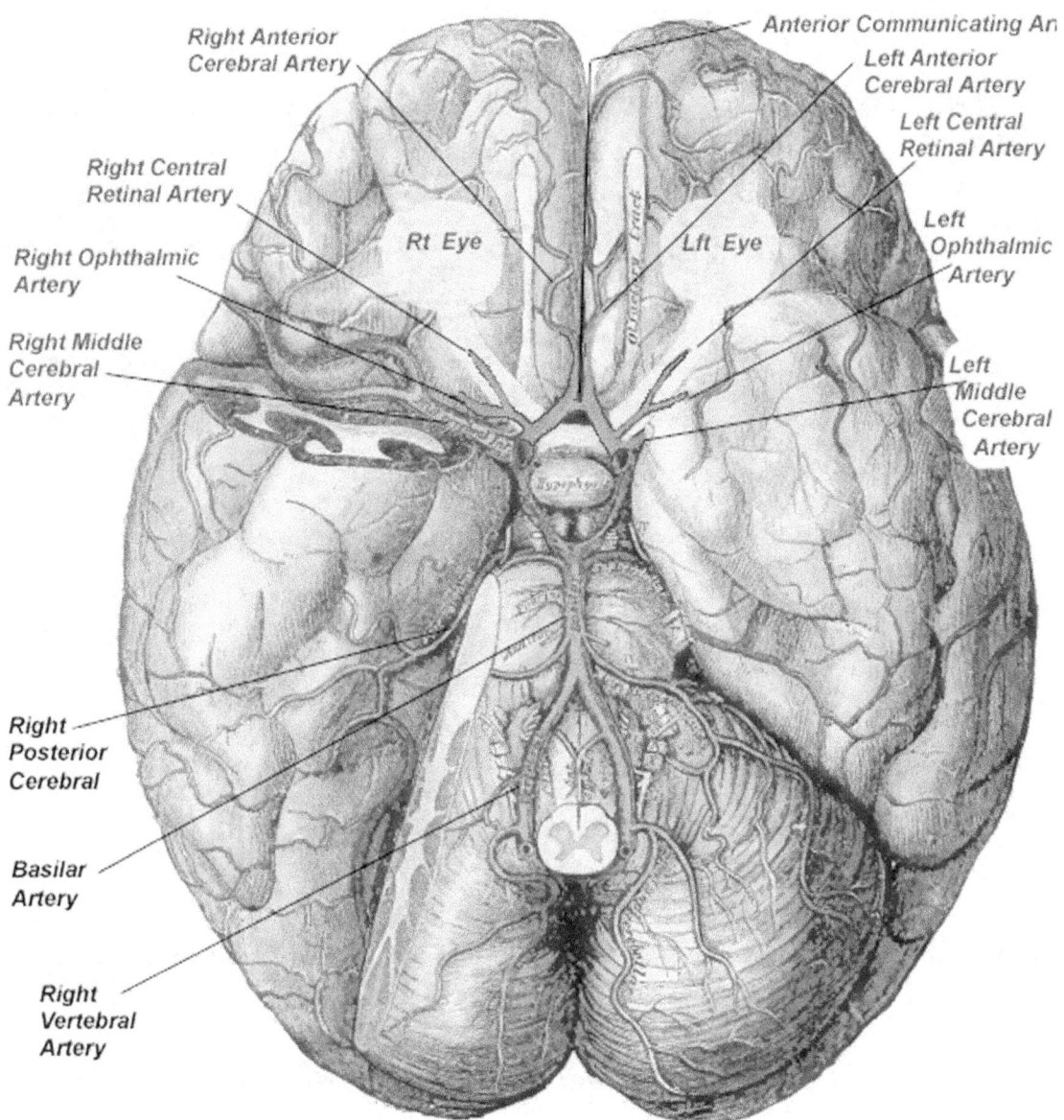

Blood Flow to the Eye as Viewed from Below

From Henry Gray, F.R.S, *Anatomy of the Human Body* 20th Edition, Lea & Febiger (Philadelphia & New York © 1918) Page 572[7] with additions & modifications by Rigney

Blood vessels of the eyeball
(The Sclera has been removed to reveal the Choroid)

Arteries carry blood to the EYE

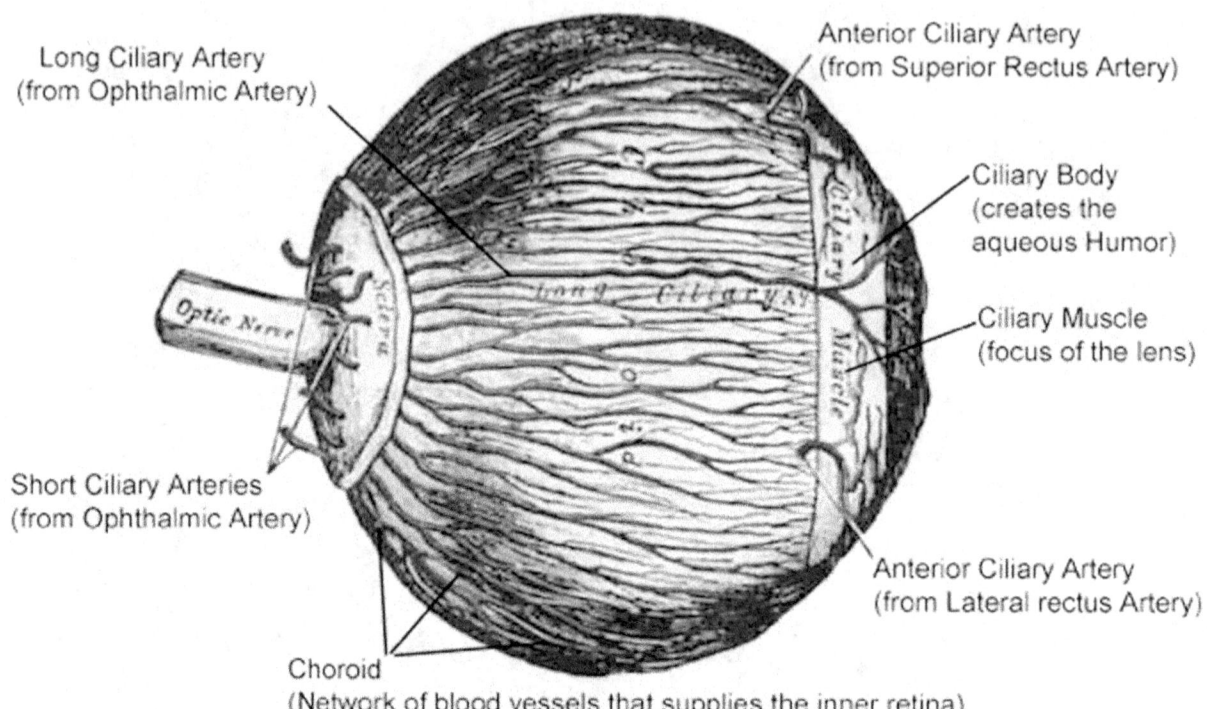

Long Ciliary Artery
(from Ophthalmic Artery)

Anterior Ciliary Artery
(from Superior Rectus Artery)

Ciliary Body
(creates the
aqueous Humor)

Ciliary Muscle
(focus of the lens)

Short Ciliary Arteries
(from Ophthalmic Artery)

Anterior Ciliary Artery
(from Lateral rectus Artery)

Choroid
(Network of blood vessels that supplies the inner retina)

From Henry Gray, F.R.S, *Anatomy of the Human Body* 20th Edition, Lea & Febiger (Philadelphia & New York, © 1918) Pages 1009[8] with additions by Rigney

All these structures are absolutely and totally dependent on the blood supply provided by the arteries and veins. If any one artery has not evolved, the tissue in that area cannot survive. ___**All**___ the arteries must be present simultaneously because all the tissues are dependent upon the blood for oxygen and nutrients; Interdependent Evidence of Creation.

After the tissues use the oxygen and nutrients in the arteries, then veins carry the blood back to the heart, lungs, and intestines to pick up and transport more oxygen and nutrients to the tissues. The Liver breaks down wastes and toxins, the Kidneys help filter out waste and toxins, the Large Intestines transport wastes out of the body[9]. Without your Liver or Kidneys you would die.

Veins carry blood back to the heart

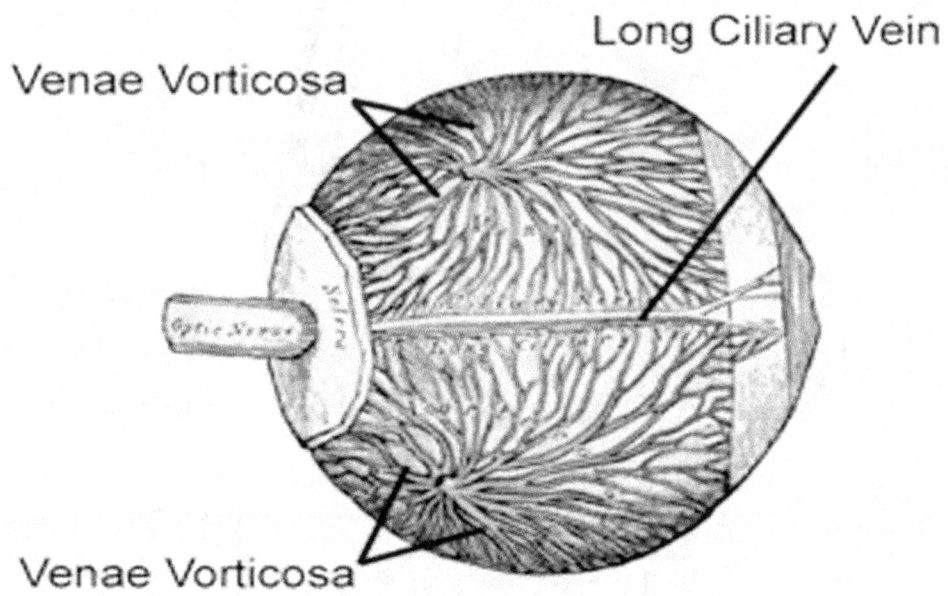

Long Ciliary Vein

Venae Vorticosa

Venae Vorticosa

From Henry Gray, F.R.S, *Anatomy of the Human Body* 20th Edition, Lea & Febiger (Philadelphia & New York, © 1918) page 1010[9] with additions by Rigney

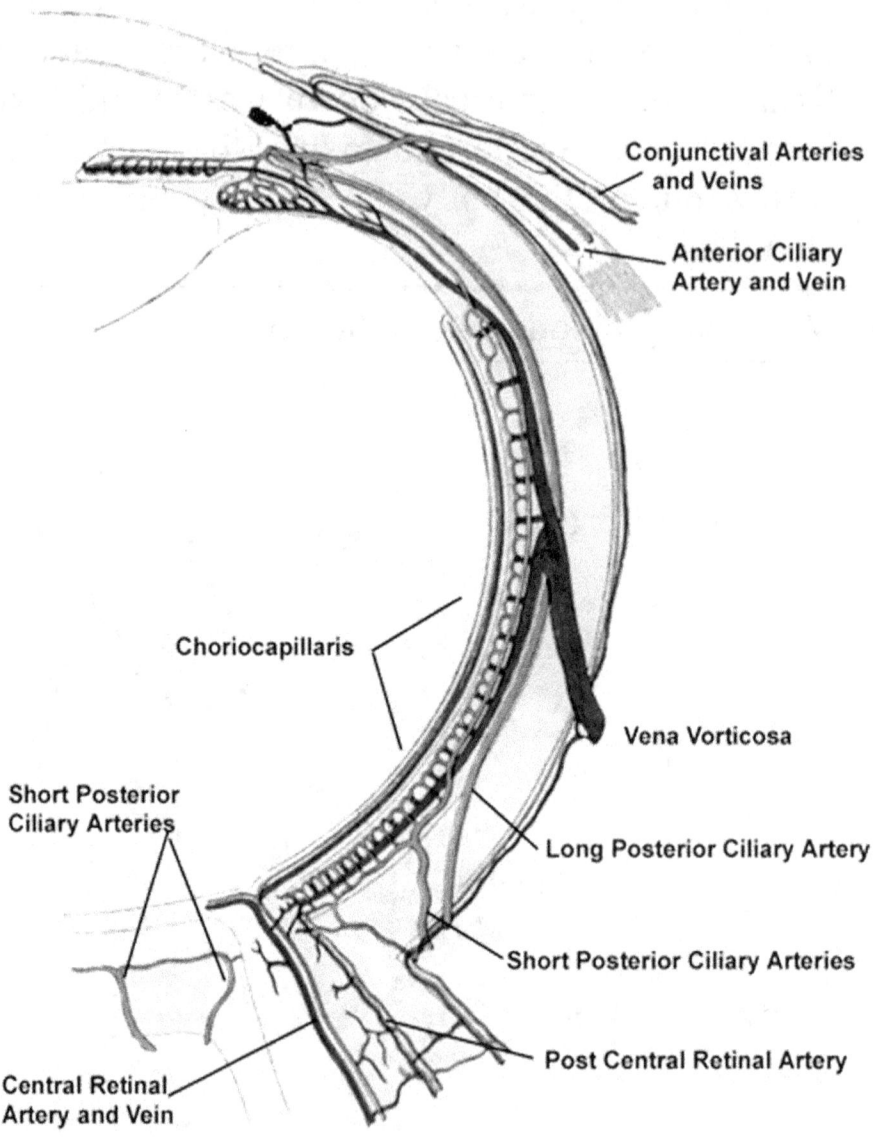

Conjunctival Arteries and Veins

Anterior Ciliary Artery and Vein

Choriocapillaris

Vena Vorticosa

Short Posterior Ciliary Arteries

Long Posterior Ciliary Artery

Short Posterior Ciliary Arteries

Post Central Retinal Artery

Central Retinal Artery and Vein

From Henry Gray, F.R.S, *Anatomy of the Human Body* 20th Edition, Lea & Febiger (Philadelphia & New York, © 1918) page 1012[10] with additions by Rigney

Veins of the Face

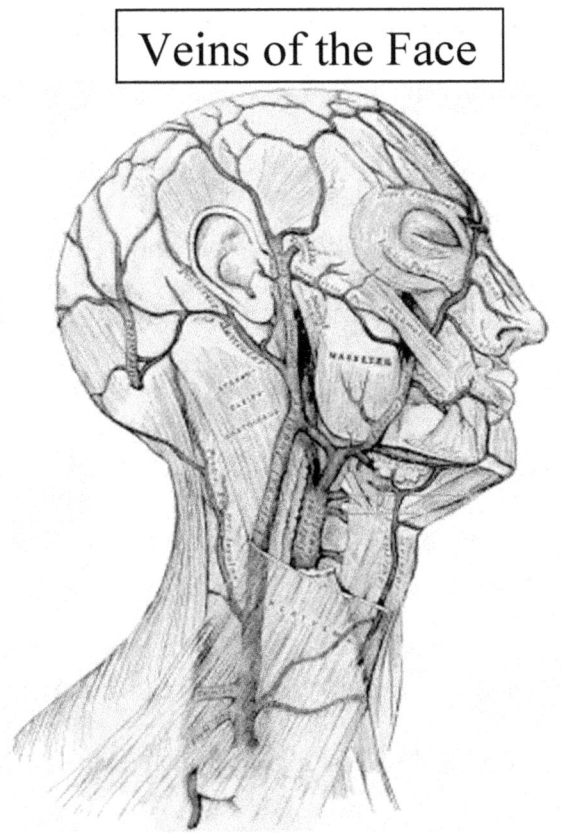

From Henry Gray, F.R.S, *Anatomy of the Human Body* 20th Edition, Lea & Febiger (Philadelphia & New York, © 1918) page 644[11]

Veins of The Eye

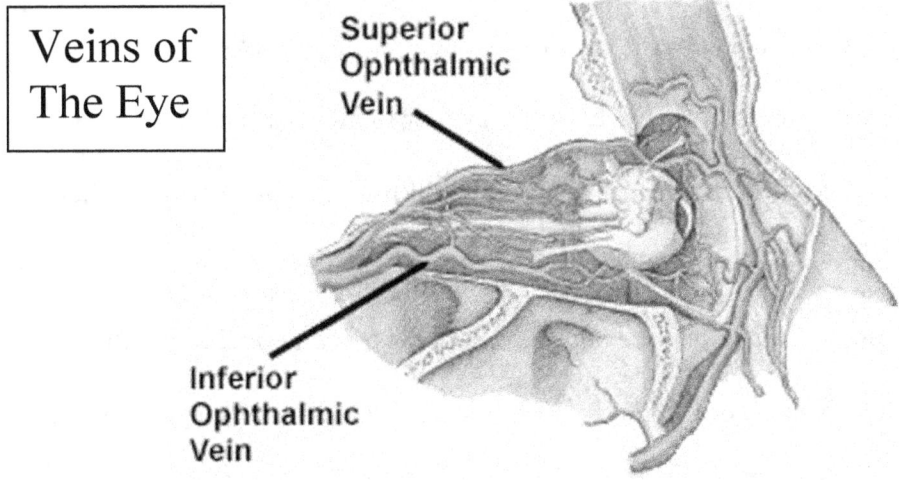

From Henry Gray, F.R.S, *Anatomy of the Human Body* 28oh Edition, Lea & Febiger (Philadelphia & New York, © 1918) page 659[12]

Look at the complexity of these last six drawings on pages 187-193. Observe and form your *own* conclusion. Do you think after observing all 56 arteries and veins, they could have evolved? The tissues cannot survive while they are waiting for the arteries and veins to evolve. Had they not evolved simultaneously those areas with no evolved arteries and veins would die. *ALL* of the arteries and veins *MUST* be present - *ALL* AT THE SAME TIME! EVOLUTION OF BLODD VESSESLS CANNOT GRADUALLY HAPPEN BY NUMEROUS SLIGHT MODIFICATIONS OVER LONG PERIODS OF TIME! THEY MUST ALL BE PRESENT SIMULTANEOUSLY.

FULLY DEVELOPED, FULLY FUNCTIONAL SIMULTANEOUS FUNCTION CAN ONLY BE EXPLAINED BY CREATION!

Observe and conclude. Someone may say, "Well that is not scientific, you can't just observe and conclude." However, this *is* science, *and* this is **_exactly_** what Darwin did when he developed his theory of evolution. Why is it acceptable to observe and conclude something evolved, but it is unacceptable to observe and conclude something was created? Just by observing the blood supply to the eye, the eye proves it could not have evolved. "Numerous, successive, slight modifications[13]" will not work when it comes to the blood supply of the eye, or the rest of the body.

I once heard Dr. R.C. Sproul Sr. teach[14], if you were walking in a forest and you happened upon a watch lying on the ground, would you conclude it was created? Yes! Did you see someone create or make it? NO! Even though you did not see anyone create it you can conclude with 100% certainty it was created - simply by observation! Would you think … "The watch evolved" or, "those pieces came together on their own" NO! The irony is the watch is much less complex than any living being. Dr. Sproul also said if you examine your shoe you can conclude it was created - just by observation[14]. How do you know your shoe was created? Could you conclude instead it evolved? You examine the evidence and after examining the evidence you conclude it was created. How do you know you are correct? ***The evidence*** points to a creator!

I once heard Dr. Stephen C. Meyer teach[15], if you find writings on a wall in a cave, would you conclude it was created? Yes, and as Dr. Meyer says, so it is with DNA and genetic code. It is written *code*, ***written information***, instructions for what the cell is and does – it is *information*[15]. Would you think the writings on the cave evolved? Did you see someone write them? NO, yet you can conclude someone did write them. They were created and *you know it* even though you did not see them write it BECAUSE THEY CONVEY INFORMATION AND INFORMATION CONVEYED MUST HAVE HAD SOMEONE TO SUPPLY OR PROVIDE THE INFORMATION[14]. It is the same with DNA and the genetic ***CODE.***[15] AS Dr. Stephen C. Meyer says, the DNA contains the genetic code which conveys information, and information conveyed is evidence of a Creator of that information.

Observe and conclude. The eye; just as the watch, just as the writings in a cave, JUST AS YOUR SHOE is evidence of a Creator! All life - *plant and animal* is evidence of ***THE CREATOR!***

21. Additional Scientific Evidence of Creation

Darwin was twenty-two years old when he set sail December 27, 1831 on a ship named the *Beagle.*[1] He was fifty years old when in his book *On the Origin of Species by Means of Natural Selection* was published in 1859.[2] Since the 1800's much has changed regarding the spectrum of knowledge in the scientific world. Science has ***proven*** at one time there was nothing and then all at once there was everything—the "big bang" has been ***proven*** in at least four different ways.

1. Edwin Hubble in 1929 at the Mount Wilson Observatory *proved* the universe is expanding.[3, 4, 5] The Hubble telescope now confirmed this and more. The Hubble telescope has been recording the movement of the stars as they are ever expanding and spreading farther and farther away from each other. Some of the stars and galaxies are moving at a speed of 90 million miles per hour![4, 5] If you then play the video of the expansion of the universe backward, in reverse you would see the movement of the stars - moving to a central point they call *the point of singularity.*[5] Therefore, in forward regular play we see the galaxies and the stars are expanding out from a central point. This alone, *proves*, everything in the universe, at one time, had a common point of origin. Cosmologists call it the "big bang"—the moment everything came into existence. Again, this fact has been ***proven***.

2. Additionally, Albert Einstein's theory of relativity also predicted a point of origin for the universe. Edwin Hubble at the Mount Wilson Observatory[5, 6, 7, 8] confirmed and proved his theory.

3. The "big bang" has also been *confirmed* and proven by observation and measurement of the *red shift* of light as it passes from galaxies of different distances.[9]

4. Additionally, just recently published[10, 11] in March of 2014, scientists at the project known as Bicep 2 or Background Imaging of Cosmic Extragalactic Polarization[10,11] (which also included researchers from Caltech, the Jet Propulsion Laboratory, Stanford University, and the University of Minnesota) have discovered the first direct evidence for the theory of cosmic inflation—the mysterious and violent expansion after the "big bang." Again, *proof* the "big bang" actually DID occur!

The fact there was a "big bang" has been ***proven*** as exampled above, in at least four different ways. What *caused* it is the current debate.

The evolutionists believe the "big bang" was the moment everything came into being from *nothing*, and *nothing* caused it to happen. *Nothing* caused *nothing* to become *everything*. The evolutionists believe and have faith nothing can become everything by nothing. However, can *nothing* create or become anything without an outside influence? Think about it. Meditate on the possibility of nothing becoming everything by nothing.

The creationists believe and have faith the "big bang" was the moment God said, "Let there be light," and He created the universe. The creationists believe and have faith an eternal, all-powerful God made everything, and when God said, "Let there be light," the universe exploded into existence.

Everything... from nothing? or *Creation?* Again, which is harder to believe? In either case it requires faith to believe? Either put your faith in nothing or put your faith in God?

Evolution? or *Creation?* The evidence overwhelmingly demonstrates Creation! If God can create the universe. This ***IS*** what the scientific evidence has ***PROVEN !*** Therefore, it actually does not require faith to believe in Creation! It does require faith to believe in evolution ! Isn't that ironic, it requires faith to believe in evolution while Creation has been scientifically ***PROVEN ! Creation is science and evolution is a religion !*** He certainly can create the *living* part of life. Speaking of life, how can elements, amino acids, proteins, come to life? The evolutionist, cosmologist, physicists, and statisticians will tell you, "by chance", given enough time, because there are billions and billions of opportunities for it to happen "by chance" - given enough time (billions and billions of years).

However, we are speaking of life; the *living* part of "all organic beings[12]"- that which Darwin was referring to when he said, "into which life was first breathed,"[12] (And, as you will learn in Chapter 24, pages 306 -307 Darwin believed in the Creator.) The living part of life is the ultimate *proof* of God. Men have attempted to create life in the laboratory.[13] Men have attempted to *cause* to happen, what they think spontaneously happened by chance, for life to *evolve,* when the earth was in its infancy. In multiple different experiments Stanley Miller took the elements they believe were present when the earth was in its infancy and added electricity (energy) to see what would happen. The most that happened was a few amino acids were formed[13,14] (amino acid groups are required to combine to create proteins. Proteins are required as the building blocks for living cells.[13, 14]) Because amino acids were shown to have formed, the evolutionists have said, "See! There are amino acids! That is the beginning of life!" However, in all the experiments these amino acids did not then go on to form proteins. Additionally, and of even more importance, they have taken this experiment several steps further. They *know* what chemicals and compounds are necessary to create amino acids, they *know* what chemicals and compounds are necessary to create proteins, they *know* the vitamins, minerals, and nutrients necessary to sustain life. Knowing this today, you can take all these ingredients necessary for life; all the chemicals, all the proteins, all the amino acids, add in vitamins, minerals, nutrients, water - whatever you think might aid in the forming or evolving of life; then add light, energy, heat, electricity, or whatever you think may be even remotely necessary in order to facilitate a reaction. Create the conditions we *know* are even ***more ideal*** than those present when, according to evolutionists, the "earth was in its early evolutionary infancy;" then put it all in a test tube, beaker, pot, pan, incubation chamber, or whatever you think would be the best conditions to aid in creating life. Establish the conditions so they are even ***more ideal*** to support life than what they believe to be present when the earth was in its infancy; try to *facilitate* life to occur in any and all ways possible, and in any combination, and of any and every kind. Do anything and everything, necessary and possible, to *cause* life to occur - but you cannot make it happen, you can't get life to happen – even when tipping the tables in a way to try to *make it happen,* and *try to help it happen,* you can't *evolve* or create life.

The evolutionist will tell you, "Well, if you have enough time, *eventually* it *will* happen." Why would it ***not*** happen on one day - especially when you are *trying to make it happen,* and *facilitate it to happen* (which means it is not actually happening by chance) yet, on another day, for some reason it *can* happen, and happen on its' own - by chance? Their answer is "Given enough time by *chance* it can happen." They have *great* faith in *chance* yet they have no faith in God! Again, evolution is a religion and Creation is science!

This is a personal note from the author to the evolutionist or atheist who may happen to be reading this book. Open your eyes and open your mind! Consider what you are *doing,* what you are putting your trust in. You *are* putting your trust in something. You *have* made a choice. You have faith and trust in *chance* and evolution. You believe everything occurred by chance with no outside intervention; it just happened all on its own - by *chance*. Look at the evidence! Have faith in and trust in God instead of chance or evolution! Choose God! It makes more sense logically and intellectually to choose God! Especially knowing that science has ***PROVEN*** Creation Look at all the evidence in the world. Look at the evidence in this book! Look at what Darwin and the evolutionists are telling you to believe, "***probably*** all the organic beings which have ever lived on this earth have descended from some one primordial form, into which life was first breathed"[15] and, "all animals and plants have descended from some one prototype"[15] Think for yourself! Don't be deceived! Who did the *"breathing into"* in Darwin's statement, "into which life was first breathed"?[15] This is a statement. This states someone "breathed life" into the first primordial form. You can't deny God exists and at the same time say "into which life was first breathed![15] Darwin is thus saying, there ***must*** have been GOD "first"[15] - in the beginning! If there was God "first", God did the "into which life was first breathed."[15]

Again, "imagination", "supposition", and "belief" - require faith. Based on these statements made by the father of the theory, evolution is really a religion. It requires faith to believe it. Evolution is really a religion and therefore should not be taught in schools. If evolution can be taught in schools so should creation, it is only fair and right. In fact, Creation makes more sense.

The goal of the evolutionist and the atheist is to eliminate God, to take God out of ***all*** equations. When you do this there is no absolute truth. There is no right and no wrong. You can do whatever you please, and to whomever, however as you please if that is what it takes for the individual to survive - or if it's just what you want to do.

Thankfully there is a God and if you choose NOT to acknowledge Him this does NOT mean He doesn't exist. If He does exist, the problem is; one day there will be a reckoning, and those who refuse to acknowledge him will suffer eternal punishment in hell[16] (Matthew Chapter 24 through Chapter 25, The Amplified Bible). The evidence leads to the conclusion there is God. HE is the GREAT I AM.[17] And consider this, if you have an eternal soul when you die, your soul will continue to exist eternally, either in heaven eternally, or hell eternally, you get to choose who you will follow and where you will spend eternity. God has revealed Himself to everyone. You now have the proof you may be looking for. And he says you have no excuse_not believing in Him.[18] We see God through His creation of the things we can see, we can also see God's wisdom, intelligence, foresight, planned prevention, exceeding abundance, beauty, wonder, and power in the creation and the working of His engineering marvel—the design and function of the eye. God has revealed Himself to us *through* our eyes and God has revealed Himself to us *in* our eyes.

Even so, there is additional scientific evidence of Creation; the following portion of this Chapter, specifically, this section, I originally removed from the book altogether. However, I had been praying, asking God whether I should keep it in or not. This Chapter, Chapter 21, while it is seventh from the end of the book, as it has happened, is the last chapter I am writing. Again, I wasn't sure if God wanted me to say what I had originally written here so I originally took this section of Chapter 21 out. But I am now writing about it again as I think the Holy Spirit has caused events to take place such that He wanted me to put it back in. As a matter of God's timing, or I should say *because of God's timing*, I am including the next section of Chapter 21 in the book. I am about to disclose a rather outlandish theory I have had, about something I think God's word, the Bible, says – but, I actually felt it was a little *too far out there* to include. Then, today, *before* I began writing, I read a headline in the news I think, confirms my rather outlandish theory; at the least, it lends much credence to my theory. Therefore, even though it is rather, "out there" in thinking, it may answer some questions many people have about Creation, and God's word, and the reliability and accuracy of scripture. Specifically, it is regarding the age of the earth. Again, this is new thinking (rather outside the box thinking) on my part, and I wasn't even going to share it; except *TODAY* as I was reading the news, I came across an article I think, confirms or at least substantiates what I had written - but felt uneasy including in this book. It is ironic that *TODAY* my plan was to leave this section out and thereby; my book would have been finished. However, because I read this particular article *today* I am thinking possibly the HOLY SPIRIT WANTS ME TO GO AHEAD AND ADD IT BACK IN THIS SECTION OF MY BOOK. (As I was typing I just *accidentally* hit the caps key - but I am leaving it that way, all in caps, as I am questioning if *that too* was divinely appointed).

I am paste-ing back-in, at this moment what I previously had cut out! (Isn't this fun – isn't this exciting!? It is *very exciting to me*, to think of the timing of how all this has happened.) Here it is, please keep an open mind to what I am suggesting in this theory as I paste it back in.

(I have just pushed the Paste☺ button. Why, or how, this smiley face has appeared I haven't a clue! I do not know how, or why, it got here. I have searched for ways I could have accidentally caused it to be placed here but have not found a smiley face icon in *any* of the Microsoft Word programs I am working in. But I like the fact I was questioning adding this chapter in, and when I was obedient to the Holy Spirit; when I pushed the "paste" command, it appeared. *And,* I like the fact it appeared by no method known to me, so I am leaving it in - because I too am smiling; happy God's timing was such He has showed himself real to me!)

I understand there is a dispute between evolutionists and creationists over the age of the earth. Evolutionists, using varying different means (carbon 14 dating, erosion rates, sedimentation rates, etc.) typically come up with a number for the age of the earth in the billions of years, while Creationists using dates, ages of individuals, and times stated in the Bible, typically come up with a number in the thousands of years.[19] As a result evolutionists, astronomers, geologists, and archeologists say, "see the Bible is in error", and therefore, they say, "what the Bible says of creation, and anything else the Bible says, for that matter, is of no account" – because they say, "the Bible is full of errors". I have heard, or read that statement many, many times in my life. "The Bible is full of errors." I DISAGREE! I believe it is without error - not even one.

I have settled on a couple of my own theories, *possibilities,* which I am comfortable with - and I now feel one of my theories has been confirmed, which may explain the difference in what the age evolutionists, astronomers, geologists, and archeologists say the earth is, and what the creationists say the age of the earth is. We can *reconcile the disagreement of the age of the earth* in these two differing views. I have not heard these ideas anywhere else so these are *my theories* and *my thinking* and I may be totally in error (so keep that in mind as you read this chapter). It is possible someone else at some other time has presented this theory or something similar – if they have I am not aware of it, but if they have great! (When more people come up with similar theories I think it only lends more credibility to the theory.)

Please remember I originally intended to leave this section of this chapter out because it is a little *out there* in thinking, but after you read it I think you will agree it is plausible. I can't say I have had any special revelation form God regarding this. I *can* say however, I pray and ask for God's Wisdom *continually, daily*. He said He will give us His Wisdom if we ask Him. Therefore, I ask Him for His Wisdom to guide me as I write this book, so I am hoping there is His input and He is inspiring me into what I am writing. Again, I originally questioned, but now believe He indeed wanted me to include this section on differences in the estimated age of the earth. *The day* I was deciding whether to include it- an article came out in the news I believe confirms what I had previously written. *Plus,* in pasting this section back in, the smiley face appeared – and I cannot find a smiley face Icon in the processing program I am using to write this book. I think this is confirmation from God that I needed to put this section back in.

My first possible theory is the theory *God's days, initially, are different, than man's days*. (I back this theory up with Biblical facts.)

One of the ways I have reasoned and feel comfortable with why there is a difference between what evolutionists, astronomers, geologists, and archeologists *think* the age of the Earth is, and what creationists *think* the age of the Earth is, the difference may possibly be due to God's *days* being different than our *days* initially. The **God's days _initially_ are different from man's days** theory is as follows.

First, I believe the Bible is without error and what is written was directly inspired by God as is stated in scripture in 2 Timothy 3:16, "Every scripture is God-breathed (given by His inspiration)"[20] So, if you read what God *Himself has* to say about it you should be able to see *His* answer. When you read what God *Himself* says about it in the Genesis account in the Bible, *I think* when God is stating evening and morning He is stating a different evening and morning than what we think about as <u>one day</u> in *our time* (we are thinking 24 hours). God Himself tells us in Genesis 1:1, "In the beginning God (prepared, formed, fashioned,) created the heavens and the earth."[21] That is, *GOD SAYS* the heavens and the earth were formed *before* the galaxies and our solar system in the beginning (the ***very*** beginning - the very first thing, the Bible says, "In the beginning God Created the heavens and the earth."[21])

Genesis 1:**1** says, "In the beginning God created the heavens and the earth."[22] So, He created the earth <u>even before he created light</u>; at least this is what GOD has written in Genesis 1:1-3.

Genesis 1:**2** says, "The earth was without form and an empty waste, and ***darkness*** was upon the face of the very great deep." [23] "The Spirit of God was moving (hovering, brooding) over the face ***of the waters***". Thus, we KNOW: 1. There was water present and 2. The earth was present 3.There was darkness on the face of the earth, and 4. God's Spirit was present.

God has yet to say, "let there be light". The earth could have been in this state… for eons of time. Therefore, He says the earth was created in the beginning – even before ***HE*** said let there be light, and His Sprit was moving (hovering, brooding) over the face of the waters[23]. It is very important to remember that water is present "in the beginning" as the presence of water on the earth "in the beginning" is ***the verifying factor*** now confirmed by science of this theory.

Genesis 1:**3** says, "and God said, let there be light; and there was light."[24]

Genesis 1:**4** says, "And God saw that the light was good and He approved it: and God separated the light from the darkness."[24]

Genesis 1:**5** says, "And there was evening and morning, <u>one day.</u>"[24]

He is calling "evening and morning "one day" before He has formed the sun and our solar system. (Watch and see, He does not form the sun until Genesis 1:**14** <u>and here we are in Genesis1: 5</u>).

Genesis 1:**6** says, "And God said, Let there be a firmament [the expanse of the sky] in the midst of the waters, and let it separate the waters [below] from the waters [above].[24]

Genesis 1:**7** says, "And God made the firmament [the expanse] and separated the waters which were under the expanse from the waters which were above the expanse, and it was so."[24]

Genesis 1:**8** says, "And God called the firmament Heavens. And there was evening and there was morning, <u>a second day.</u>"[24]

Genesis 1:**9** says, "And God said, Let the waters under the heavens be collected into one place [of standing], and let the dry land appear. And it was so."[24]

Genesis 1:**10** says, "God called the dry land Earth, and the accumulated waters He called Seas. And God saw that this was good (fitting, admirable) and He approved it.[24]

Genesis 1:**11** says, "And God said, Let the earth put forth [tender] vegetation: plants yielding seed, according to their own kinds and trees bearing fruit in which was their seed, each according to its kind, upon the earth. And it was so."[24] (Remember, He Created light in Genesis 1:3 so the plants had light to grow.)

Genesis 1:**12** says " The earth brought forth vegetation: plants yielding seed according to their own kinds and trees bearing fruit in which was their seed, each according to its kind. And God saw that it was good (suitable, admirable) *and* He approved it."[24]

Genesis 1:**13** says, "And there was evening and morning, a third day."[24] ***Three*** of what God is calling "evening and morning", lets call these ***God days*** have passed, (but *we*, (man) are thinking; *a day; our solar day* has now passed) but according to God's Word, ***our sun has not yet been created***, so… Three ***God days*** have passed and our sun has yet to be Created!

Genesis 1:**14** says, "And God said, Let there be lights in the expanse of the heavens to separate the day from the night, and let them be signs and tokens [of God's providential care], and [to mark] seasons, days, and years."[24]

Genesis 1:**15**: "And let them be lights in the expanse of the sky to give light upon the earth. And it was so."[24]

Genesis 1:**16**: "And God made the two great lights-the greater light (the sun) to rule the day and the lesser light (the moon) to rule the night. He also made the stars."[24]

Genesis 1:**17**: "And God set them in the expanse of the heavens to give light upon the earth."[24]

Genesis 1:**18**: "To rule over the day and over the night, and to separate the light from the darkness. And God saw that it was good (fitting, pleasant) *and* He approved it."[24]

Genesis 1:**19**: "And there was evening and morning a ***fourth day***."[24]

So, **Three** of what God is calling "days" have passed, *before* on the **fourth** "day" God creates what *we* use to determine "seasons, days, and years" (our sun) has even been created.

Therefore, in summary, He says the first thing He did was form the Earth (and His Sprit was moving over the face of the ***waters***). And then He says there was evening and morning one day, yet there was <u>at that time</u> **not yet** a sun or stars. In fact, He has yet to say,"Let there be light." Then on the fourth of what *<u>He</u>* is calling days, He says He formed the stars and the greater light - our sun. Therefore, He *<u>can't</u>* be talking about *<u>our</u>* days, what we call the solar day based on the solar calendar (how long it takes for the Earth to rotate on its axis one time around), because He formed the Earth (with its water) -*before* He had even formed our sun. I know this is totally new thinking - but GOD SAYS THIS IS THE WAY IT HAPPENED! I believe what He said, and it makes sense if you think about it! He formed the Earth ***<u>with all the water present on it first</u>***, just as He says in *Genesis 1:1.* Then, four "God days" later, on the fourth "*God day*" He formed the stars[24] (which would mean the galaxies and *our* solar system were formed <u>much later</u> in *Genesis 1:14* [24].

I think, after he created our sun and our solar system, He then threw or *launched* the earth into orbit into our solar system around our sun at the precise place needed to sustain life –that is, *after* he created our solar system on the fourth ***God day***. Just as we launch satellites into orbit around the earth and they continue to orbit the earth even as I write. (We know you can *place* an object into orbit when you balance the speed of the launched object trying to escape from the gravitational pull, to the gravitational pull on the object at the distance and speed it is traveling. Therefore, I think God formed the earth and water *first* just as He says. Then, on the fourth of what *<u>He</u>* calls "evening and morning", ***<u>God days</u>***, have passed, He formed the stars- which would include our sun and our solar system – just as He says. I think He then launched the pre-made earth into orbit around our sun at just the right speed and just the right distance from our sun to place it in orbit around our sun. He knew just the right place in our solar system to sustain life, so He ***<u>placed</u>*** the earth in orbit at that the exact spot needed, just as we ***<u>place</u>*** satellites into orbit.

Who knows how much longer three of what God is calling evening and morning, a "day" **_is_** – *before* what we call "days" or the way we measure days has *even been created*. This would account for the difference in what the Bible refers to as *days, and what we call days,* and why the *physical* age of the Earth **_is_** actually older than what the solar calendar days, or what counting dates and times listed in the Bible would indicate! According to Genesis, the earth, **_with its water_**, was present before –possibly LONG, L O N G, L O N G before what we think of as solar days came into existence. (Because, three of what God is calling *days* has passed before what we use to call "seasons, *days* and years" has even been created. It is on the <u>fourth</u> of what God calls evening and mornings He then Creates our sun.[24])

He had Created Light in Genesis 1:3[24] (also before he created our solar system) so *His* light or some form of light was present to sustain the earth before it was placed into orbit around our sun, but this light was coming from the light he uses to call "days" which again was present three of what He calls "days" before what we call "days" even existed. (Who knows, it may be sometime in the future we may discover the entire galaxies are orbiting around some structure or super black hole, and possibly galaxies orbiting around some structure which may have "galaxy years". Just as we know there are solar years, there may be galaxy years or something such as that. When we speak of years, the years we are referring to are relating to *solar years.* There may be *galaxy years* or *universe years* we know nothing about.) With all that having been said, *we do know* God is discussing "DAYS" before what we call " DAYS" can even occur. He may be discussing *universe days* or galaxy days, or **_"God days"_.** We are thinking he is talking about what we would call "solar days". It *can't* be solar days, as our sun (which is a star) was not *created* until three of what God is calling His "days" had already passed! This would explain why the age of the earth *is* older than what the calculations based upon dates and times listed in the Bible would suggest. He created the earth ***with its' water first*** just as He says in Genesis 1:1! Then He placed (or launched, or threw) earth with its water, into orbit at the perfect place to sustain life.

Then published *TODAY*, September 28, 2014 in the Journal *"Science"*[25]

The Headline reads:

"50% of Earth's water is older than the sun and came from interstellar ice, research says[25]"

... it then goes on to say...

"A recent research published in the journal *Science* estimates that about 30% – 50% of the water in the solar system like the water on Earth, the discs around Saturn, and the meteorites of other planets was already around even before the birth of the sun."[25]

Wow, thank you God! ISN'T THAT GREAT! RESEARCH CONFIRMS WHAT THE BIBLE SAYS!

One way this, "50% of Earth's Water is Older Than the Sun" - **which is what this article states,**[25] could have occurred, is the earth was made first just as Genesis 1:1 says and *with the water* just as it says in Genesis 1:2.[26] Then on the fourth of what God calls *days*; He then creates the sun and the stars.[26] So, if this *IS* the way it happened ***(which IS what the Bible says)*** the earth's water *would be* older than the sun! Which is what science is now saying! Thus, my rather *out there* theory is confirmed! Therefore, I included this section *reconciling the age of the earth* back in! Thank you, God, Your Word is true! Now science has ***PROVEN*** your word is true!

A second theory I have that may also explain why there is a difference between what evolutionists, astronomers, geologists, and archeologists *think* the age of the Earth is, and what Creationists *think* the age of the Earth is could be due to two different spin rates of the earth. This **different spin rate** *theory* is as follows:

The earth as we know it now (in its place within our solar system) may have been rotating the sun is in our solar system, but rotating at a much slower speed on its' axis, than it is now. What we call a day now, could have been much longer. For example, if the earth were rotating on its axis 100 times slower, our days and nights would be 100 times longer. If at some time later an asteroid or meteor strikes the earth with so much force and speed, at a glancing angle rather than a perpendicular angle, it would cause the earth to now spin faster creating shorter days and nights. (This would look/work/be the same as when a basketball player is spinning the ball on the tip of his finger and as it starts to slow down he pushes the side of the ball, at a glancing angle, to create more spin to keep it spinning on his finger.) This scenario would also account for how the earth received its tilt on its axis, which we know is what causes seasons which according to Genesis 1:14[26] was also later than when the earth was formed. Genesis 1:14 was when God created the sun and the stars,[26] possibly at that time a piece from that creative event formed an asteroid or meteor that hit the earth causing the different spin rate.) Also, there is much evidence to suggest an asteroid or meteor at some time struck the earth and caused the ice age or a solar winter to occur (which is what they *think* caused the dinosaurs to die). This **different spin rate** theory causing a difference in the time of *days* possibly would agree with that also. So, when the earth was spinning slower - what the Bible is calling days could have been much longer. Personally, I think the first theory I explained regarding ***God's days <u>initially</u> are different than man's days*** fits the Genesis account better, but they both do fit, and they both do support this recent aritcal published September 28, 2014 that the "earth's water is older than our sun."[27] ***<u>Science</u>*** has confirmed in five different ways; Creation. And now, ***<u>science</u>*** has confirmed what the Bible says, and has science has *reconciled the disagreement of age of the earth*. Thank you, God,! Your Word is True! And now science has confirmed your Word, the Bible, is true.

"In the beginning God created the heavens and the earth, "The earth was without form and an empty waste, and ***<u>darkness</u>*** was upon the face of the very great deep. The Spirit of God was moving (hovering, brooding) over the face of the waters."[28]

22. Vision

Light + Eye + Brain = Vision

Without vision, everything would look dark.

Close your Eyes and mediate on the darkness. Think about spending the next hour in absolute darkness.

Think about spending the rest of the day in absolute darkness.

Think about spending the next week in the dark.

Think about living and surviving in absolute darkness.

Think about surviving while waiting on your vision to evolve.

Think about how your eye and brain could *evolve* to see… It can't -even with millions and millions of years. The eye cannot function until EVERYTHING is present, and everything *MUST* - ***ABSOLUTELY MUST*** be present SIMULTANEOUSLY, and that can ***ONLY*** happen by CREATION !

Darkness: Without God everything would be dark.

Think about the world without God. There would be no right, there would be no wrong. Lying, stealing, killing would be rampant. The world would be dark.

Without God everything would be dark. But God gave us His Gift of Light, Jesus.

We see God through His Creation of the things we can see, and we can see God's wisdom, intelligence, foresight, planned prevention, exceeding abundance, beauty, wonder, and power in the Creation and the working of His engineering marvel—the design and function of the eye. God has revealed Himself to us *through our eyes and in our eyes*.

Observe with your eyes open, what is actually happening for you to see. *Look* at what *all* must happen. Look closely, and as you are looking, think about it. Think about it *exactly*. Ponder the miracle of your vision! Think about *all* that is happening for you to see.

First, light MUST be present. Where did the light come from? It is evident that God created the light. It didn't just *happen*. Man has **_PROVEN_** the "Big Bang" in numerous ways already referenced in this book (see references 3-14 of Chapter 21 of this book, "Additional Evidence of Creation"). There was a moment in time when there was nothing, and then suddenly there was everything—the "big bang", the moment when the universe suddenly came into being or in other words, "the moment the universe was created." This is a **_proven FACT!_**

Then, light from the sun must strike an object here on earth. Where did the sun come from? Did the sun just one day on its own become something from nothing? We **_KNOW_** Science has **_PROVEN_** the BIG BANG is **_FACT_**. Therefore, what caused the BIG BANG ? Do you really think, "everything from nothing- caused by nothing"? If you really **_believe_** it happened that way, your **_FAITH_** is GREAT, and you **_DO_** believe in religion. Only it is the religion of evolution.

Then the light travels from the sun to the earth. The light must then bounce off an object here on earth; (where did the earth come from?) and after the light bounces off all the many shapes, angles, and colors of the object, the light must pass through the living tissues of the eye as it enters the eye. The tissues then must change the light rays and focus the light rays exactly on the retina. At the same time, the tissues must not deflect, distort, or negatively alter the light rays.

The eye then must convert the light rays into electrical impulses. Those electrical impulses must then pass from the eye in the front of the head through the optic nerve to a place in the rear lower portion of the brain, *the visual cortex*, where the centers for vision processing are located, which again is in the back lower portion of the head. The brain's visual cortex then somehow converts those electrical impulses into a picture in the brain as it *is* in the real world. But your eye and brain do not just create a single picture. To see in a way which allows you to work and play, your eye must create these electrical impulses, your brain must process and *create* the picture, then the eye and the brain erases the now old picture and *creates* the next *new* picture. Your eye and brain can make a new picture in one one-thousandth of a second! Think about the miracle of your vision! Light is converted by the eye into an electrical impulse. The eye sends this electrical impulse to the brain. The brain creates a picture, erases the old picture and continues to update and create new pictures! Thank you, God, for making a way for us to see and *keep on seeing!*

How can the eye and the brain turn light into a picture in our brain? How does the brain create a picture in our mind, as it is in the real world! Darwin did not know,[1] man still does not know![2] Only God knows!

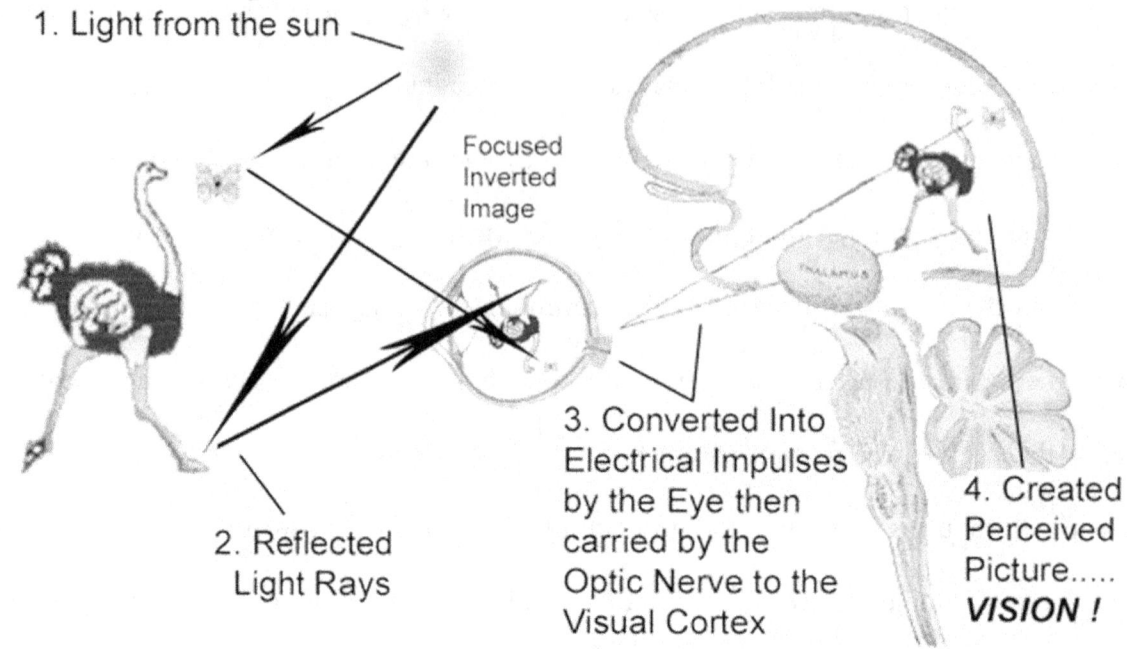

1. Light from the sun

Focused
Inverted
Image

3. Converted Into
Electrical Impulses
by the Eye then
carried by the
Optic Nerve to the
Visual Cortex

4. Created
Perceived
Picture.....
VISION !

2. Reflected
Light Rays

Eye and Brain: Henry Gray, F.R.S, *Anatomy of the Human Body* 20th Edition, Lea & Febiger (Philadelphia & New York, © 1918) Pages 766, 1006[3]. Additions by Rigney

1.The sun provides the light. The light shines on an object. 2.The light reflects off the object. All the different angles and colors of light are focused on the retina. 3.The retina converts all the different light rays with all the different colors, angles, intensities, shadows and any movement by the object into electrical impulses. The impulses are carried to the brain by the optic nerve. Then the impulses are split into appropriate fields of vision at the optic chiasm. The impulses proceed to the optic radiation where some preprocessing occurs for pupil function and interaction with other cranial nerve nuclei within the brain stem. 4.The impulses then proceed to the visual cortex where they are somehow converted into real time pictures that are perceived by the brain as it is in the real world – VISION! Thank you GOD!

How could the eye, as Darwin supposes, go from blindness to seeing clearly *gradually* over time?[4] How could the being function and provide for itself and protect itself while it is waiting for its' vision to - at some point in a future generation, evolve to a point of clarity. Any of the preceding generations that had dark or blurry or cloudy vision would not be able to survive.

Furthermore, you would see this problem with *evolving vision* in each species - and therefore, no species could have evolved because no species could have survived without being able to see while it was waiting on its' vision to evolve. In each different species, vision had to be fully functional *all at once* otherwise each species could not survive while it was waiting for its' vision to evolve.

Another thing that we know today that Darwin did not know in the 1800s. We know that when an eye is out of focus at birth, and remains out of focus from birth through about three years, the visual processing center in the brain (the visual cortex) does not fully develop and the eye is permanently blind.[5] That is, for the vision-processing center (the visual cortex) to fully develop after birth, it requires stimulation- and, that stimulation, *MUST,* be clear. When it is not clear, a weak, distorted signal is sent to the visual processing centers of the brain. As a result, the vision processing cells (neurons) do not develop properly in the brain's vision processing center, the visual cortex, and *permanent* visual impairment results. This is what is known as amblyopia – or a "lazy" eye.[5] A "lazy" eye means the person does not see -not only because the eye does not see, but also because the neurons, the nerve cells in the vision processing center of the brain, do not develop.[5] When the neurons in the brain do not develop, the person is not only visually blind but also neurologically blind. Not only does the eye have a bad focus, but also the brain cannot process or *form* vision because the neurons in the brain did not properly develop resulting in neurological blindness. We know that when a person has amblyopia[5] or a "lazy" eye, if we do not detect it and attempt to treat it and correct it by age three the weakness is permanent. That is, there is a *critical period* of neurological development from birth through about age three that _requires_ a clear focus on the retina. If a clear focus is not obtained during that time, the _critical period of development_, the persons' processing centers in the brain do not fully develop and the person is permanently visually impaired.[5] This too proves that the eye could not go from not seeing, to seeing clearly gradually over time as Darwin has supposed.[6] This proves his theory as he himself has stated, "absolutely breaks down."[7] Fully functional vision can only happen through creation! We know this because fully developed, fully functional vision-processing neurons of the visual cortex **_MUST_** have a clear focus to develop properly and fully. If a clear focus is not present from birth to age three you miss the neurological development boat, and the neurons responsible for creating vision, the visual perception cells in the visual cortex do not develop fully and the person cannot see clearly.[8]

In studies (experiments) on cats and monkeys where one eye has been deprived seeing by monocular lid closure, (that is, at birth, for the experiment, one eyelid is sewn completely closed so that eye does not see.) The results found in these experiments is that the eye has a permanent visual impairment due to underdevelopment of the neurons in the brain's visual processing center, the visual cortex, and that eye, and brain, is blind[8].

Raymond Records states, "There are well established postnatal morphologic changes in the retina and retinal projections relative to the presence or absence of visual stimuli. In the lateral geniculate nucleus (LGN), neuronal cell growth is greatly reduced (by 25% - 40%)[8, 9, 10, 11] in the laminae with connections to visually deprived eye as compared with a non-deprived eye."[8, 9, 10, 11] (The lateral geniculate nucleus is an area within the visual system, located between the retina and the visual cortex which is somewhat of a preprocessing center for binocular vision within the visual system.)

"Profound defects in retinal imagery from birth (for example, dense cataract) affect the visual system so strongly that removal of the cataract after two years of age, even given a clear image and after occlusion therapy of the good eye (if unilateral), is insufficient to restore vision above 20/200."[12] And, "If the eye is deprived between 6 and 30 months of age, finger counting is the best visual acuity achieved."[13] ("Finger counting" level of vision means the person could not see the very largest letter on the eye chart. To record the level of vision present, the next measure for recording the level of vision is noted by recording the distance at which fingers can be counted when holding your hand up and displaying one, or two, or three, or four fingers up. The distance at which fingers can be counted is then recorded as CF at 3 feet, or CF at 2 feet, or CF at 4 feet. This level of vision is the equivalent of less than 1% of normal vision.[14] By today's vision standards this would be considered legally blind because the person does not have functional vision.) If the eye is visually deprived from birth through 30 months the best vision is less than 1% and by today's vision standards, would be considered legally blind. This *critical period of development* is "either it does or doesn't develop." For example, if a baby is born with a cataract which obscures vision, it's not that the eye cannot see and is black, but no form or shapes can be detected because the neurons in the visual cortex do not develop. If the cataract is not removed by 30 months, the best vision obtained *even after removal of the cataract and vision therapy* is less than 1%.

The drawings below show the typical physical underdevelopment of the cells, 25% - 40% underdevelopment, ("morphologic changes[15]"), physical changes, in the development of the cells within the Lateral Geniculate Nucleus of a cat's eye that had been visually deprived by sewing the eyelid closed at birth. The cells in picture "A" did not develop fully because they lacked visual stimulus. You can see that; lack of stimulation of the neurons causes them to become underdeveloped.

Renditions of Right and Left Lateral Geniculate Nucleus of a cat reared with monocular lid closure. Because the occluded eye (A) was deprived visual stimulation, the processing neurons did not develop properly. There is physical under-development (A)as compared to the processing neurons of the unoccluded, normally developed eye (B).If clear stimulation of the retina is not achieved before age 3 normal development as exhibited in (B) does not occur, and cannot be made to occur. The impairment is permanent.

This "critical period of development"[8, 9, 10, 11, 12, 13, 14, 15] is a problem for "evolving vision" as Darwin supposes[16], because *all* the parts must be fully developed for the brain to receive the proper stimulus for the processing centers of the brain to fully develop.

Furthermore, when we attempt to treat the eye that is amblyopic, "the traditional and most widely used method of amblyopia treatment is occlusion of the healthy eye."[17,18] The traditional treatment of amblyopia has been to patch the strong or good eye to force the lazy, amblyopic eye to work and thereby stimulate the neurons of the visual cortex and hope to promote activation and development. Studies have shown that if the strong eye is patched or occluded too much "reverse amblyopia" will occur in the strong - or once good, eye.[17, 18, 19] That is, patching the strong eye too much makes the strong eye weak and lazy and induces amlyopia or a lazy eye in the previously healthy, strong, non-lazy eye. Thus, *we know*; prolonged non-use ***causes*** amblyopia; a visually impaired or "lazy" eye. (This is why I thought the halibut did not "evolve" two eyes on one side of it's head as discussed in Chapter 2 reference #38 and #39. (However, we discovered evolution is not at work what-so-ever in the placement of the Halibut's eye; the DNA alone is determining this trait.) This we know; non-use of an eye causes the eye to become amblyopic ("lazy" or visually impaired), it then cannot see clearly, and the amblyopia or "lazy" eye is permanently amblyopic; permanently "lazy." (The laziness or amblyopia is permanent - assuming there is no subsequent therapeutic intervention with glasses, patching, or eye exercises.[17, 18, 19])

With this in mind; when Jesus healed the blind man at Bethsaida; if you look up the passages in the Bible regarding this healing event you will see Jesus had to heal him twice before the man's vision was completely restored and only then could he see clearly and truly; Mark 8:22-25.[20] It took two times to heal him! Why would God want us to know that Jesus had to heal the man two times? Why is it recorded this way? I think it is recorded this way because God wanted to make a point. *Possibly*, (and this is solely *my* opinion) the first time Jesus healed him, his eye and nerve were healed. The second time Jesus healed him, Jesus then healed the man's processing centers (and underdeveloped neurons) in the man's brain. That is, the man still could not see clearly when Jesus <u>first</u> opened his eyes. The man's *eyes* could see but his *brain* could not see. Only when the man's *mind* (brain) was healed, *from Jesus' <u>second</u> healing*, could the man then see clearly. I think God, Jesus and the Holy Spirit allowed, performed, and recorded it, respectively, this way to illustrate a similar parallel example. This is the parallel He may intentionally be using this two-step healing to emphasize this principle. I think He intentionally did it this way for us to question His intent and thereby emphasize this principle. God has revealed himself to you, to me, and to everyone. We all see Him, that is, our eyes receive the evidence of His existence every moment they are open - through His Creation *every moment of every day.*[21] Romans 1:19 &20 says, "For that which is known about God is evident to them and made plain in their inner consciousness, because God Himself has shown it to them. For ever since the creation of the world His invisible nature and attributes, that is, His eternal power and divinity, have been made intelligible and clearly discernible in and through the things that have been made. So men are without excuse."[21]

Therefore, ***He*** has shown ***Himself*** to **EVERYONE!** Everyone sees Him. But, until we accept and acknowledge Him ***with our mind*** we do not perceive Him. I think therefore (at least *it is possible*) this is why this healing event was Performed in this *TWO-STEP* manner. It certainly was inspired by the Holy Spirit to be <u>recorded</u> in this manner. If Jesus has the Power to open the eyes of the blind, (which He did on numerous accounts) with just One Touch, or even just with His Word, He could have healed this man's blindness all at once, with one event, ***IF*** He had wanted to. I think He did it this way as an illustration, ***FOR A SPECIFIC REASON***. He wanted to emphasize a point to you and me. Unless you open your mind and realize that He is revealing Himself to you through His creation[21] you are blind to Him. That is, you see Him with your eyes. In fact, He says you *know* - without question, that you see HIM through your eyes[21]. Because HE, "Himself" has shown you.[21] But, until you perceive HIM in your mind as GOD, you are blind to HIM. You are Spiritually blind. Open your mind to HIM. Your brain is where *perception* takes place. Your eye is where light is *converted* to impulses. That is, the eye sees but the eye does not perceive. The brain perceives! He has shown HIMSELF to you. He has shown HIMSELF to everyone, with our eyes. We ALL see

HIM! Don't just see him, see HIM *and perceive HIM*. Acknowledge HIM with your brain and be *converted* out of darkness into LIGHT!

In another recorded instance of healing in the Bible,[22] Jesus took two steps for the healing to be accomplished; John 9:1 a man was born blind from birth. Everyone knew him as a blind beggar. All he could do was sit and beg.[22] When Jesus healed him, again, He did it in two steps. First, Jesus spat upon the ground and made clay (mud) and spread it on the man's eyes. Secondly, Jesus said to the man, "Go, wash in the Pool of Siloam. So, he went and washed, and came back seeing.[23]

Why would Jesus heal this different man, but again in a two-step manner? He could have healed him with just His Word, or by Touching him as He did with so many others. Why a second time, a different blind man, in a different two-step healing? Again, Jesus may be *illustrating, and at the same time emphasizing* a principle. Possibly, (again this is my theory- my opinion) the mud represents the earth or the world. Just as Jesus first put the mud on the man's eyes, Jesus has put the world (all of creation) before us for us to see <u>Him</u>.[24] He then *told* the man to go and wash in the Pool of Siloam. ***Only*** when the man <u>listened and obeyed *Jesus*</u>, and actually <u>*did*</u> what ***Jesus*** said to do, was the man then healed.[25]

Likewise, we too must <u>listen and obey Jesus</u>. *The man* washed off the earth. Jesus didn't wash it off *of* him. Jesus could have washed it off but He didn't. <u>*The man*</u> had to listen to Jesus *and* obey him; ***the man*** had to wash off the mud. ***THE MAN*** HAD TO LISTEN AND OBEY JESUS! ONLY THEN DID HE RECEIVE HIS HEALING! When Jesus talks to ***you*** through the Holy Spirit ***you*** must listen and obey Him and accept Him, just as this man had to listen and obey Jesus and ***accept*** the healing that Jesus provided him. Only when the man obeyed Jesus' instructions was he healed. ***You*** too must *listen to Jesus*, ***you*** must *obey Jesus*, and ***you*** must ***accept JESUS***!

I am writing now to the atheist and the evolutionist, and the non-Christian: The mud – that is, *the world*; (in this instance I am relating the mud, or the world's thinking, to man and Darwin's theory of evolution). The world's theory of evoluition; man's thinking; man's "vain imaginations, foolish reasoning, and stupid speculations" (Roman's 1:21, *The Amplified Bible*) is before your eyes and you cannot see. You are Spiritually blind. *You* too must listen and obey Jesus. *You* must wash off the world's theory of evolution. ***You*** take it off from before your eyes. *YOU* must see that <u>Jesus Himself has revealed HIMSELF to *YOU through His Creation.*</u> ***YOU*** Accept HIM as this man accepted his healing which was given but only received by listening and obeying Jesus' word. Listen and obey Jesus' word, and see. See HIM *and **PERCEIVE HIM***, and accept HIM, as this man accepted the healing Jesus provided when he <u>listened and obeyed</u>. Jesus didn't make him be healed, ***the man*** had to listen and obey to ***accept*** the healing Jesus ***offered***. Similarly, Jesus ***offered*** His perfect life as a sacrifice for your sins. He took the punishment you deserve for your sins. ***Accept his free gift of salvation***. Accept Jesus as your Savior and ask Him to be Lord of your life.[26] (Romans 6:23 and Romans 10:13, *The Amplified Bible, Expanded Edition.*[26])

Jesus was without sin, He never sinned so He is the only one able to take our punishment for our sins- because He didn't deserve ANY punishment. Jesus was the ***perfect*** sacrifice to God for your sins. Accept His free gift of reconciling us back into a relationship with God. Why would God do that? If we sin and disobey God, why would He want to provide a way for us to have a relationship with Him even though we sin and disobey Him? Think about that. God didn't have to make a way for restoration of our relationship with Him. Mankind sins and is disobedient to God. Mankind doesn't want to be in submission to God. Mankind rebelled against God. Mankind denies God. Mankind attempts to do away with God. Mankind doesn't acknowledge Him as God even when He has revealed Himself as God; and He has revealed Himself so effectively that He says mankind is "without excuse". Why would God want to provide a way for us to have a relationship with Him especially considering all the grievances He should have against mankind? Why? …It just doesn't make sense. Why?……. Because He loves us! "For God so greatly loved and dearly prized the world (mankind) that He gave up His only begotten Son, that whoever believeth in Him shall not perish, but have everlasting life. John 3:16[27]

Jesus healed the blind to fulfill the prophesy about HIMSELF in the Old Testament written by the prophet Isaiah[28], (Isaiah 35:5,42:7,). Fulfillment of Prophecy is one of the ways we ***know*** that the Bible is true. (Isaiah 9:6 and Isaiah 53:1-12[29] are also prophesies of Jesus that were fulfilled). Jesus healed the blind at least 5 different times in the Bible. It was witnessed and documented by at least 4 different people in the Bible, Matthew, Mark, Luke and John – plus probably the other 8 disciples witnessed Jesus healing the blind, and many others present witnessed Jesus healing the blind. John11:37 says, "But some of *them* said, "could not He who opened the eyes of the Blind man have kept this man from dying?" (The Amplified Bible) They were at that time referring to Lazarus who had died but Jesus raised Lazarus from the dead, John 11:14-45.[30]

Healing of the blind is fulfillment of prophecy of Jesus. This is one of the ways we know the Bible is true; events were witnessed and documented. Those events were written down, recorded by witnesses. The Bible is not a book of fables; it is a book of eyewitness accounts. Matthew records at least six different people being healed of blindness, (Matthew 9:27, 12:22, and 20:29). Mark records two; Mark 8:22, 10:46. Luke records one specific event; Luke18: 35-42. But also Luke 7:21 says, "In that very hour Jesus was healing many [people] of sicknesses and distressing bodily plagues and evil spirits, and to <u>many</u> who were blind He gave [a free, gracious, joy-giving gift of] sight (The Amplified Bible). John records one specific event, John 9:1-25.

God made your eyes so you could see. God made your eyes so you can see *Him* through His wonderful creations. Through the eye, <u>everyone</u> *knows* He is God.[31]

Furthermore, He states that He has made Himself *so plainly evident* that when we stand before Him one day, NO ONE will be able to say things such as, "I didn't know." Or, "No one ever told me." Or, "I asked You, God, for proof and you never answered me." Or, "If you only would have shown me." Or, "But everyone was telling me to believe in evolution." Or, "All the experts said there was no creation, only evolution." Or, "all they taught me in school was evolution and they said it was fact and true," Or, …. whatever excuse you may attempt to make.

However, Romans 1:20 says, "men are without excuse." He will say, "<u>I did reveal myself to you.</u>" "You knew. You do *know. I have made it plain to everyone.*"[31]

> "For that which is known about God is evident to them and made plain in their inner consciousness, because God Himself has shown it to them. Romans1:19&20[31]

He says: ***"Made Plain"***[31]

> "For ever since the creation of the world God's invisible nature and attributes, that is, His eternal power and divinity, have been made intelligible and clearly discernable in and through the things that have been made. So that men are without excuse". Romans 1:20[31]

He says, "so men are, ***Without Excuse"***[31]

23. The DNA Dictates the Species

"For that which is known about God is evident to them and made plain in their inner consciousness, because God Himself has shown it to them. For ever since the creation of the world His invisible nature and attributes, that is His eternal power and Divinity, have been made intelligible and clearly discernable in and through the things that have been made. So men are without excuse. Romans 1:19&20 Amplified Bible[1]

(For the picture references used on pages 222 and 223; See *References:* Chapter 23 The DNA Dictates the Species, references #2--#12.)

God made your eyes so you can see *Him* through His wonderful creations. Through the eye,

EVERYONE ***knows*** He is God.13

Darwin's theory of evolution would have you believe *ALL* these "organic beings" had the same point of origin. Darwin states, (p. 380, *The Origin of Species*, "probably all organic beings which have ever lived on this earth have descended from some one primordial form into which life was first breathed."[14] (Pictures: References; Chapter 23, #2--#12.)

Darwin states, (p. 380, *The Origin of a Species.* **"probably all organic beings which have ever lived on this earth have descended from some one primordial form into which life was first breathed."**[14] That is, the deer, the grass; the frog, the dragonfly; the geese, the water-grasses; the wolf, the deer; the fish, the butterfly; the trees according to Darwin, all "have descended from some one primordial form into which life was first breathed". Darwin knew plants and trees are "organic living beings", and if you think he didn't mean to include plants in this statement - just to make sure you know that this statement does include plants, he in the same paragraph makes this clarifying statement, Darwin says, (I am quoting him word for word) **"Analogy would lead me one step further, namely, to the belief that all animals and plants have descended from some one protype."**[14] And you; man or woman reading this right now, you are an "organic being" -Darwin states, (p. 380 *The Origin of Species*) "That probably all organic beings which have ever lived on this earth have descended from some one primordial form into which life was first breathed. According to Darwin, you ***and a tree***, you ***and a flea***, have a common ancestor, the "some ***one*** prototype."[14]

Just as the 25 plus systems are necessary *simultaneously* for the eye to function and for you to see because they are ALL interdependent, I have written these "organic beings"[14] in interdependent relationship to each other. The deer is dependent on the grass, the wolf is dependent on the deer, the frog is dependent on the dragonfly, the fish is dependent on the butterfly, and the geese are dependent on the water-grasses. The turtle is dependent on the land grasses and insects. Evolution cannot explain all these interdependent "organic beings" surviving while the species it depends on for survival has yet to evolve. They too ALL must be present simultaneously, and again only in

CREATION can they ALL survive. Think about it. Really think about everything. Think about the entire process. Logically, evolution doesn't make sense. What did they eat if they all evolved from "some one primordial form into which life was first breathed" [14]? And, if they all evolved from "some ONE primordial form,"[14] how did they reproduce?

Keep in mind that in Darwin's *theory*, the same processes that caused the grass, flower, and tree to evolve into different species, caused the dragonfly, goose, deer, wolf, and man to evolve into different species. Well, if you think about grass evolves to flower evolves to tree- that is *somewhat* logical, but according to Darwin they "all evolved from some one primordial form."[14] When you think dragonfly evolves to goose evolves to deer evolves to wolf evolves to man- that is NOT even *somewhat* logical. Yet Darwin states, **"Analogy would lead me one step further, namely, to the belief that all <u>animals and plants have descended from some one protype</u>."**[14] Mr. or Ms. evolutionist, Mr. or Ms. atheist; do you really believe that? *Why would you want to? With God you are a reason; you have purpose! If evolution is true you are an accident that happened by chance.*

The problem is when Darwin said, (p. 380, *The Origin of Species.*) "That probably all <u>organic beings</u> which have ever lived on this earth have descended from some one primordial form into which life was first breathed."[14] Darwin *fully understood* that a bacteria, a fungus, a virus, a plant, a tree, a fish, a butterfly, a turtle, a bird, a mammal, and also the dinosaurs, are ALL "organic beings" - and he knew about dinosaurs[15] from fossils when he wrote his book; (reference #15 are paleontology references Darwin writes about in his book) references Darwin writes about in his book) yet he believes - therefore His theory teaches, that probably all of them "descended from one primordial form."[16] In Darwin's *theory* a dragonfly and a goose had a common ancestor at one time![16] The tree and the deer ultimately came from the same organism![16]. How did Darwin in the 1800's come to that conclusion? How do the evolutionists in the 2000's still come to that conclusion? When anatomical parts of different species look similar, Darwin and evolutionists call it evolution! After all, a goose and a dragonfly both do have wings- therefore- (at least according to Darwin) they must be related (they are analogous).[16] And look at the branches of a tree and the antlers of a deer- they look similar (they are analogous)![16] According to his theory of evolution, at one point, they ultimately evolved from the same ancestor and the same organism![16] I intentionally used these as examples; the dragonfly and goose; the tree and the deer, because these examples do have *some* points of similar appearance; the wings of a dragonfly are similar to the wings of a goose; the branches of a tree are similar to the antlers of a deer, just as a chimp and a man have *some* points of similar appearance.

I know it sounds funny that I would relate the tree and the deer together, and the dragonfly and the goose together. *I am* somewhat joking and poking fun at what the theory of evolution teaches. However, it is actually rather sad and at the same time frightening to *know* that Darwin's theory would have you believe that not only the deer and tree, and the goose and dragonfly had a common point of origin, but Darwin and evolutionists would also have you believe that the tree and the goose, the deer and the dragonfly also had a common ancestor at some point in time- even though they physically have very little in common with each other. That is, there are few points of similar appearance. In fact, Darwin's theory of evolution would have you believe that the same processes that caused the flea to *evolve* caused the elephant to *evolve* and caused the lily of the field to evolve, and caused the dinosaurs to evolve; and the flea, the elephant, the lily of the field, and the

Tyrannosaurus Rex are related, and that they all four; insect, mammal, plant, and reptile have a common point of origin.[16] (p.380, *The Origin of Species*)

Just because you can create an image and draw lines from one picture to the next does not mean; it actually occurred. It is ***theory.*** Darwin himself stated, "If it could be demonstrated that any complex organ existed, which could not possibly have been formed by numerous, successive, slight modifications, my *theory* would absolutely break down.[17] He states, ***"THEORY"!*** *Where is the* ***"proof" of his theory?***

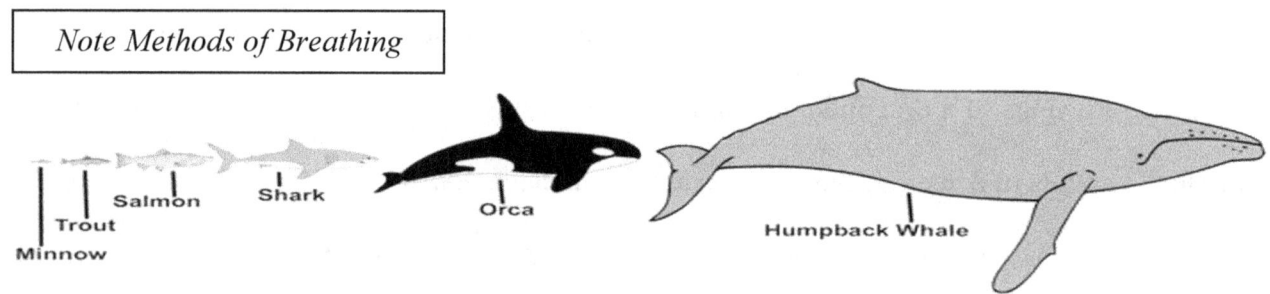

Note Methods of Breathing

IS there ANY evidence that these different species evolved? Where are the links? Why do we NOT see any transit6ional forms of any kind? The minnow, trout, and salmon are fish. The orca, and whale are mammals. How could they have evolved? (The shark bears live young as a mammal, but does not breathe air. Is it a fish or a mammal? Where are the links?) **For the Picture references used on pages 228 and 229, see:** *References:* **Chapter 23, The DNA Dictates the Species, references #18--#30.**

Where is the "proof" of his theory in this illustration? Just because you can draw something and connect the drawings with a bunch of lines does not _prove_ the theory. It only _proves_ you can draw some pictures!

Where are the links? I do realize the bat is a mammal, whereas the others are birds, but what is the predecessor of the bat, the hummingbird, and the ostrich? *Evolution???*

Where is the proof of evolution in this sequence. What proof is there of this illustration being reality? Just because you can draw it this way, doesn't mean it happened this way!

For Picture references see; References: **Chapter 23, The DNA Dictates the Species, reference #18-30.**
Picture #30 Chimp to man source is from Answers in Genesis, https://cdn-assets.answersingenesis.org https://cdn assets.answersingenesis.org /img/articles/ee / v2/ape-to-man-evolution.jpg[30]

The point is: just because you can make a drawing it does not make it true. Just as the fish and

bird drawings I made to illustrate the absurdity of the idea that these fish and birds evolved - this

illustration of a chimp evolving to man[30] is an illustration of absurdity! It has no basis; it is based on imagination ONLY- not FACT! And just because you can make a drawing with some lines on it, it does NOT make it true! I KNOW I will be criticized and ridiculed for making these preceding statements but contemplate this evolutionary fact and the following statement; somehow, somewhere Darwin and the evolutionists have a drawing with lines on it depicting evolutionary sequences that show the minnow, the hummingbird, the whale, the bat, the dragonfly, the deer, the flea, the elephant the monkey, and the man are **_ALL_** related!

Therefore, if we are going to talk about "the origin of species," what species was the predecessor to the monkey? From what species was the monkey's origin? Where are the links? And, why is it that we still have non-evolved monkeys today? Why do we not see any transitional forms (somewhat monkey-somewhat man) today? Where did the first monkey come from? Again, what was the first monkey's predecessor? Was the first monkey male or female? How did the monkeys mate while they were waiting on their reproductive parts to evolve? Did monkeys somehow evolve into different sexes, male and female, and evolve the parts exactly necessary to reproduce; - at exactly the same time - for them to be able to reproduce? (Oh, I forgot, I should not say, "*somehow*" I should say, did monkeys "*by chance*" evolve into different sexes, male and female and evolve the parts exactly necessary to reproduce; -at the same time for them to be able to reproduce? While at the same time - because it would absolutely, without question, require <u>numerous generations</u> to satisfy Darwin's statement of "numerous, successive, slight modifications[31]" even though the <u>numerous generations</u> *required* to accomplish the "numerous, successive, slight modifications" for the evolution of different sexual parts – could not happen because they could not yet reproduce! That is, they could not mate or have sexual relations because the parts necessary did not evolve because there could be no "numerous, successive, slight modifications[31]", if there are no numerous generations.

I know this sounds as if I am talking in circles, and I know this is difficult to follow. But that is exactly the point I am trying to make! If you follow the line of thinking, how can evolution apply to sexual differentiation. IT CAN'T! The species, cannot undergo the "numerous, successive, slight modifications[31]" necessary to evolve different male and female sexual parts, <u>and at the same time</u> be able to reproduce in order to satisfy Darwin's statement "numerous, successive, slight modifications[31]!

Therefore; how is it that some monkeys evolved male parts and some evolved female parts - and they so happen to exactly evolve in a way, and evolve in exactly the same time, such that they exactly work together to be able to create baby monkeys? How did the female monkey evolve ovaries with eggs, and the male monkey evolve testicles with sperm? How is it that they evolved separately, and independently yet so exactly perfectly that everything exactly fits together and joins together so that they ~~so happen~~ (oops-) *by chance* combine to create a baby monkey? How did the female monkey evolve the many, many different hormones necessary to cause reproductive cycles, and evolve the hormones necessary to sustain a pregnancy- and the hormones necessary to cause labor and delivery? How did they reproduce while she was waiting on all the hormones to evolve? If we are to believe in evolution, how did the female monkey *evolve* mammary glands in order to feed her babies once they were born? Baby monkeys cannot eat they <u>*MUST*</u> nurse! What happened to all the baby monkeys that were born before the female monkey evolved mammary glands? How did the female monkey nourish the baby in the womb while she was waiting on the umbilical cord to evolve? How did the baby fetus breathe while it was inside the womb waiting on the umbilical cord to evolve?

Everything had to be present, fully and completely functional - all at the same time. Just as evolution cannot explain the function of the eye, evolution cannot explain the reproductive system. As with the eye, **_AGAIN_**, Darwin's "**_theory_** of evolution" "absolutely breaks down"[31] when we examine the reproductive system. We see that evolution cannot happen, evolution cannot explain the reproductive system because everything not only MUST be present and fully functional, but it has to happen exactly and completely for two different sexes **_AND_** at exactly the same time. **_ALL_** the parts **_MUST_** be fully developed. ALL the functions MUST be fully functional. We observe totally interdependent parts and systems **_ALL_** with totally interdependent function, and **_ALL_** **_MUST occur_** **_simultaneously_**. This is evidence of creation, Interdependent Evidence of Creation, (I.E.C.). Think about the whole process! Don't just believe what someone has told you, really think about it. Logically, evolution doesn't even make sense. Think about it wholly and honestly. Don't allow yourself to be deceived; think for yourself! Be honest to yourself! Think about the entire process, Darwin's theory states, "all animals and plants have descended from some one prototype." [32] Think about that statement! Evolution doesn't make sense. Creation makes sense. Don't be deceived!

As Darwin has "imagined" [33], "supposed" [33], and "believed" [33] regarding the "evolution" of the eye,[33] (page 158, *The Origin of Species*) let's "imagine"[33] how evolution would apply to sexual reproduction. To follow is my "Ode to Evolution" as I "imagine"[33] and "suppose"[33] it may "by chance" have happened. (I certainly don't believe"[33] it to be true.) ……. Let's evolve a baby!

LET'S EVOLVE A BABY!

YOU EVOLVE THE WOMAN PARTS!
I'LL EVOLVE THE MAN PARTS!
LET'S EVOLVE THEM AT EXACTLY THE SAME TIME!

MAKE SURE TO EVOLVE YOUR PARTS
SO THAT MY PARTS FIT YOUR PARTS.
I WILL EVOLVE MY PARTS SO THAT
THEY EXACTLY FIT YOUR PARTS!

BE SURE TO TOTALLY EVOLVE *ALL* YOUR PARTS
AT THE EXACT SAME TIME
I TOTALLY EVOLVE *ALL* MY PARTS!

I'LL MAKE *MY* HALF OF THE BABY…
YOU MAKE *YOUR* HALF OF THE BABY…
AND LET'S MAKE OUR TWO HALVES
MATCH *EXACTLY*!

I'LL GIVE YOU MY HALF…
YOU PUT IT IN YOUR HALF.
YOU MIX IT ALL TOGETHER.

YOU *MAKE* OUR PARTS COME TO LIFE!
YOU CARRY IT INSIDE YOU AND YOU *MAKE* IT GROW
UNTIL IT'S ALL THE WAY DONE!
DON'T MAKE IT COME OUT TOO SOON OR TOO LATE!

YOU EVOLVE A WAY TO FEED IT WHILE IT IS IN YOUR BELLY.
YOU EVOLVE A WAY TO LET IT LIVE INSIDE YOU
EVEN THOUGH IT CANNOT BREATHE.

YOU MAKE SURE TO EVOLVE YOUR PELVIS
AND THE REST OF YOUR BODY TOO
SO THAT YOU CAN ACTUALLY *MAKE* THE LIVING BABY COME OUT!
YOU PUT ALL THE PARTS TOGETHER EXACTLY RIGHT!

TEACH IT HOW TO CLOSE ITS' HEART
SO THAT IT CAN BREATH AIR WHEN IT COMES OUT!
AND WHEN IT'S DONE… DON'T YOU FORGET TO MAKE THE MILK!!

By Jan Jay Rigney O.D. ©2016

(To purchase a copy of ***Let's Evolve a Baby*** go to www.letsevolveababy.org ALL proceeds derived from www.letsevolveababy.org are used to help unexpected babies and their mothers, or to help unwanted babies to get adopted rather than aborted.)

Reproduction cannot happen while one of the sexes is waiting for any one part, or any one system, or any one hormone to evolve! And even more mind staggering is that IF evolution WERE to occur simultaneously for the two different sexes it would have to have occurred simultaneously, exactly, and completely for each and every different species! WOW! If you believe THE SEXES DIFFERENTIATED AT EXACTLY THE SAME TIME FOR EACH AND EVERY SPECIES you have **GREAT FAITH** in evolution. Evolution requires <u>Faith</u> to believe in it. It is odd that evolutionists fault Christians for our faith-based beliefs. Yet, evolutionists must have Faith to believe in evolution.- especially to believe that "all <u>animals</u> and <u>plants</u> have descended from some one prototype."[34] If there is evidence of a Creative event (remember, science has PROVEN in four different ways that there was a creative event), and if science has confirmed that the water on earth is older than our sun (remember science has proven that also) and, if the evidence ***proves*** creation by proof of the theory; Interdependent Evidence of Creation I.E.C. Then, realize that there is a Creator. *Furthermore*, if there is a Creator there too is the Devil. Know this, The Devil does not want you to think he is real, yet he is! He wants to deceive you ! Don't be deceived. The goal of the devil is to cause you to question if there is God and to separate you from God. He is the Great Deceiver. He is a liar. Think about what he is attempting to get you to believe. **"all animals and plants have descended from some one prototype."[34] And, "probably all the organic beings which have ever lived on this earth have descended from some one primordial form, into which life was first breathed"[34]** THINK! Darwin says, "animals and plants have descended from some one prototype.[34]" And, "probably all the organic beings which have ever lived on this earth have descended from some one primordial form, into which life was first breathed."[34] If you believe that then you certainly can believe in creation ! Especially since there _**IS**_ actual evidence of Creation ! Where is the evidence of **"all animals and plants have descended from some one prototype" OR, "probably all the organic beings which have ever lived on this earth have descended from some one primordial form, into which life was first breathed"**[34]? He Believes, (and again thankfully he actually wrote it down so we…

THINK! Darwin says, *"animals and plants have descended from some one prototype.*[34]" And, "probably all the organic beings which have ever lived on this earth have descended from some one primordial form, into which life was first breathed."[34] *Think about those statements! Do you not have a problem with them?*

He believes, (and again, thankfully, he actually wrote it down so we do not have to assume, or infer, or guess, "What did he *really* think?" Or, "I wonder what he *really* meant?" Or, "What was he trying to say?" He actually wrote it down. _We know_ what he thought, _we know_ what he meant, _we know_ what he is saying. It is written on page 380 of his book *The Origin of Species*[34])**… probably all the organic beings which have ever lived on this earth have descended from some one primordial form into which life was first breathed**[32]. And, *Everything…* **"from some one primordial form"!** [34] *Everything* from *one!*

I ask you to be truthful to yourself. Honestly consider; which one is more difficult to believe; an all powerful creator, *creating* everything… or _everything_ evolving from "some one primordial form"[34], and *"plants __and__ animals, from some one prototype"!*[34] There actually is *VERY LITTLE DIFFERENCE.* You have to have *faith* in each theory. You have to have faith to believe in either one. *THEY BOTH ARE A RELIGION! "All organic beings which ever lived from one primordial* form"[34] *and* "animals and plants have descended from some one prototype."[34] *Where is ther any proof of that?* Where is there ***ANY*** evidence of that? It requires *faith* to believe it. It is truly a religion.

One of the tools evolutionists use as a key factor in determining if two species evolved from a common ancestor is jawbone structure, function and development.[35]

Using this method of jawbone analysis, evolutionists today say (and this is current thinking as of the date of the writing of this book, that the elephant, the sea manatee, and the hyrax are all directly related.[35, 36, 37] (The hyrax is a rodent-like mammal about the size of a badger; some call it the *rock badger.)* Because the hyrax has teeth and jawbone structure that is *somewhat* similar to an elephant they say it is related to the elephant through evolution. For similar reasons they also say the manatee and the elephant, and therefore the hyrax and the manatee are related.[35, 36, 37]

Evolutionists Say the Elephant Manatee and the Hyrax are directly related.[35, 36, 37]

For picture references; See References: Chapter 23, #38, #39, and #40

In addition, on June 26, 2014 an article appearing in the Journal of Mammalogy 95(3):443-454,2014 states the African *Sengi Mouse* has genes that are similar to the "elephant, manatee, hyrax, aardvarks, golden moles, and tenrecs which collectively belong to the supercohort Afrotheria."[41, 42] This means that the sengi mouse and the elephant are evolutionally related. The article uses the words "distant phylogenetic affinities"[41, 42] to say that they are genetically related. In fact, the article references DNA similarities[41, 42]. That is, the article says the elephant, hyrax, sea cow (manatee) and the sengi mouse are distant relatives by way of evolution and "all collectively belong to the supercohort Afrotheria." [41, 42] Consider that! They say a MOUSE and an ELEPHANT are ***directly*** related (distant phylogenetic affinities[41, 42]) - yet we see *NO* transitional forms! That is unless you consider the sea manatee and hyrax transitional forms! This particular mouse does indeed have a nose that protrudes and is elongated, but nothing to the extent as that of an elephant! But because the DNA of this mouse has some similarities as that of an elephant they (the "experts") say this sengi mouse and the elephant are distant but directly related. Not that they are related because (as according to Darwin and the theory of evolution) the sengi mouse *evolved* a nose different from other mice through "multiple numerous, successive, slight modifications"[43] in order that it had a competitive advantage over other mice; and due to *"the struggle for existence"[44]* or *"natural selection"[44]* and *"the vigorous, the healthy and the happy survive and multiply"[45]* (survival of the fittest) the sengi mouse evolved an elongated nose becoming a different species of mouse. On the contrary, evolutionary processes DO ***NOT*** show to be at work at all when the experts say they are related[46]. It is only the physical traits exhibited, by its' DNA - and the DNA dictates those traits exhibited. Again, we observe; that is, the conclusion reached is; the DNA dictates the similarities. The DNA dictates the differences - not evolution. If the sengi mouse did evolve through the evolutionary processes espoused by Darwin and the evolutionists it would not have DNA similar to an elephant it would have DNA more similar to other mice. So we are not seeing evolution at all. It is PURELY the DNA!!!

And, did this mouse *"evolve"* from the elephant, or did the elephant *"evolve"* from this mouse? Whichever the case, the *DNA* of the mouse is similar enough to the *DNA* of the elephant that they are saying *this mouse* and the elephant are distantly, but directly related![46] No "numerous successive slight modifications"[47]- NOT EVOLUTION!

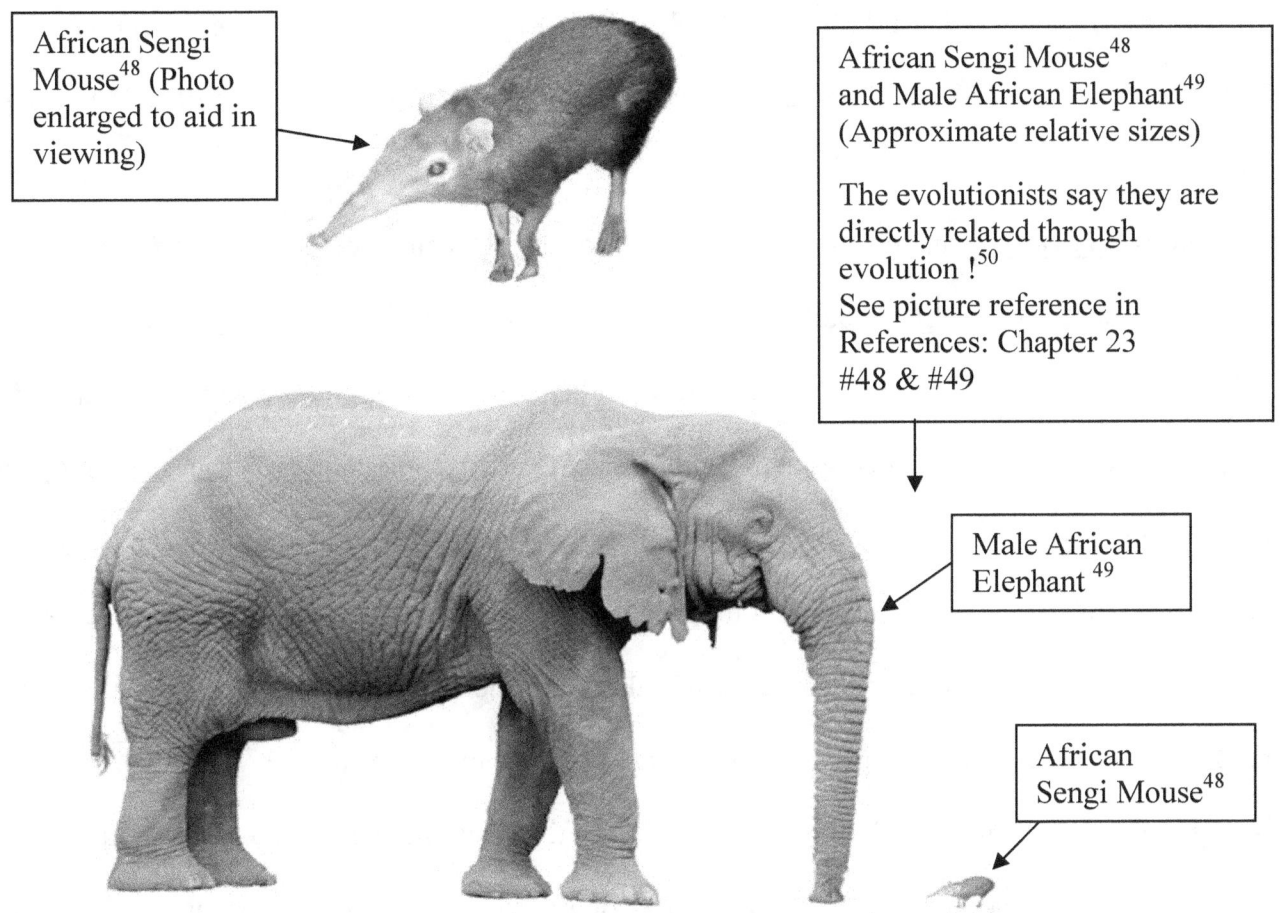

African Sengi Mouse[48] (Photo enlarged to aid in viewing)

African Sengi Mouse[48] and Male African Elephant[49] (Approximate relative sizes)

The evolutionists say they are directly related through evolution ![50]
See picture reference in References: Chapter 23 #48 & #49

Male African Elephant[49]

African Sengi Mouse[48]

Could it be that the sengi mouse and the elephant DNA are telling us, in fact *SCREAMING* at us!

"IT IS THE GENETIC INFORMATION – THE DNA, THAT DICTATES THE SPECIES! NOT EVOLUTION!" THE SENGI MOUSE DNA AND THE ELEPHANT DNA ARE TELLING US- GIVING US EVIDENCE THAT THE CREATOR DESIGNS THE DIFFERENT SPECIES UTILIZING THE DNA (THE GENETIC INFORMATION WHICH INSTRUCTS LIVING CELLS HOW TO DIFFERENTIATE, MULTIPLY AND ASSEMBLE). HE THEN MANIPULATES OR PROGRAMS THIS GENETIC CODE (THE DNA) TO CREATE DIFFERENT SPECIES! EVOLUTION HAS NOTHING TO DO WITH IT! THE SENGI MOUSE DNA AND THE

ELEPHANT DNA ARE SCREAMING TO US! *"SIMILARITIES DON'T INDICATE EVOLUTION -THE DNA DICTATES THE SIMILARITIES. THE DNA DICTATES THE SPECIES - NOT EVOLUTION!"*

There is <u>no evidence</u> that either the sengi mouse or the elephant evolved- as Darwin states, "by numerous, successive, slight modifications"[51] from a common ancestor. Pure logic, rational evaluation, observation - any *one* of these *alone* would tell you the sengi mouse and the elephant did not evolve from the same predecessor or common ancestor, <u>it is the similarities in the DNA that causes them to come to that conclusion</u>[52]. Yet even in the face of this <u>evidence</u> they say it is evolution. They say the sengi mouse and the elephant *evolved* from the same distant predecessor. I find it hard to believe that they could conclude the sengi mouse and the elephant had a common ancestor! As I continued to think about this, and meditate on these facts, I wondered, how can **<u>anyone</u>** believe such an outlandish statement. It is not that they *theorize* that they are related, they say it as a matter of fact, that is, it is a statement, not a theory. They say the elephant and the sengi mouse and the Hyrax and Sea Cow (Manatee), and the Aardvark, and the Golden Mole, and the Tenrecs are related through evolution.[52] (They arrive at that conclusion based upon the DNA, and structural similarities such as jawbone structure dictated by the DNA.) As I was trying to reason how anyone could actually think an elephant and a mouse are *directly* related and at some point in their past they had a common ancestor, it suddenly occurred to me; Darwin and the evolutionists believe *<u>everything</u>*, and again I quote Darwin himself, "probably <u>all</u> the organic beings which have ever lived on this earth have descended from some one primordial form, into which life was first breathed"[53] and, "animals and plants have descended from some one prototype"[53]. Considering *those* statements, I guess it *is* somewhat *easy* for the evolutionist to *THINK* that the sengi mouse and the elephant are related and have a common ancestor –because at least there is *some* similarity- ___the nose!___ The fact that there are no transitional forms, however, tells you they are NOT. Think about it, meditate on these facts in your mind and what the evolutionists are saying. Think about what they expect you to believe. Do you really believe that a mouse and an elephant are directly, evolutionarily, related? Where are the transitional forms? Again, it takes faith to believe it! Instead, could it be the genetic code, the DNA- the blue print for building each species, is where the similarities come from? Is it that the DNA is dictating the similarities? YES!

YOU DO NOT REALLY NEED THE FACT THAT THERE IS NO EVIDENCE OF TRANSITIONAL FORMS FROM THE ELEPHANT TO THE SENGI MOUSE TO SUPPORT THE THOUGHT THAT THEY ARE **NOT** TWO SPECIES THAT *EVOLVED* FROM ONE! Look at the physical traits, habits, diet, habitat, instincts, struggle for survival, etc. that Darwin uses to attribute *causes* of evolution. What of these evolutionary processes that supposedly *CAUSE* evolution can be found similar in the sengi mouse and the elephant? Where are they??? What are they??? *WHY did the sengi mouse evolve into an elephant? Or, is it that the elephant evolved into the sengi mouse??? In either case, again... WHY? What evolutionary processes could have been at work? What evolutionary processes are evident? NONE!* It is the similarities in the DNA that is causing the "experts" to say this mouse and the elephant are related. ___It's the DNA, and the DNA alone___ – not the habits, the struggle for existence, survival of the fittest, feeding habits, etc. that Darwin says <u>*causes*</u> evolution. Nothing supports that they evolved – ___except the DNA___ - *yet they are saying they did evolve.* Similarities don't Indicate Evolution – The DNA Dictates the Similarities!

Could it be that the similarities are present because the basic blueprints for organic life are based upon a similar plan and design in the coding – the written information - in the DNA? The basic blue print is retained throughout the DNA of all species. We observe today the same phenomena in different types of vehicles. The DNA Dictates the Species as a basic blue print for life, just as the same basic design principles dictate vehicular transportation design - therefore vehicles ALL follow the same basic blueprint of design, and that design is retained in different vehicles. You don't think vehicles evolved on their own <u>by chance</u>. Yet, they are not nearly as complex as "organic life".

Also, it is because of the similarities in the DNA that the evolutionists say that man evolved through the monkeys, chimpanzees, or apes. And, just as the sengi mouse and the elephant are telling us… it is the manipulation of the DNA by the Creator that creates these two <u>different</u> species *AND THAT THEY DID NOT EVOLVE,* the chimp, ape, and man species are telling us "it is the manipulation of the DNA by the Creator that creates these three <u>different</u> species AND THAT THEY DID NOT EVOLVE! Again, He (GOD, THE CREATOR) takes the genetic code - the DNA, and manipulates it to create different species. (Man does this today, we have many genetically modified crops; GMO corn.) The genetic information was modified or altered to create two different species; a mouse and an elephant, also; a chimp and a man. The similarities in appearance are therefore similar because of the similarities in the DNA, not due to evolution. Just as there are no intermediary or transitional forms from the sengi mouse to the elephant because they did NOT evolve from a common ancestor, (that is, there is no evidence of that- again ***<u>except the DNA!</u>*** It is the DNA that dictates the similarities, and the DNA dictates the species!) there are no transitional or intermediary forms from the chimp, or monkey, or ape to man because they too did not evolve from a common ancestor either.

Evolutionists say the elephant and the sengi mouse are directly related![48]
Picture References; See references: Chapter23, #54 & 55

Evolutionists say the ape, chimp and man are directly related!
(Picture from: Answers in Genesis, https://cdn-assets.answersingenesis.org
https://cdn-assets.answersingenesis.org/img/articles[56])

Furthermore, if you were to relate these species in a line in the evolutionary "tree of life" what is the common predecessor to the sengi mouse? What was the common predecessor to the elephant? What is the common predecessor of ape and man? Why did some evolve into a man and others not?

Additionally, Published Jul 31, 2014 in *Reuters,*[57] a report on an Article in the journal <u>Science:</u> August 1, 2014[58] states, and I quote; <u>"massive, meat-eating, ground-dwelling dinosaurs evolved into agile flying birds: they just kept shrinking and shrinking, for over 50 million years."</u>[57, 58] It further states, I quote; **<u>"Being smaller and lighter in the land of giants, with rapidly evolving anatomical adaptations, provided these bird ancestors with new ecological opportunities, such as the ability to climb trees, glide and fly. Ultimately, this evolutionary flexibility helped birds survive the deadly meteorite impact which killed off all their dinosaurian cousins."</u>**[57, 58]

This study said that through "12 <u>shrinking transitional forms</u>" the great dinosaurs evolved into birds and were ultimately the only survivors of a meteor strike on the earth that caused atmospheric changes to occur too fast for the large dinosaurs to adapt and that these birds were the only survivors of the asteroid strike.[57, 58]

The problem stating that birds were the only survivors of the meteor strike that wiped out the dinosaurs is; that statement is equivalent to saying, all life that exists on earth today had to have evolved from birds.

I am NOT making this up! These articles *are for real* ! Yet people do not question it! They just accept it. Think about it. Do you really believe it ? A mouse evolved from an elephant.[59, 60] And, the dinosaurs kept shrinking and shrinking for over 50 million years until they have now evolved into the birds we see today, and that after the meteor strike only the birds survived.[61, 62] If this is true, all life now on earth evolved from birds. For some reason, as of the date of this writing, I have seen little to no further discussion about these two articles; the sengi mouse having DNA similar to an elephant, and all dinosaurs having evolved into birds. I personally think it is because the evolutionists are still trying to figure out how to bend or spin these *stories* to fit their *theory* so that they can still deceive as many as possible to believe in evolution. Their problem is, it just doesn't make sense! It is amazing to me that somehow people can believe these types of things -yet they cannot believe in creation. Why is that ? I personally think it is because people are all about ***self***. People are self-willed, and self-motivated and do not want to answer to, or submit to, and therefore acknowledge that there is, a higher authority.

To follow is my representation of how an evolutionary tree of life might be illustrated if life were to have evolved pre-meteor strike and post-meteor strike.

As you look at this illustration think this through a little further. They say the birds today evolved from the dinosaurs "shrinking and shrinking over 50 million years."[57, 58] If evolution were true, what competitive advantage was taking place over those 50 million years that would cause the dinosaurs to keep "shrinking and shrinking"? You would think that the bigger and stronger would have competitively eliminated the smaller dinosaurs over the 50 million years if the tenets of evolution were true? Why did they instead keep "shrinking and shrinking"[61, 62]? And if they "evolved", why did they first start off **small** "from some one primordial form" as Darwin states,[63] and **then evolve to become bigger**, to be the size we know that they were (because of fossils evidence), and **then, over 50 million years evolve to become smaller?**[64] Wow! I laugh out loud when I think about that! Supposedly they evolved from one primordial form; *small to big*, then over the process of 50 million years; *big to small.*[64] It just doesn't make sense! And if they did evolve, how did the dinosaurs evolve? What were their predecessors? What did the evolutionary tree look like before they were as we see their fossils? I have attempted to duplicate what their *"evolutionary tree"* may have looked like, in my illustration to follow of the *"Supposed Evolutionary Tree of Life Before, and After the Meteor Strike"*. In my rendition, we see the dinosaurs at the base of the tree in their completely "evolved" state prior to the meteor strike. Again, what did they "evolve" from? What were *their* predecessors? Where is the fossil evidence of Dinosaur evolution from "some one primordial form"[65] as Darwin theorizes. (He knew of, and wrote of dinosaurs in his book.[66]) Therefore, in my illustration of *the "Supposed Evolutionary Tree of Life Before and After the Meteor Strike"* I just illustrated the dinosaurs in their assumed Created form at the base of the tree because how they evolved "from some one primordial form"[67] is not known. However, I think this "*Supposed Evolutionary Tree of Life - Before and After the Meteor Strike*" illustrates exactly what this article is stating.

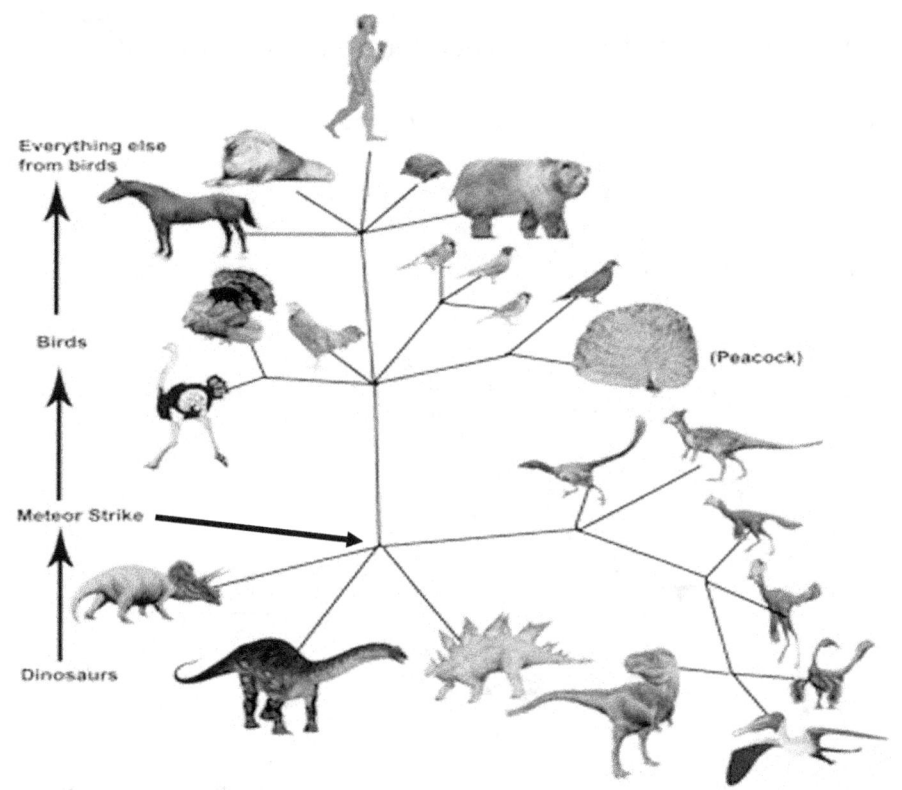

"Supposed Evolutionary Tree of Life Before and After the Meteor Strike"
(Picture References see Chapter 23 #68-85)

"Supposed" evolutionary tree of life before and after the meteor strike

*This is **my interpretation** of what a supposed tree of life would have to look like before and after the meteor strike. The quadrapeds (bottom left) died and the theropods bottom right survived the meteor strike. Then...somehow....all life as we see it today evolved from the theropods. OOPS!.., I should say, Then .... **by chance**.... all life as we see it today evolved from the theropods. (Somewhere in here we would need to make room for insects, fish, and, ... well, ... every life form we see on earth today).*

Making evolutionary assumptions based upon physical similarities of body parts such as jaw bone structure would be similar to looking at the wheel and saying, "Well, a bicycle has a wheel, a car has a wheel, a big-rig semi-truck eighteen-wheeler has wheels; they all must have *evolved* from the same entity on their own and of their own power by *chance* because they had millions of years for it to happen." Or; in trying to be more in line with Darwin's theory I should say, "we must *imagine*"[81] that there is a greater advantage to having more wheels. Then as Darwin says, *"we must suppose"*[86] that having more wheels leads to better survival. *"May we not believe"*[86] that there became more wheels on these vehicles over time so that they had a competitive advantage over vehicles with only one or two wheels, and that these changes in the number of wheels occurred by "numerous, successive, slight modifications"[87] on their own *by chance!*

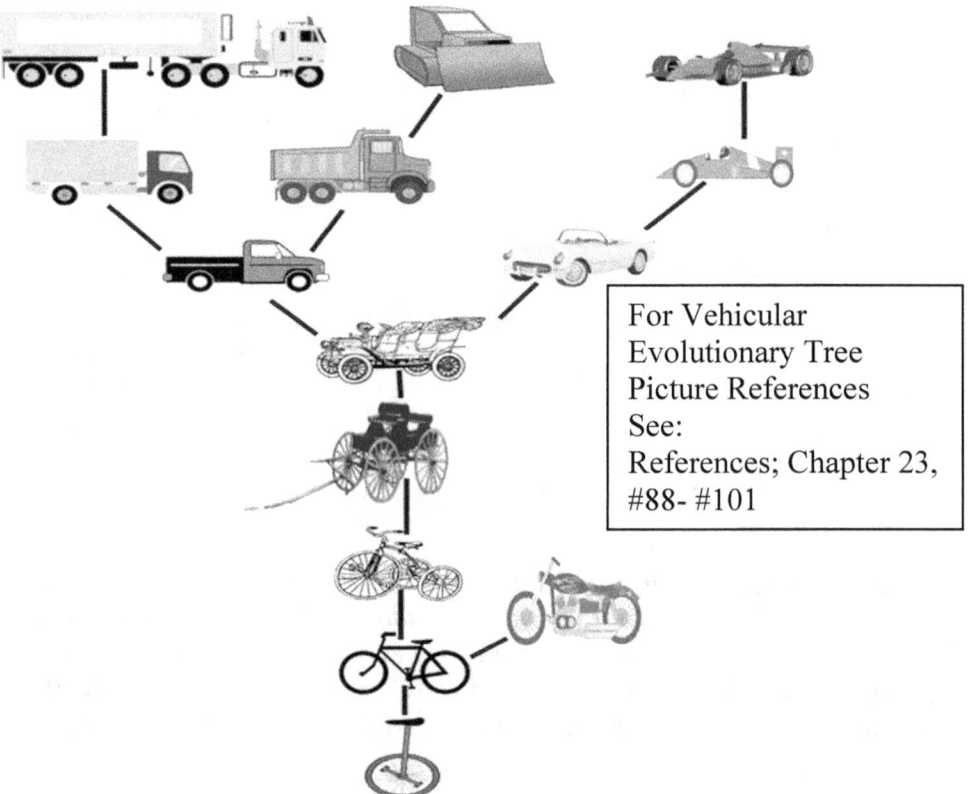

For Vehicular Evolutionary Tree Picture References See: References; Chapter 23, #88- #101

Is this Evolution ? Did any of these just happen? NO ! And *none* of these examples are *nearly* as complex as any living creature. You can observe the copmlexity of these examples and **KNOW** that someone created them even though you did not see someone create them. When we observe creation, we too can **KNOW** someone created it even though we did not see them create it.

When observing these vehicular examples, no one would question if they were created or if they evolved by chance. It is obvious they were created and everyone would agree they were created even though they did not see them be created. It is the same but infinitely more so when observing life. You ***KNOW*** these vehicles are created Can you not ***KNOW*** life was created?

KNOWING that none of these vehicular examples are nearly as complex as even a common housefly, let alone a human being; evolutionists still make these kinds of assumptions and conclusions regarding evolution every day! They use similarities in structure, to attempt to determine if they ***think*** different species are directly related. (Remember evolution would have you believe ***ALL*** species –plants and animals are related.)

Simply by *looking* at these vehicles (observation), everyone *knows* that they did not evolve by *chance*. Yet even while *knowing* that organic beings are ***far more intricately made*** - people still believe organic beings; life; occurred by "chance". Only God can make organic beings come to life. Life; the *living* part of life is the ultimate ***PROOF*** of God the Creator. Life proves there is God. Life cannot just happen. Even when man tries to ***MAKE*** it happen, man can't make life happen. Life proves God is.

We see God through His creation of the things we can see, we can also see God's wisdom, intelligence, foresight, planned prevention, exceeding abundance, beauty, wonder, and power in the creation and the working of His engineering marvel—the design and function of the eye. God has revealed Himself to us *through* our eyes and God has revealed Himself to us *in* our eyes.

When you look at someone's eyes you are seeing 25 different systems ALL working together! All 25 systems are totally INTERDEPENDENT!

At the same time; when you look at someone's eyes you yourself are using 25 different systems simultaneously. All 25 systems must be present and all 25 systems must be working totally, completely, and simultaneously

If one system is missing the eye cannot function and in some cases if only one system is missing the eye cannot remain alive.

When we look someone in the eye we see 25 different systems all working together in order for *their* eyes to function properly.

When we look at someone, 25 different systems are at work *in us* in order to see them. We see the following…

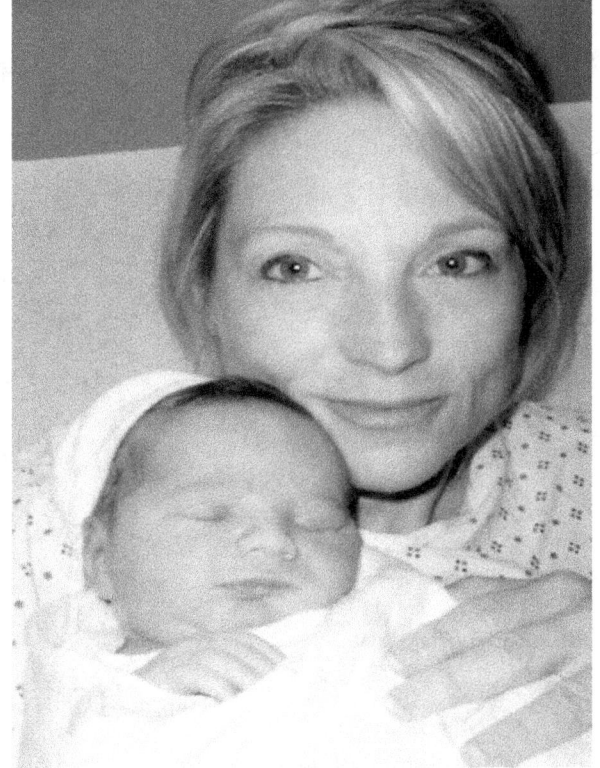

Interdependent Systems

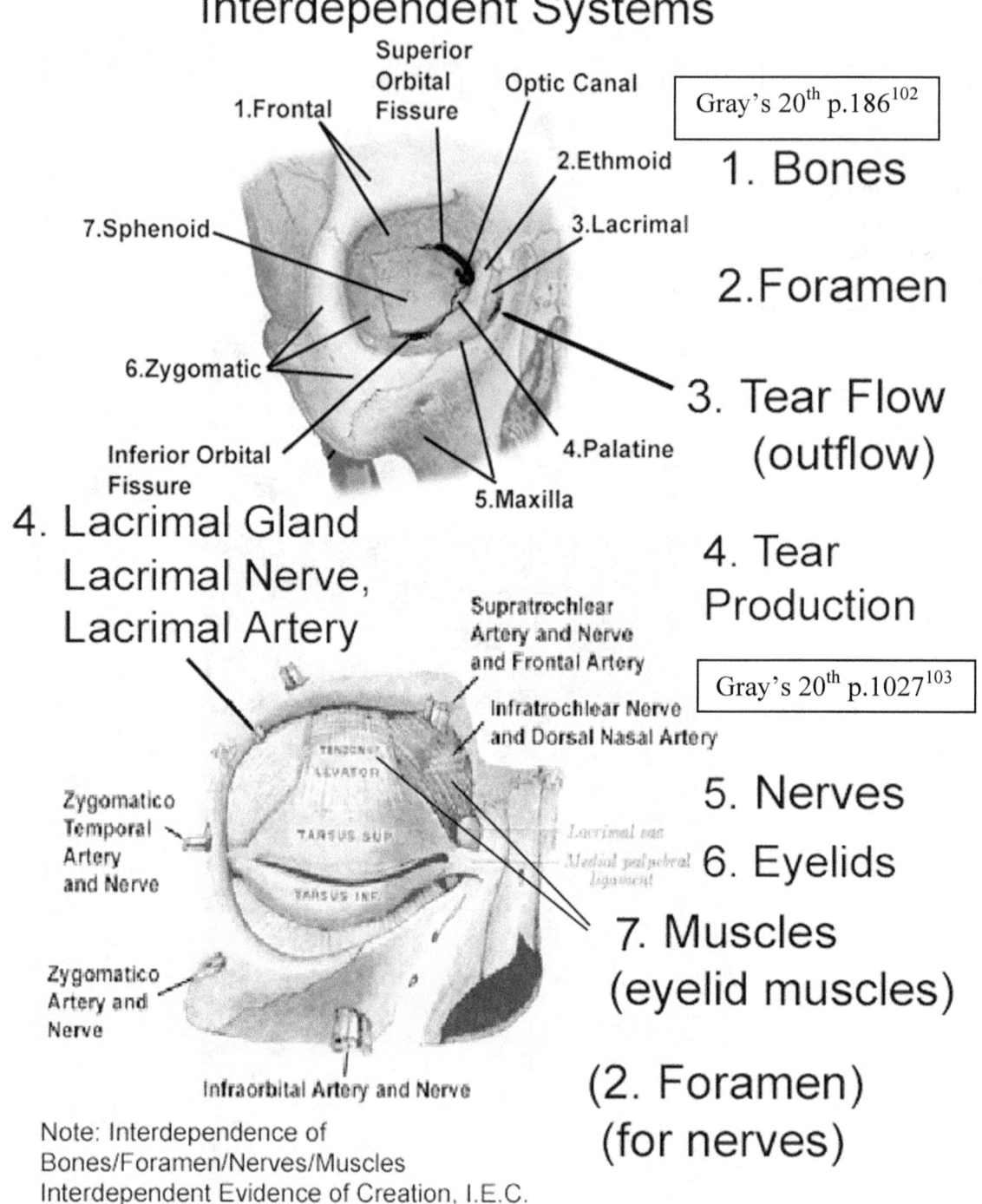

Superior Orbital Fissure

Optic Canal

1.Frontal

2.Ethmoid

3.Lacrimal

7.Sphenoid

6.Zygomatic

Inferior Orbital Fissure

4.Palatine

5.Maxilla

Gray's 20th p.186[102]

1. Bones

2.Foramen

3. Tear Flow (outflow)

4. Tear Production

4. Lacrimal Gland Lacrimal Nerve, Lacrimal Artery

Supratrochlear Artery and Nerve and Frontal Artery

Infratrochlear Nerve and Dorsal Nasal Artery

Gray's 20th p.1027[103]

5. Nerves

6. Eyelids

7. Muscles (eyelid muscles)

(2. Foramen) (for nerves)

Zygomatico Temporal Artery and Nerve

Zygomatico Artery and Nerve

Infraorbital Artery and Nerve

TENDON of LEVATOR

TARSUS SUP

TARSUS INF

Lacrimal sac

Medial palpebral ligament

Note: Interdependence of Bones/Foramen/Nerves/Muscles Interdependent Evidence of Creation, I.E.C.

Interdependent Systems

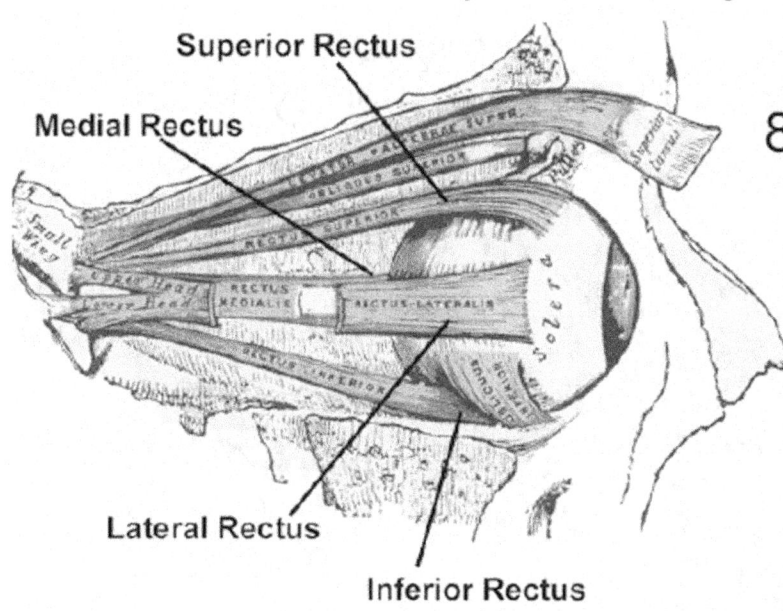

Superior Rectus

Medial Rectus

Lateral Rectus

Inferior Rectus

8. Extraocular Muscles
1. Bones

Gray's 20th p. 1022[104]

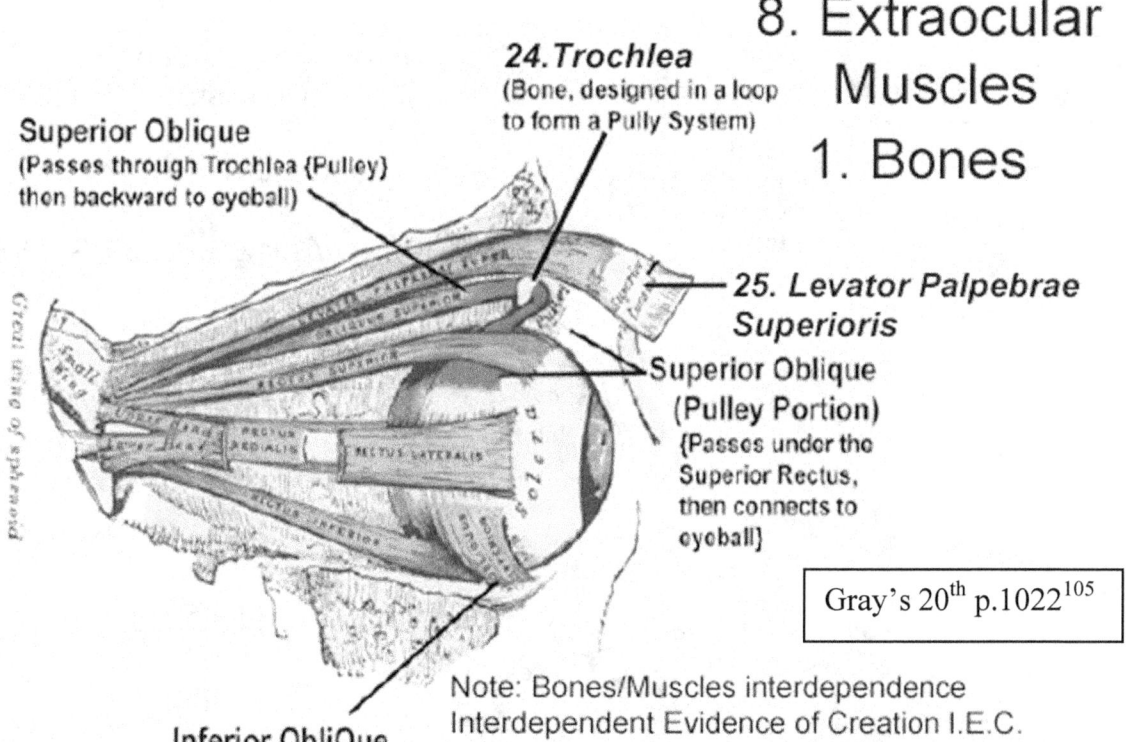

Superior Oblique
(Passes through Trochlea {Pulley}
then backward to eyeball)

24. Trochlea
(Bone, designed in a loop
to form a Pully System)

Inferior ObliQue

8. Extraocular Muscles
1. Bones

25. Levator Palpebrae Superioris

Superior Oblique
(Pulley Portion)
{Passes under the
Superior Rectus,
then connects to
eyeball}

Gray's 20th p.1022[105]

Note: Bones/Muscles interdependence
Interdependent Evidence of Creation I.E.C.
(Nerves must be present also - additional I.E.C.)

Interdependent Systems

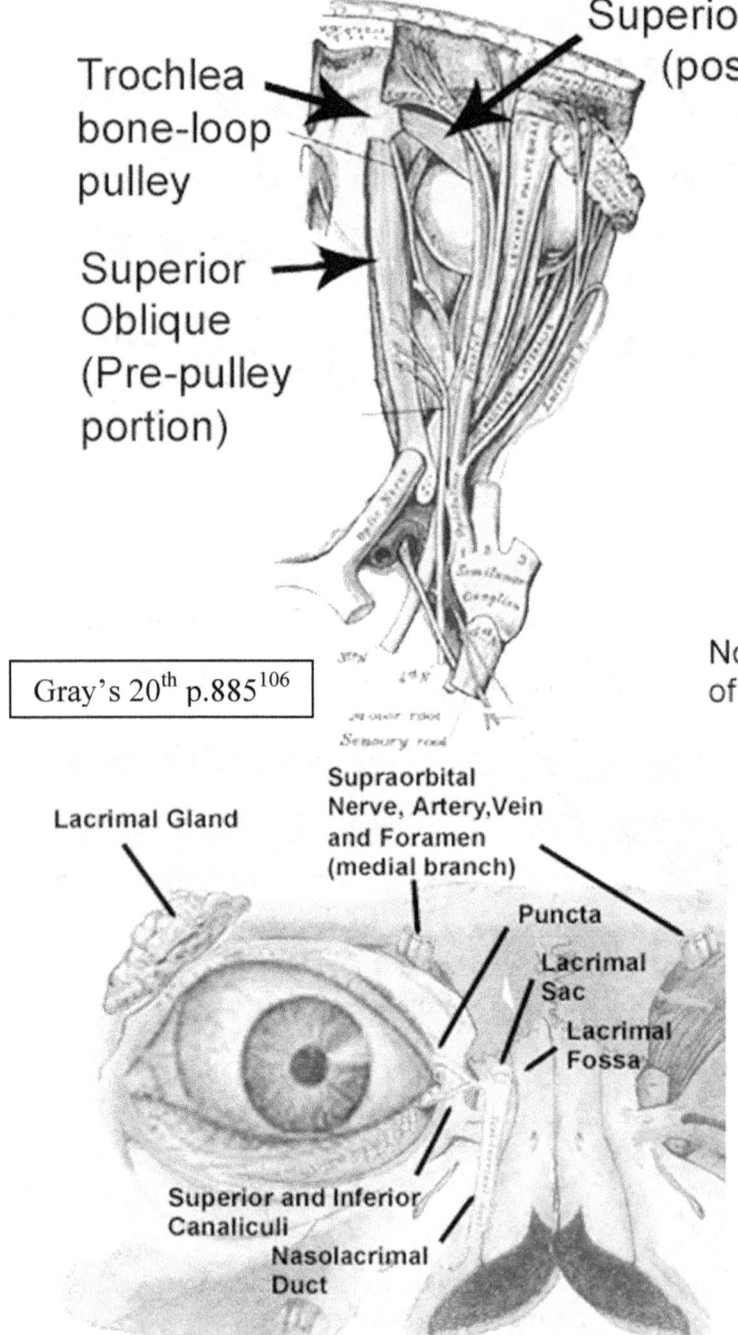

Trochlea bone-loop pulley

Superior Oblique (Pre-pulley portion)

Superior Oblique (post-pulley portion)

1. Bones
2. Foramen
4. Tear Flow (tear production)
5. Nerves
8. Extraocular Muscles
9. Arteries
10. Veins

Note interdependence of all these systems; I.E.C.

6. Eyelids

7. Facial Muscles

3. Tear Flow (tear outlow)

4. Tear Flow (tear production)

5. Nerves

2. Foramen

Gray's 20th p.885[106]

Lacrimal Gland

Supraorbital Nerve, Artery, Vein and Foramen (medial branch)

Puncta

Lacrimal Sac

Lacrimal Fossa

Superior and Inferior Canaliculi

Nasolacrimal Duct

Gray's 20th p.1026, 1027[107]

Interdependent Systems

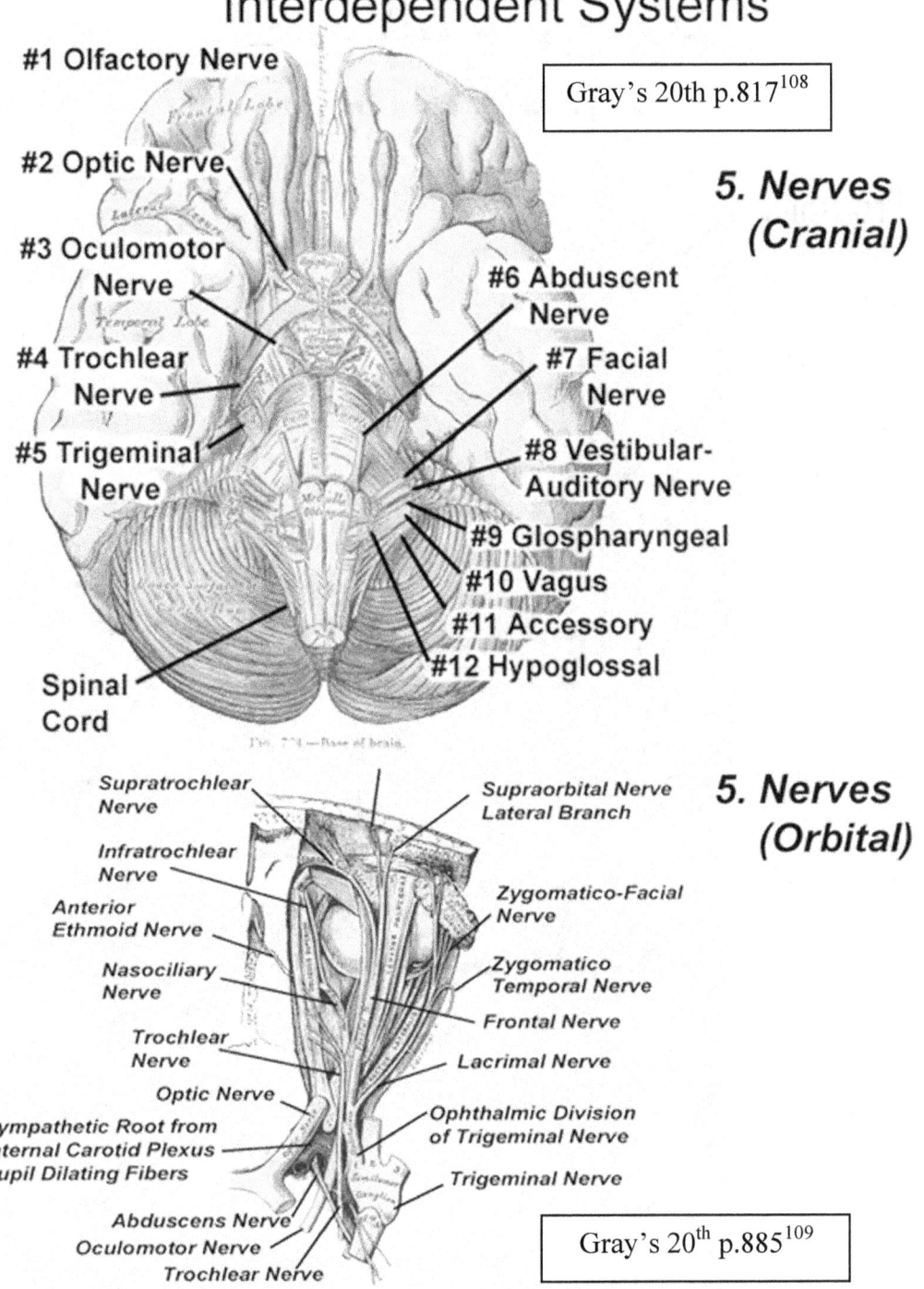

#1 Olfactory Nerve

#2 Optic Nerve

#3 Oculomotor Nerve

#4 Trochlear Nerve

#5 Trigeminal Nerve

Gray's 20th p.817[108]

5. Nerves (Cranial)

#6 Abduscent Nerve

#7 Facial Nerve

#8 Vestibular-Auditory Nerve

#9 Glospharyngeal

#10 Vagus

#11 Accessory

#12 Hypoglossal

Spinal Cord

Fig. 774 —Base of brain.

5. Nerves (Orbital)

Supratrochlear Nerve

Infratrochlear Nerve

Anterior Ethmoid Nerve

Nasociliary Nerve

Trochlear Nerve

Optic Nerve

Sympathetic Root from Internal Carotid Plexus Pupil Dilating Fibers

Abducens Nerve

Oculomotor Nerve

Trochlear Nerve

Supraorbital Nerve Lateral Branch

Zygomatico-Facial Nerve

Zygomatico Temporal Nerve

Frontal Nerve

Lacrimal Nerve

Ophthalmic Division of Trigeminal Nerve

Trigeminal Nerve

Gray's 20th p.885[109]

6. Facial Muscles
(Orbicularis Oculi)

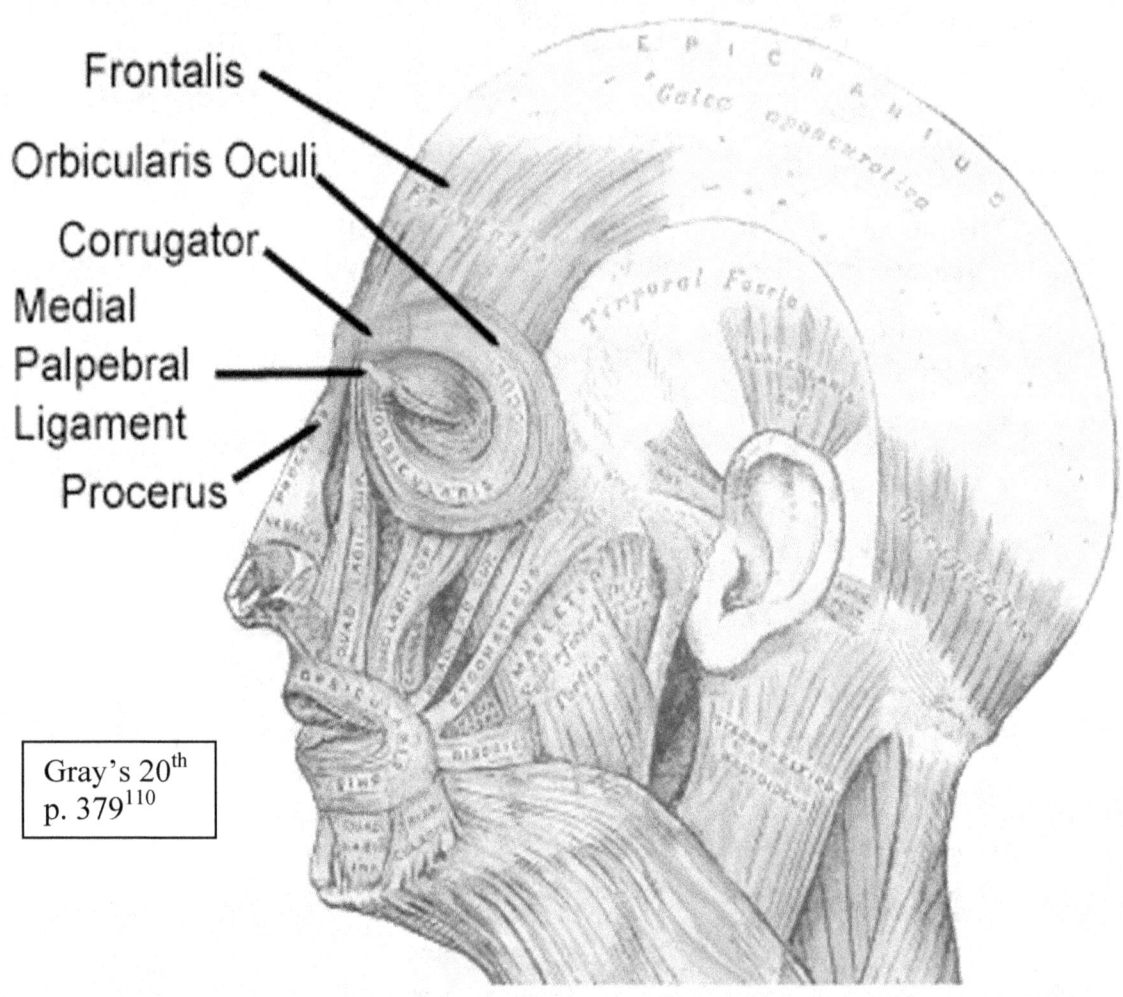

Frontalis

Orbicularis Oculi

Corrugator

Medial
Palpebral
Ligament

Procerus

Gray's 20th
p. 379[110]

Interdependent Systems

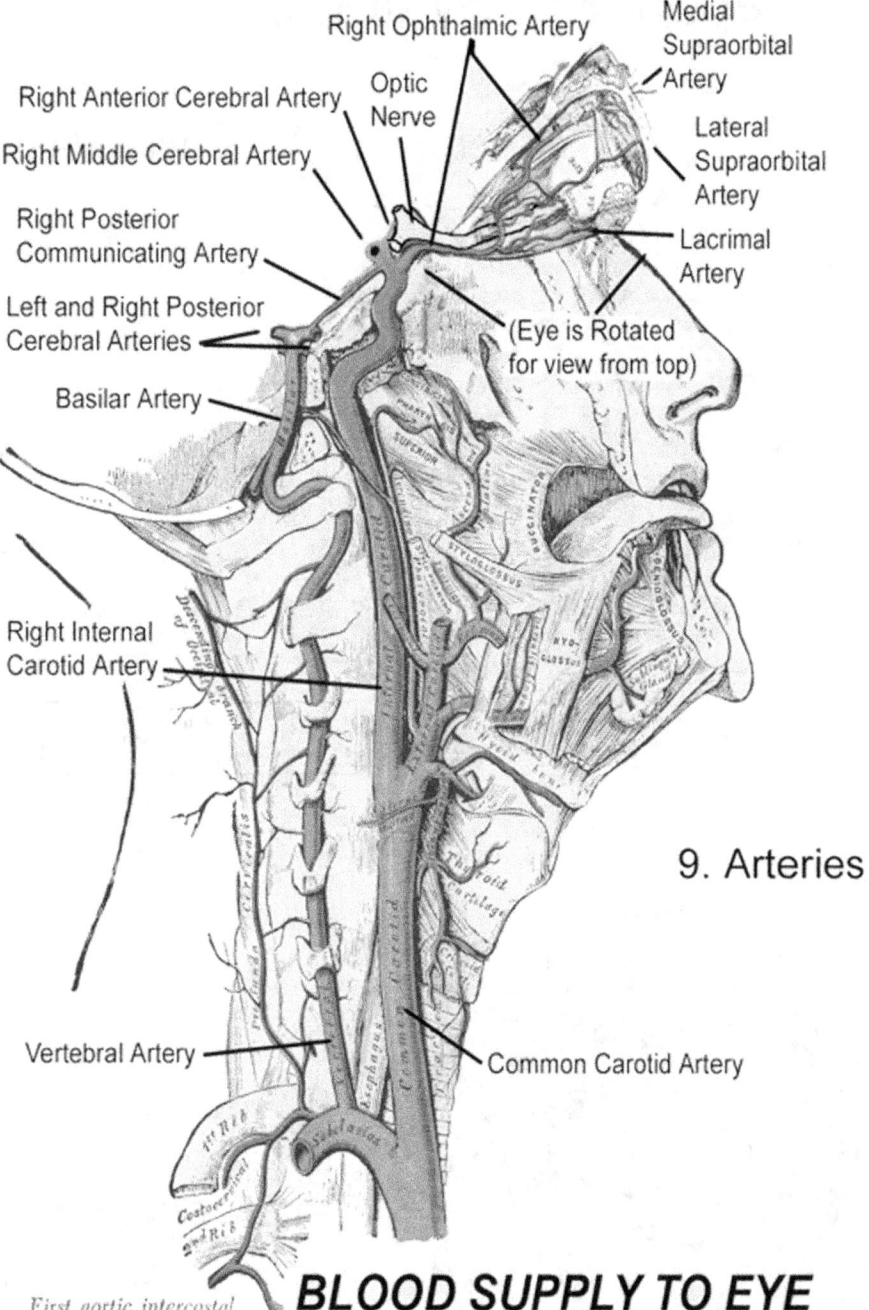

Right Ophthalmic Artery

Medial Supraorbital Artery

Optic Nerve

Right Anterior Cerebral Artery

Lateral Supraorbital Artery

Right Middle Cerebral Artery

Right Posterior Communicating Artery

Lacrimal Artery

Left and Right Posterior Cerebral Arteries

(Eye is Rotated for view from top)

Basilar Artery

Right Internal Carotid Artery

9. Arteries

Vertebral Artery

Common Carotid Artery

First aortic intercostal

BLOOD SUPPLY TO EYE

Gray's 20th p. 566 and p.569[111] combined by Rigney

Interdependent Systems

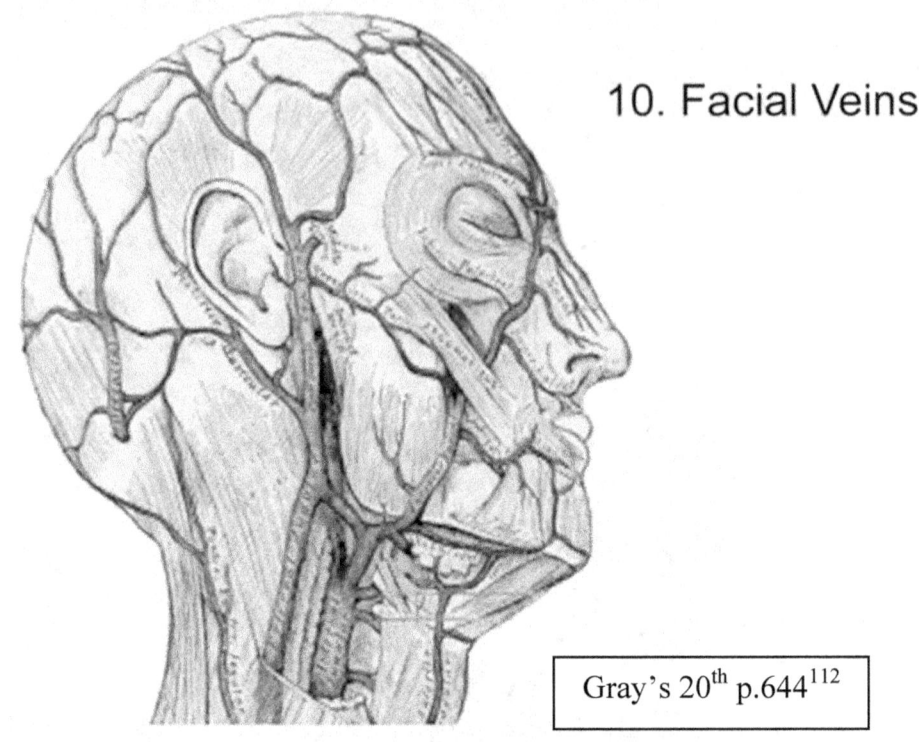

10. Facial Veins

Gray's 20th p.644[112]

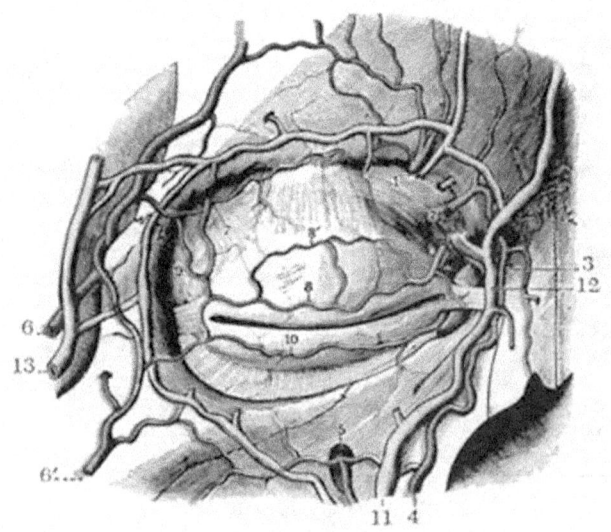

9. Arteries
10. Veins
(of Lid and Adnexa)

Gray's 20th p.570[113]

Interdependent Systems

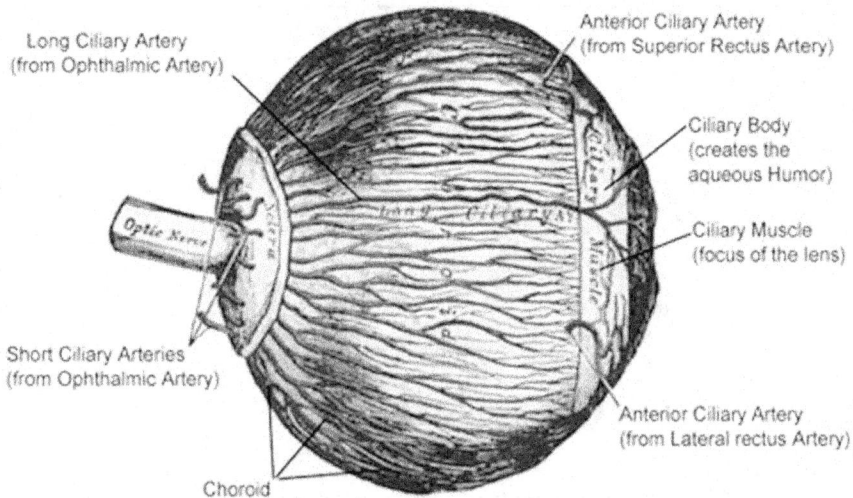

Long Ciliary Artery
(from Ophthalmic Artery)

Anterior Ciliary Artery
(from Superior Rectus Artery)

Ciliary Body
(creates the
aqueous Humor)

Ciliary Muscle
(focus of the lens)

Optic Nerve

Short Ciliary Arteries
(from Ophthalmic Artery)

Anterior Ciliary Artery
(from Lateral rectus Artery)

Choroid
(Network of blood vessels that supplies the inner retina)

Gray's 20th p.1009[114]

9. Ciliary Arteries

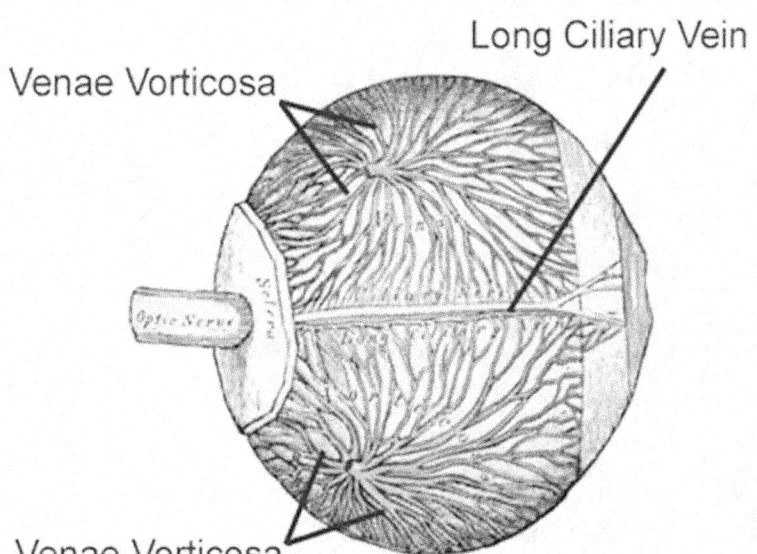

Long Ciliary Vein

Venae Vorticosa

Optic Nerve

Venae Vorticosa

10. Ciliary Veins

Gray's 20th p.1010[115]

Blood vessels of the eyeball
(The Sclera has been removed to reveal the Choroid)

Interdependent Systems: 11. Pupil Function

Pupil Constrictor Nerve Route

Gray's 20th p.972[116]

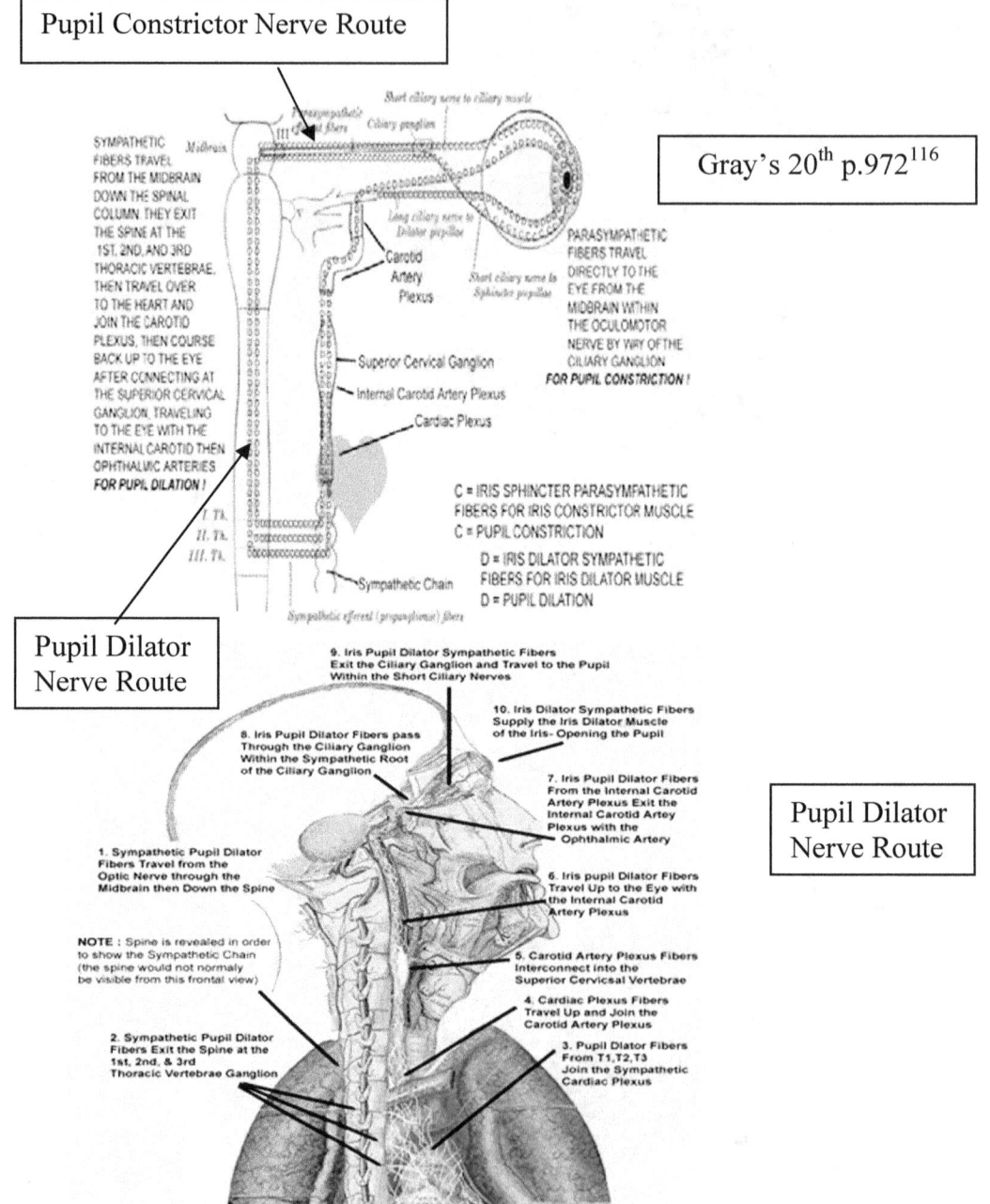

SYMPATHETIC FIBERS TRAVEL FROM THE MIDBRAIN DOWN THE SPINAL COLUMN. THEY EXIT THE SPINE AT THE 1ST, 2ND, AND 3RD THORACIC VERTEBRAE. THEN TRAVEL OVER TO THE HEART AND JOIN THE CAROTID PLEXUS, THEN COURSE BACK UP TO THE EYE AFTER CONNECTING AT THE SUPERIOR CERVICAL GANGLION, TRAVELING TO THE EYE WITH THE INTERNAL CAROTID THEN OPHTHALMIC ARTERIES *FOR PUPIL DILATION !*

PARASYMPATHETIC FIBERS TRAVEL DIRECTLY TO THE EYE FROM THE MIDBRAIN WITHIN THE OCULOMOTOR NERVE BY WAY OF THE CILIARY GANGLION *FOR PUPIL CONSTRICTION !*

C = IRIS SPHINCTER PARASYMPATHETIC FIBERS FOR IRIS CONSTRICTOR MUSCLE
C = PUPIL CONSTRICTION

D = IRIS DILATOR SYMPATHETIC FIBERS FOR IRIS DILATOR MUSCLE
D = PUPIL DILATION

Pupil Dilator Nerve Route

Pupil Dilator Nerve Route

9. Iris Pupil Dilator Sympathetic Fibers Exit the Ciliary Ganglion and Travel to the Pupil Within the Short Ciliary Nerves

10. Iris Dilator Sympathetic Fibers Supply the Iris Dilator Muscle of the Iris- Opening the Pupil

8. Iris Pupil Dilator Fibers pass Through the Ciliary Ganglion Within the Sympathetic Root of the Ciliary Ganglion

7. Iris Pupil Dilator Fibers From the Internal Carotid Artery Plexus Exit the Internal Carotid Artey Plexus with the Ophthalmic Artery

1. Sympathetic Pupil Dilator Fibers Travel from the Optic Nerve through the Midbrain then Down the Spine

6. Iris pupil Dilator Fibers Travel Up to the Eye with the Internal Carotid Artery Plexus

NOTE : Spine is revealed in order to show the Sympathetic Chain (the spine would not normally be visible from this frontal view)

5. Carotid Artery Plexus Fibers Interconnect into the Superior Cervicsal Vertebrae

4. Cardiac Plexus Fibers Travel Up and Join the Carotid Artery Plexus

2. Sympathetic Pupil Dilator Fibers Exit the Spine at the 1st, 2nd, & 3rd Thoracic Vertebrae Ganglion

3. Pupil Dlator Fibers From T1,T2,T3 Join the Sympathetic Cardiac Plexus

From Henry Gray, F.R.S, *Anatomy of the Human Body* 20th Edition, Lea & Febiger (Philadelphia & New York, © 1918) p.566, 569, 527, 766[117] combined and modified by Rigney

Interdependent Systems

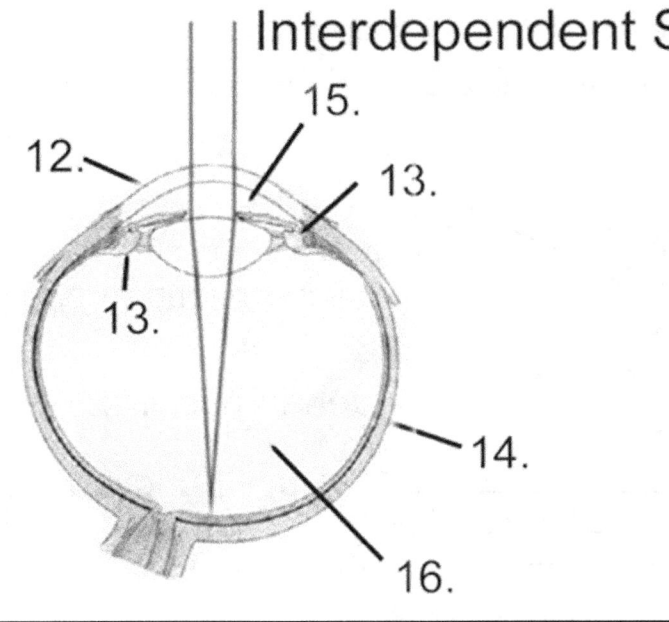

12. Cornea

13. Focusing Control

14. Sclera

15. Aqueous

16. Vitreous

(Above and Below)
Gray's 20th p. 1006[118] with modifications by Rigney

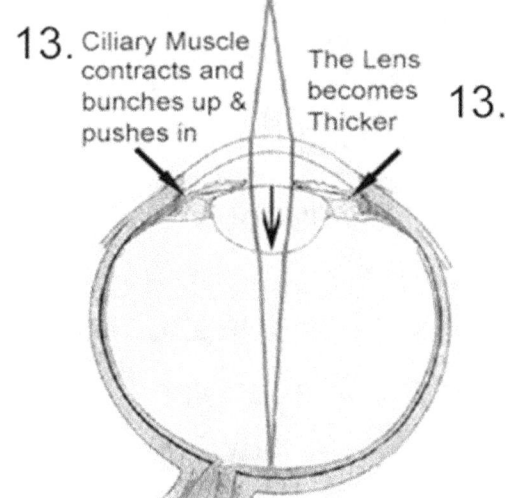

13. Ciliary Muscle contracts and bunches up & pushes in
The Lens becomes Thicker
13.

13. Ciliary Muscle

13. Lens

13. Lens Zonules

5. Oculomotor Nerve to Ciliary Muscle

Note Interdependence of **ALL** these <u>systems</u> in order to achieve focusing control Interdependent Evidence of Creation, I.E.C.

J. Jay Rigney, O.D.

Note Interdependence of **All** the Systems to
maintain eye pressure. **ALL MUST BE PRESENT; I.E.C.**

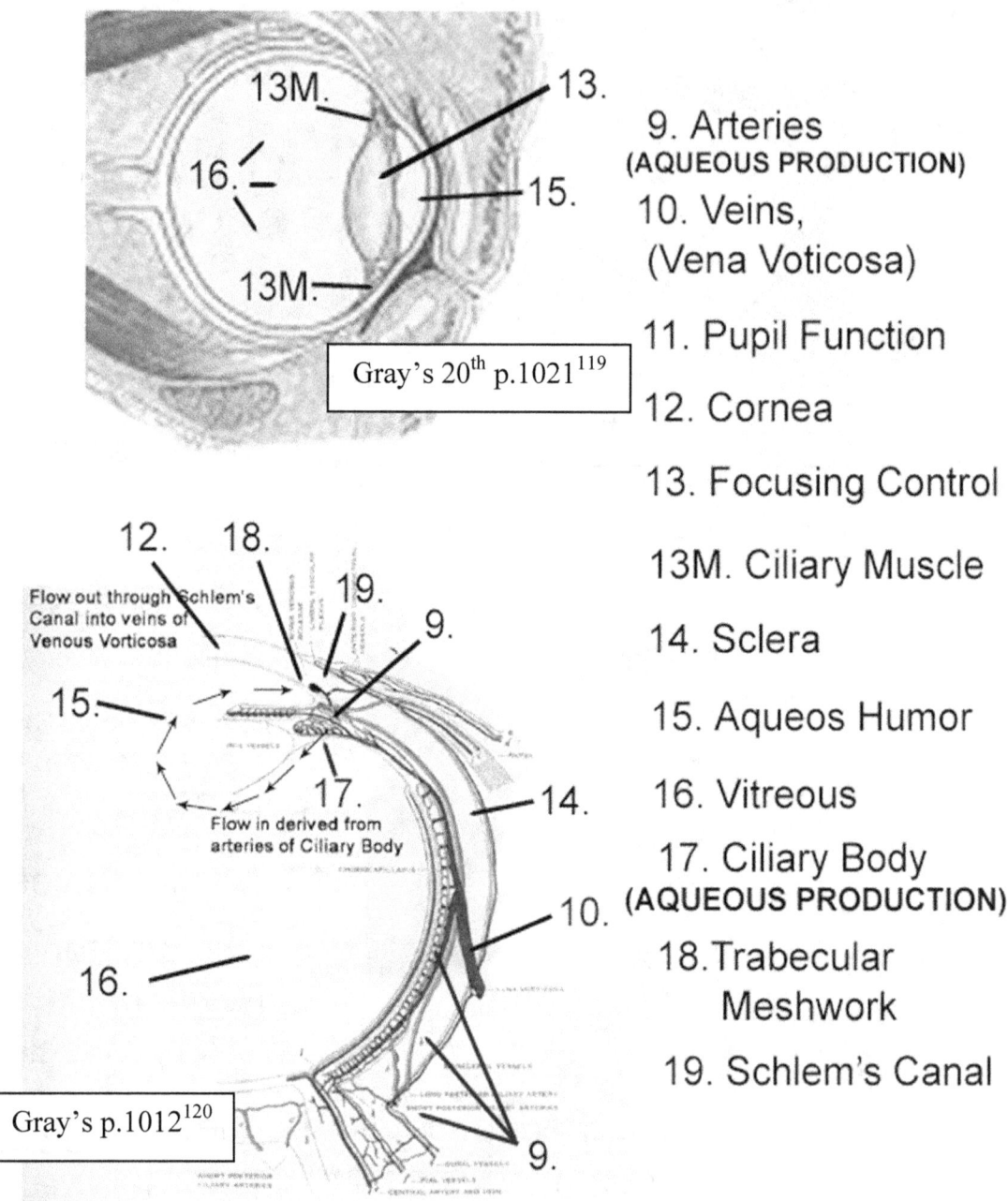

Gray's 20th p.1021¹¹⁹

Gray's p.1012¹²⁰

9. Arteries
(AQUEOUS PRODUCTION)
10. Veins,
(Vena Voticosa)
11. Pupil Function
12. Cornea
13. Focusing Control
13M. Ciliary Muscle
14. Sclera
15. Aqueos Humor
16. Vitreous
17. Ciliary Body
(AQUEOUS PRODUCTION)
18. Trabecular Meshwork
19. Schlem's Canal

I.E.C. INTERDEPENDENT EVIDENCE OF CREATION

Interdependent Systems

20. Retina:

(Rods, Cones, Additional Retinal Cells, Retinal Piment Epithelium)
I.E.C.: Function of rods and cones is dependent upon nourishment from 9. arteries, 10 veins, plus function is also dependent on 21 Optic Nerve. Interdependent Evidence of Creation

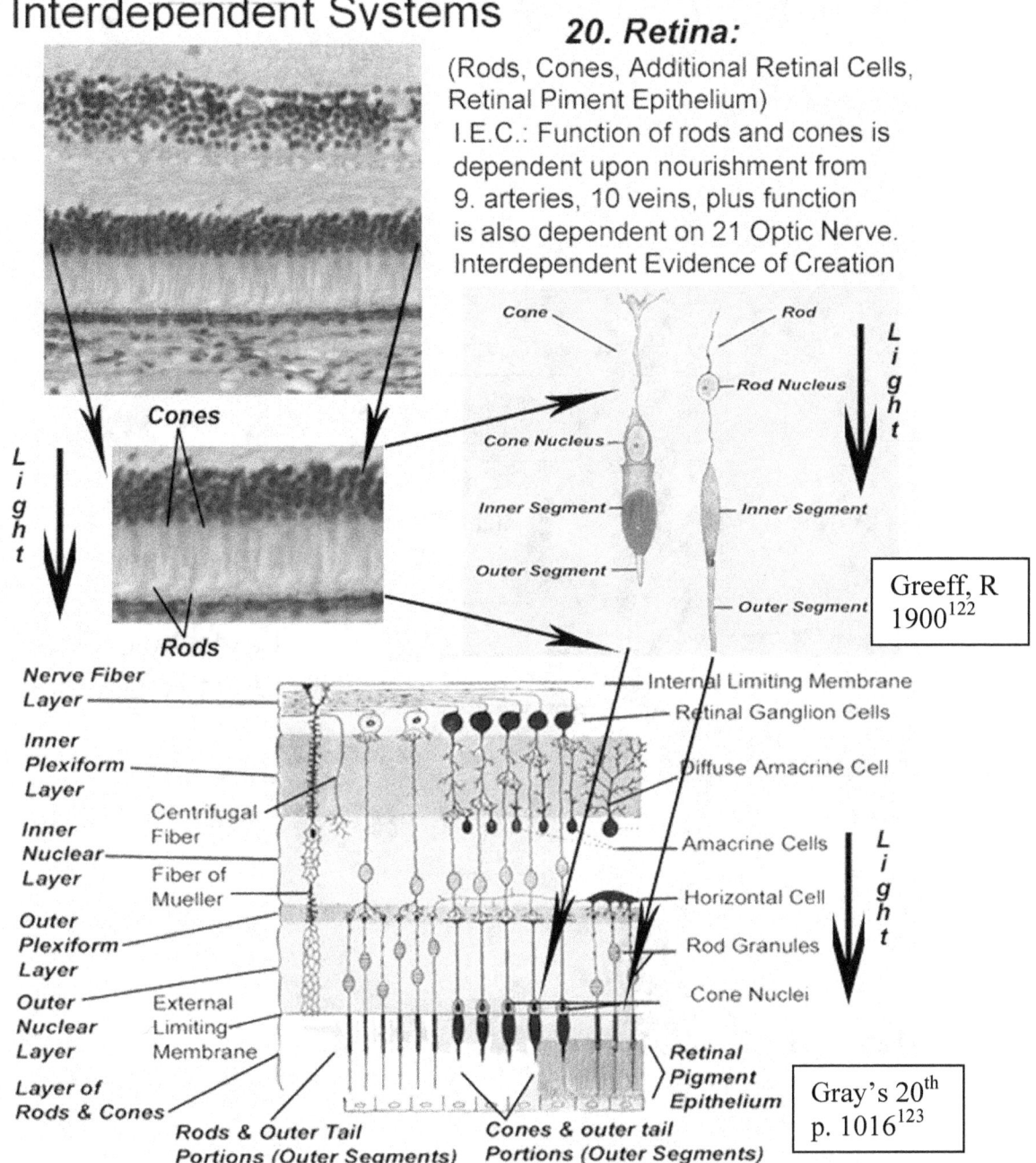

Greeff, R 1900[122]

Gray's 20th p. 1016[123]

Interdependent Systems

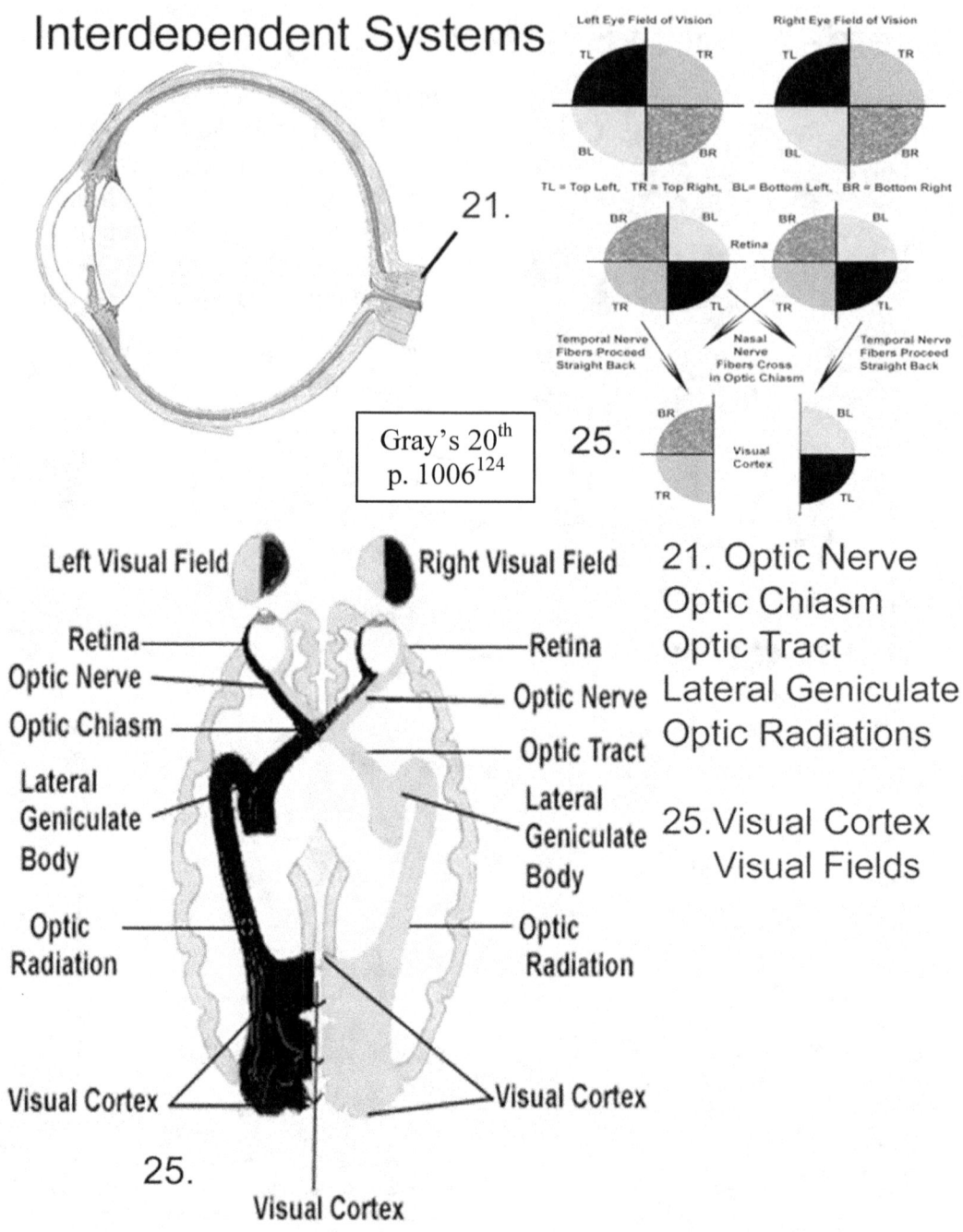

21. Optic Nerve
Optic Chiasm
Optic Tract
Lateral Geniculate
Optic Radiations

25. Visual Cortex
Visual Fields

Gray's 20th
p. 1006[124]

25.

Left Eye Field of Vision | Right Eye Field of Vision

TL = Top Left, TR = Top Right, BL= Bottom Left, BR = Bottom Right

Retina

Temporal Nerve Fibers Proceed Straight Back | Nasal Nerve Fibers Cross in Optic Chiasm | Temporal Nerve Fibers Proceed Straight Back

Visual Cortex

Left Visual Field — Right Visual Field

Retina — Retina
Optic Nerve — Optic Nerve
Optic Chiasm — Optic Tract
Lateral Geniculate Body — Lateral Geniculate Body
Optic Radiation — Optic Radiation
Visual Cortex — Visual Cortex

25. Visual Cortex

Interdependent Systems

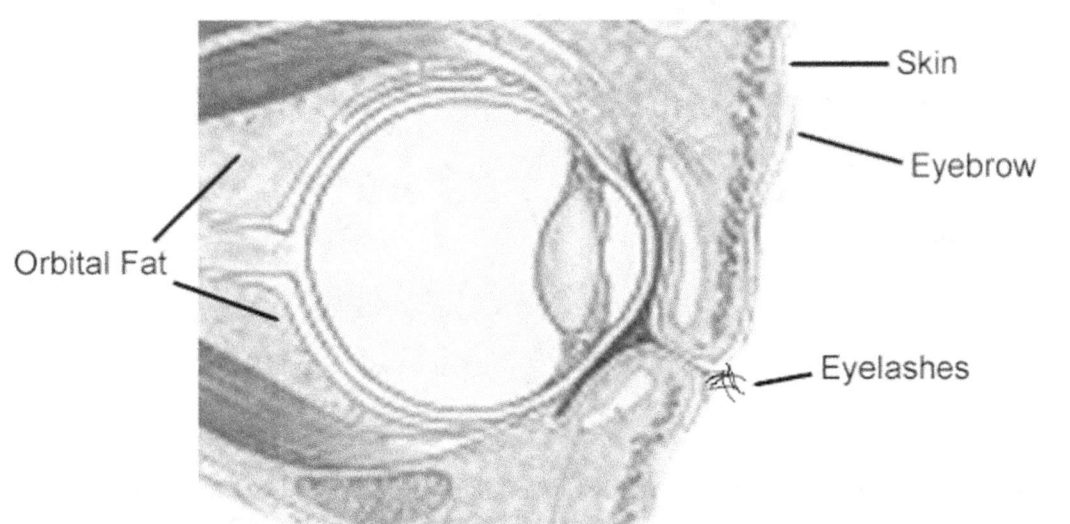

Eyebrow

Supraorbital Nerve, Artery,Vein and Foramen (medial branch)

Orbicularis Oculi

Levator Palpebrae Superioris

Supraorbital Nerve, Artery, and Foramen (lateral branch)

Lacrimal Nerve
Lacrimal Artery

Frontal-Zygomatic Bone suture

Eyelids

Eyelashes

Lateral Palpebral Ligament

Zygomatico-Facial Foramen, Nerve, Vein, and Artery

Medial Palpebral Ligament

Infraorbital Foramen, Nerve, Artery, & Vein

Maxillary Bone

22. External Ocular Adnexa
Skin of Eyelids, Eyebrows, Eyelashes, Orbital Fat

Skin

Eyebrow

Orbital Fat

Eyelashes

7. Eyelids; Facial Muscles   23.Eyelid (Levator) Muscle

Top: Gray's 20th p.1026 & p.1027[126] combined by Rigney. Bottom Gray's 20th p.1021[127]

23. Levator Palpebra Superioris

Eyebrow

Supraorbital Nerve, Artery, Vein and Foramen (medial branch)

Orbicularis Oculi

Supraorbital Nerve, Artery, and Foramen (lateral branch)

Lacrimal Nerve
Lacrimal Artery

Frontal-Zygomatic Bone suture

Eyelids

Eyelashes

Lateral Palpebral Ligament

Zygomatico-Facial Foramen, Nerve, Vein, and Artery

Medial Palpebral Ligament

Infraorbital Foramen, Nerve, Artery, & Vein

Maxillary Bone

23. Levator Palpebrae Superioris

Bulbar Conjunctiva Clear Skin on Eyeball

Superior Fornix
Pink eyelid skin
(Palpebral Conjunctiva)
becomes clear
eyeball skin
(Bulbar Conjunctiva)

Inferior Fornix
Pink eyelid skin
(Palpebral Conjunctiva)
becomes clear
eyeball skin
(Bulbar Conjunctiva)

The Levator Palpebrae Superioris (eyelid muscle) is not an Extraocular Muscle, it is not a Facial Muscle, The Levator Palpebrae Superioris is a separate system in and of itself. Evoluition cannot explain the formation of this muscle by "numerous, successive slight modifications". Darwin's theory "absolutely breaks down.".

Top: Gray's 20th p.1026 & p.1027[128] combined, with additions by Rigney
Bottom; Gray's 20th p.1021[129]

Interdependent Systems

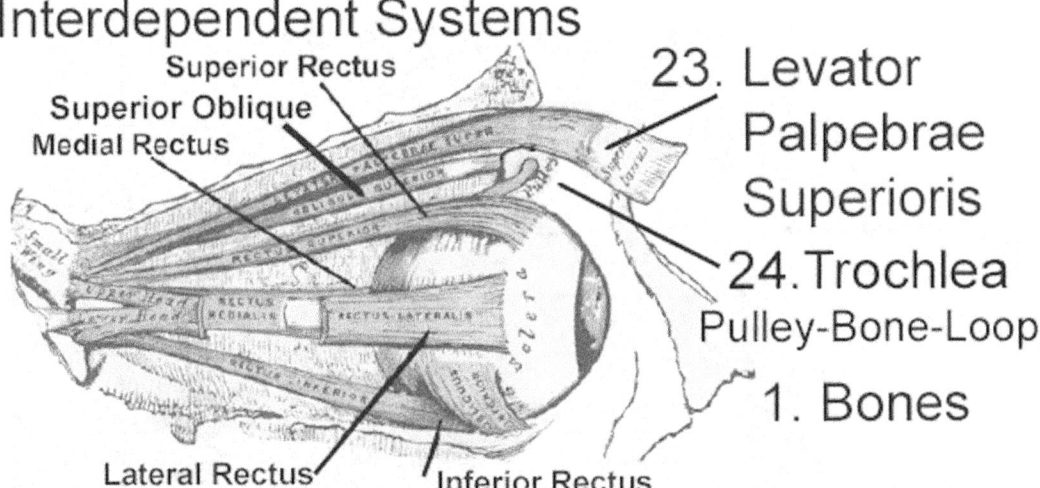

Superior Rectus
Superior Oblique
Medial Rectus

23. Levator Palpebrae Superioris

24. Trochlea
Pulley-Bone-Loop

1. Bones

Lateral Rectus

Inferior Rectus

The Trochlea (Pulley-Bone-Loop) is not a Foramen. It is a system in and of itself. Evolution cannot explain its formation and the Superior Oblique Muscle evolving to pass through it, then evolving to proceed back to the eye, attaching on the back of the eyeball underneath the Superior Rectus Muscle;The trochlea and superior oblique muscle system cannot be explained by "numerous, succssive, slight modifications", as Darwin himself anticipated, "my theory ... "absolutely breaks down".
We observe; Interdependent Evidence of Creation, I.E.C.

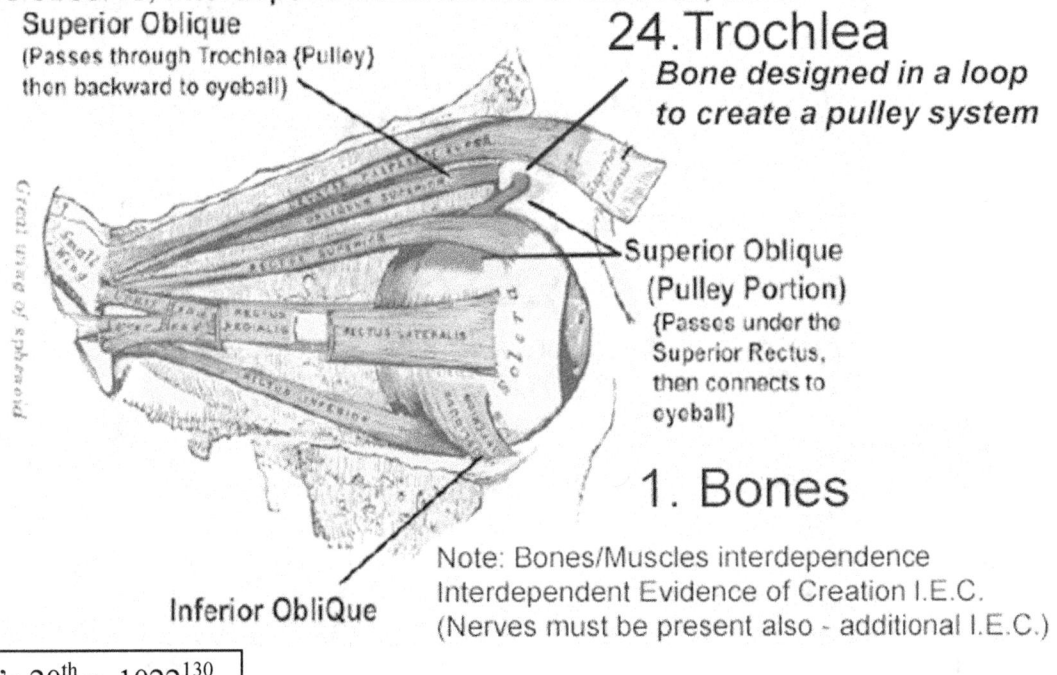

Superior Oblique
(Passes through Trochlea {Pulley}
then backward to eyeball)

24. Trochlea
Bone designed in a loop to create a pulley system

Superior Oblique
(Pulley Portion)
{Passes under the
Superior Rectus,
then connects to
eyeball}

1. Bones

Note: Bones/Muscles interdependence
Interdependent Evidence of Creation I.E.C.
(Nerves must be present also - additional I.E.C.)

Inferior ObliQue

Gray's 20th p. 1022[130]

Interdependent Systems

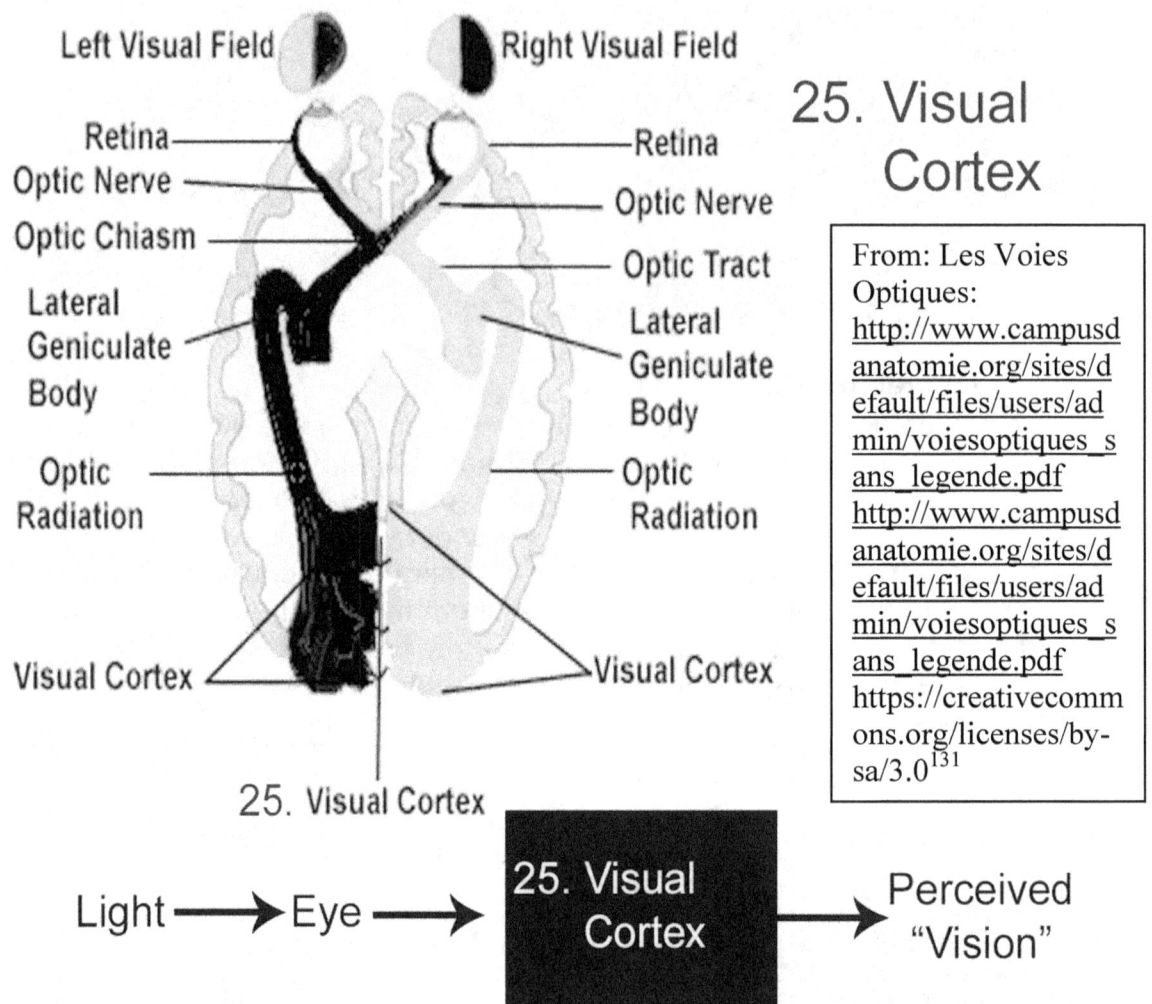

25. Visual Cortex

Left Visual Field   Right Visual Field

Retina   Retina

Optic Nerve   Optic Nerve

Optic Chiasm   Optic Tract

Lateral Geniculate Body   Lateral Geniculate Body

Optic Radiation   Optic Radiation

Visual Cortex   Visual Cortex

25. Visual Cortex

From: Les Voies Optiques: http://www.campusdanatomie.org/sites/default/files/users/admin/voiesoptiques_sans_legende.pdf http://www.campusdanatomie.org/sites/default/files/users/admin/voiesoptiques_sans_legende.pdf https://creativecommons.org/licenses/by-sa/3.0[131]

Light ⟶ Eye ⟶ **25. Visual Cortex** ⟶ Perceived "Vision"

Only God Knows how the Visual Cortex converts electrical impulses into "*VISION*". THINK ABOUT IT ! MEDITATE ON THIS FACT: WHAT IS PERCEIVED IS AS IT IS IN REALITY. *THANK YOU GOD !*

I.E.C.: ALL 25 SYSTEMS MUST BE PRESENT. ALL 25 SYSTEMS MUST BE FULLY FUNCTIONAL. ALL 25 MUST BE PRESENT AND FULLY FUNCTIONAL *ALL AT THE SAME TIME.* NO ONE SYSTEM CAN WAIT ON ANY OTHER ONE SYSTEM TO EVOLVE BECAUSE THEY ARE ABSOLUTELY AND FATALLY INTERDEPENDENT. ANY ONE SYSTEM'S ABSENCE IS FATAL TO THE FUNCTION AND SURVIVAL OF THE EYE. **DARWIN'S THEORY "ABSOLUTELY BREAKS DOWN".** WE OBSERVE: *INTERDEPENDENT EVIDENCE OF CREATION*

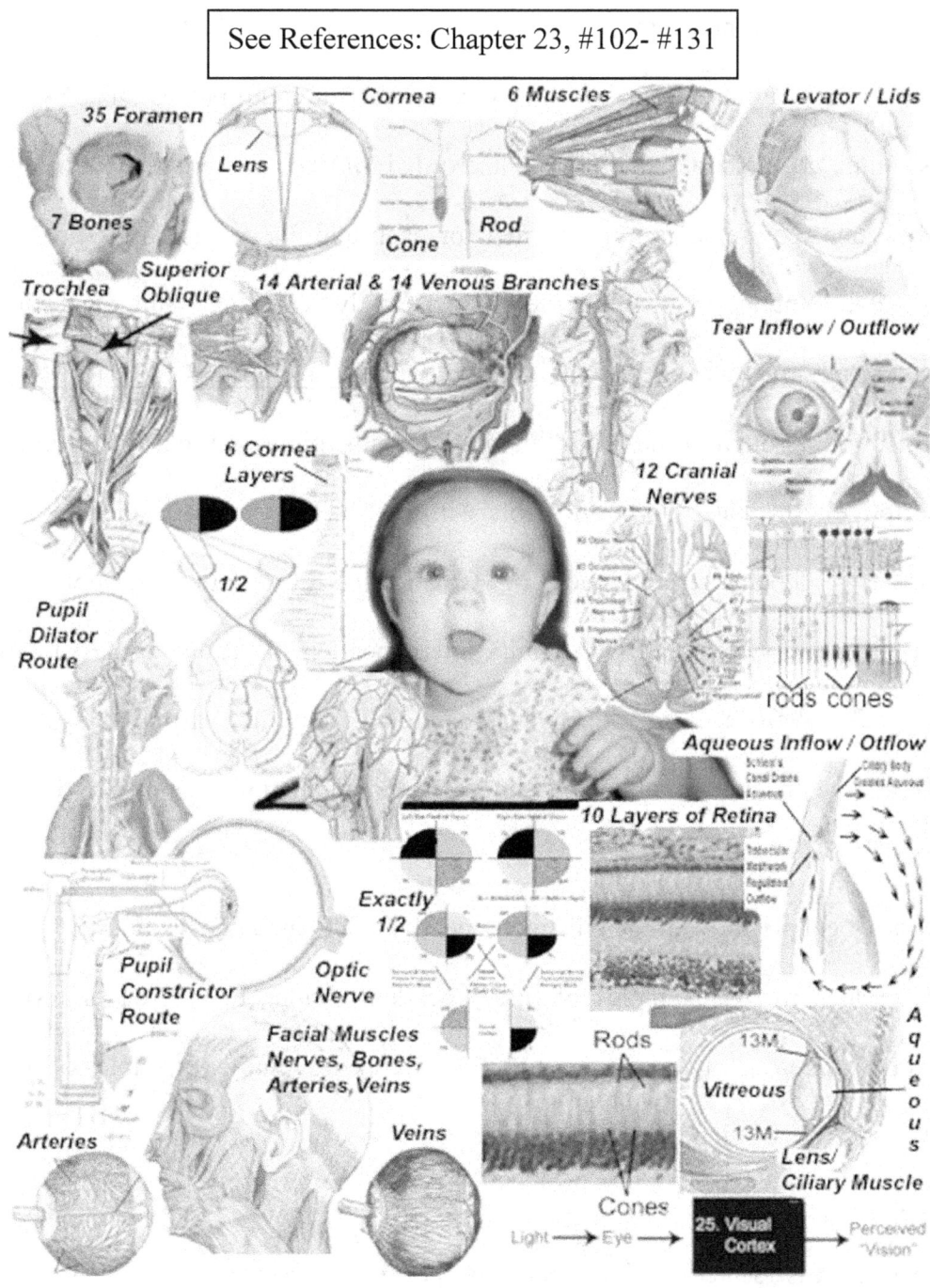

See References: Chapter 23, #102- #131

When you look at someone's eye, you are observing 25 different systems at work. Observe that each system must be fully formed and fully functional. Observe how they are formed and fashioned together. Observe that each system is dependent on all the other systems. We observe Interdependent Evidence of Creation, I.E.C.

All the parts have to be present, *all* working properly, *all* fully functional, and *all* at the same time. If any *one* part were missing or not fully functional, you could not see. The eye cannot work and will not see until *all* the parts are present—and *only* if *all* the parts are fully developed and fully functional. The same requirements must be met for each and every part, organ, or system in the body. *All* the different parts, organs, or systems of the body have totally interdependent parts and systems all with totally interdependent function, ***and* All MUST OCCUR SIMULTANEOUSLY**. I have struggled for an acronym to describe this but have not found one that I am comfortable with. The best descriptive term that I am comfortable with is "Interdependent Evidence of Creation" or I.E.C. Perhaps Interdependent Evidence of Creation, I.E.C. could be defined as totally interdependent parts and systems all with totally interdependent functions **and** ALL ***MUST Occur Simultaneously***. Simultaneous, interdependent function of multiple different systems can only happen by ***Creation*** ! In fact; simultaneous, interdependent function of multiple different systems describes CREATION.

To follow are a couple of examples of Interdependent Evidence of Creation as it pertains to the eye. There are many, many, examples of Interdependent Evidence of Creation, written in each chapter of this book. Here are two examples; at least twenty-three other examples of Interdependent Evidence of Creation are outlined in other chapters of this book.

I.   The Lacrimal System:

The tear gland is located on the upper outer portion of the eye. The tear drain duct is in the inner corner. The lacrimal system must have 1.The *lacrimal nerve* to 2.the *lacrimal gland*. The lacrimal nerve must pass from the brain by way of 3.the *ophthalmic division* of 4.*the efferent potion of the trigeminal nerve* (the portion of the nerve that goes from the brain to the organ) through the 5.*bones* of the eye socket, by way of 6.the *foramen* of the *superior orbital fissure*. The blood supply to the lacrimal gland from the 7.*Ophthalmic artery* and then the 8.*Lacrimal artery* must be present. 9.The *lacrimal vein* and 10. *ophthalmic vein* must also be present. The 11. *puncta* (two of them) must be present in both the upper and lower lids. The 12. *canalliculi* must pass from the lid margin through the lid, then pass through the 13. *lacrimal fossa* (the opening or *foramen*) in the inside of the eye socket and side of the nose. The foramen must form and the canalliculus must pass through it. The 14. *Valve of Hasner* must be present inside the nose to prevent debris from the nose getting into the eye and causing recurrent infections. Plus, the 15. *afferent portion of the trigeminal nerve* (the part that goes from the organ to the brain) must be present in the 16.*cornea* to sense dryness and tell the lacrimal gland to make tears. The 17.*eyelids* must be present to distribute the tears. (The lids must also have all the necessary muscles, nerves, blood supply, foramen etc. also but for simplicity we will not even include all those required interdependent functions and systems - also Interdependent Evidence of Creation.) So there are at least seventeen (17!) different systems all working together ***interdependently***, and ***simultaneously*** to ***create*** just the Lacrimal System. You can observe their structure, location and function and at the same time observe all the other simultaneously involved structures, their location and their function and conclude that they had to be created. When observing the eye, just as Dr. R.C. Sproul Sr. relates in observing a shoe,[132] you may not have seen someone make your shoe, but by examining the evidence, you can conclude that your shoe was created. You observe the different parts, that they are stitched together, glued together, cut to specific matching parts –dependent and interdependent parts, dependent and interdependent functions all working together to perform a completed entity, in this case a *non-living* "being" a shoe. By simple observation it is easy to conclude that the shoe and likewise the lacrimal system were Created.

II. The Intraocular Pressure System:

The 1.Sclera and 2.Cornea must form the watertight globe of the eyeball. The 3.Ciliary Body must be present to produce the 4.Aqueous Humor, the fluid that fills the eyeball. (Precisely how the Ciliary Body ***creates*** the aqueous humor and all the additional Interdependent Evidences of Creation (I.E.C.) associated with how the ciliary body creates the aqueous, we will not even include at this time.) The 5.Short Ciliary arteries, the 6.Long Ciliary arteries and the 7.Anterior Ciliary artery and 8.Posterior ciliary arteries must be present to provide the blood flow in order for the Ciliary Body to create the Aqueous Fluid that fills the eye. The 9.Ophthalmic artery must be present to supply the Ciliary arteries. The 10.Short Ciliary veins and the 11.Vena Vorticosa must be present to return the aqueous fluid from the Ciliary Body to the Circulatory System. The 12.Trabecular Meshwork must be present to regulate the flow of fluid out of the eye. 13.Schlem's canal must be present to carry the aqueous from the eye back into the venous blood supply.

So there are at least thirteen (13!) different systems all working together ***interdependently***, and ***simultaneously*** to ***create*** just the Intraocular Pressure System. You can observe their structure, location and function and at the same time observe all the other simultaneously involved structures, their location and their function and conclude that they had to be created. Again, just as in observing a shoe. You may not have seen someone make it, but by examining the evidence you can conclude that it was created.[132] You observe the different parts, that they are stitched together, glued together, cut to specific matching parts – dependent and interdependent parts, dependent and interdependent functions all working together to perform a completed being, in this case a *non-living* "being" a shoe. By observation it is easy to conclude that the shoe and likewise the Intraocular Pressure System were created.

Observe and conclude. That is exactly what Darwin did when he hypothesized his theory. Why is it that when Darwin and the evolutionists observe and conclude- it is science, but when Creationists observe and conclude it is OPINION?

I now am addressing the atheist or evolutionist that may be reading this book, Take off your shoe. Now look at your shoe. Observe and conclude! Did it evolve? No! Did you see it be created? No! How do you *know* that it was created? By logical observation and conclusion. Do you not realize that the structures and functions of the eye and the body are much more intricate and much more complicated than your shoe. If you can ___**KNOW**___ that your shoe was created, even though you did not see it be created- simply by logical observation and conclusion; how much more so can you ___**KNOW**___ that the eye and the body- ___*and all of life;*___ ___**WERE CREATED !**___ You ___**KNOW**___ the shoe was created and it does not display the ultimate evidence of ___**THE**___ CREATOR …….. ___**LIFE !**___ (Darwin's statement, "into which ___*life*___ was first breathed"[133] IS ACTUALLY A STATEMENT, *"THERE WAS A CREATOR!"*)

Observe and conclude. Be truthful to yourself. Don't allow yourself to be deceived! Don't fall for Satan's deceit! And don't deceive yourself! It requires more faith to believe in evolution than Creation. How did LIFE – the **_LIVING BEING_**, become alive? The answer is; the "into which life was first breathed" that Darwin himself[133] describes! Believe in God and be saved! Your very soul is in peril! Matthew 13:49, The Amplified Bible; So it will be at the close and consummation of the age. The angels will go forth and separate the wicked from the righteous (those who are upright and in right standing with God) and cast them (the wicked) into the furnace of fire; there will be weeping and wailing and grinding of teeth.[134]

Now; since Darwin says, "we ought in imagination"[135]… "we must suppose"[135]… and, "may we not believe"[135]; Now "imagine[135]" variations of how the eye might have evolved, **_IF_** the eye actually did evolve. That is, if the eye did evolved why did it not evolve differently?

1. What did beings look like before they evolved eyes? How did they survive without eyes?

   "Imagine"[135] all beings without eyes - before they "evolved" eyes.

2. Why are our eyes as they are, two, in front, recessed, with lids?

   "Imagine"[135] beings with eyes that have not evolved lids.

3. Why is it that we don't have one eye, or three eyes or four eyes?

   "Imagine"[135] beings with one, three, four or five eyes.

4. Why is it that we do not have eyes on the sides or in back of our heads?

   "Imagine" [135] beings with eyes on top or on the sides of the head, or in the back of the head.

   Or, "imagine"[135] beings with four eyes, two in front and two in back.

5. Why did we not evolve compound eyes such as the insects?

"Imagine"[135] beings with bug eyes

6. If both "animals and plants have descended from some one prototype"[136] as Darwin states – if *beings* evolved eyes, why did the same processes that caused animals to evolve eyes not cause plants to evolve eyes?

"Imagine"[137] plants with eyes. Will plants one day evolve eyes?

7. How does evolution explain growth? "Imagine"[137] if there was no growth process. *"Imagine"*[137] either only children or only adults. How does evolution explain growth? What evolutionary advantage was at work to cause growth?

8. How does puberty "evolve"? How would all the necessary hormones evolve? What did beings look like while they were waiting on all their growth hormones or puberty hormones to evolve? How could the species survive while they were waiting on their reproductive hormones to evolve? Again, we observe I.E.C. *"Imagine"*[137] what all the species would look like while waiting on their puberty or reproductive hormones to evolve. There would be NO LIFE what-so-ever!

Remember, the reason I am asking you to "imagine"[137] these possibilities is in direct response to Darwin' statements, "all animals and plants have descended from some one prototype"[138] and, "probably all the organic beings which have ever lived on this earth have descended from some one primordial form, into which life was first breathed."[138] Now, *"imagine"*[139] everything –plants and animals, from one. If "all animals and plants have descended from some one prototype"[140] and if, "probably all the organic beings which have ever lived on this earth have descended from some one primordial form, into which life was first breathed"[140] were true, it would mean that first a bacteria would have to come to life *by chance*. Then, that one form of bacteria evolved to both plants and animals and once it came to life it then evolved its' digestive, nervous, cardiovascular, respiratory, immune, endocrine (horemones), metabolic, muscular, skeletal urinary, and reproductive systems. And, in addition to all of that "absolutely" having to evolve - it had to happen for both sexes of each species and cooperatively and simultaneously. ***And***; that means that the same processes caused a bacteria to evolve to animals - and at the same time to evolve plants! So; animals evolved all their anatomical parts from one being - a bacteria. So, a bacteria was just sitting around being a bacteria then, legs evolved, arms evolved, feet evolved, hands evolved, heads evolved, the heart evolved, the lungs evolved, the intestines evolved, the brain evolved, and they all evolved for all the different species! It had to happen that way if, "all animals and plants have descended from some one prototype!"[140] And then, the ultimate question is, where did the "ONE"[140] come from??? Did it just happen? How did the "some one"[140] come to life? Who "breathed" life "into" it?[140] And, why do we TODAY not observe bacteria evolving parts?

Think long and hard about these two statements. Darwin says, "animals and plants have descended from some one prototype." [140] And, "probably all the organic beings *which have ever lived* on this earth have descended from some one primordial form, into which life was first breathed"[140]. Then after thinking about these two statements, I challenge you to do what Darwin has suggested; "imagine"[141] now how the eye may have evolved. You now know the components and systems of the eye. Now show me how it could have happened ***IF*** the eye evolved. Illustrate some of the "imagined"[141], "numerous, successive, slight modifications"[142] that Darwin espouses to evolution. See if you can do it. Draw some "imagined"[143] intermediary or transitional "imagined"[143] evolutionary states, or forms of the eye. Or, produce by way of illustration some reasonable simulation of how evolution of the eye possibly could have occurred - IF you can "imagine"[143] it! I tried- specifically, at the end of this book after presenting the review of the 25 systems that make up the eye, to come up with some way of ordering of the 25 systems that could possibly make some sense of an evolutionary process. I tried to "imagine"[143] how the eye could have evolved as Darwin has "supposed"[143] and "believed"[143], but I could not even "Imagine"[143] it.

Yes, I know it is difficult to imagine the eye evolving, so let's make it real easy. Instead, try now to "imagine"[143] and illustrate the evolutionary processes that occurred for arms and legs. Or, try to illustrate the evolutionary processes that occurred for reproductive parts of the male and female <u>of each species</u>. Illustrate some 'imagined"[143] intermediary transitional forms. Think about the process of evolving from non-walking, no-foot, no-leg, beings; to walking, footed, legged beings. Think about it. Think about the processes involved. Think for yourself. Don't be deceived.

Now think about, and try to *"imagine"*[143] or illustrate; evolution of male and female sexes. *"Imagine"*[143]; no sexual parts, evolving to simultaneously differentiated sexual parts -and all the associated internal components necessary for sexual reproduction, (that is differentiation of egg, sperm, and hormones) for each and every species! Evolution doesn't even come close to making sense! It is no wonder why Darwin had doubts himself[144]. It is no wonder he left open and anticipated and prompted for a challenge when he said, "If it could be demonstrated that any complex organ existed which could not possibly have been formed by numerous, successive slight modifications, my theory would absolutely break down."[145] Thus, with the writing of this book

Darwin's challenge, as Darwin himself anticipated, is answered. The eye proves Darwin's "theory absolutely breaks down"[145]. The eye proves the theory ***Interdependent Evidence of Creation***, I.E.C. And if the eye did evolve why did it not evolve differently? Why do we observe such consistency from being to being? The answer is to all of these "Why" question is: it is because of the information evident within the basic blueprint for life; the Information and code in the DNA. The DNA; information about how a cell is made and what it does, is evidence in and of itself of a Creator.

The following illustrations on pages 277 – 283 are not "imagined"[146], they are actual photographs, illustrating what happens when things go wrong with the eye itself, or, with one of the 25 systems required for the eye to function properly.

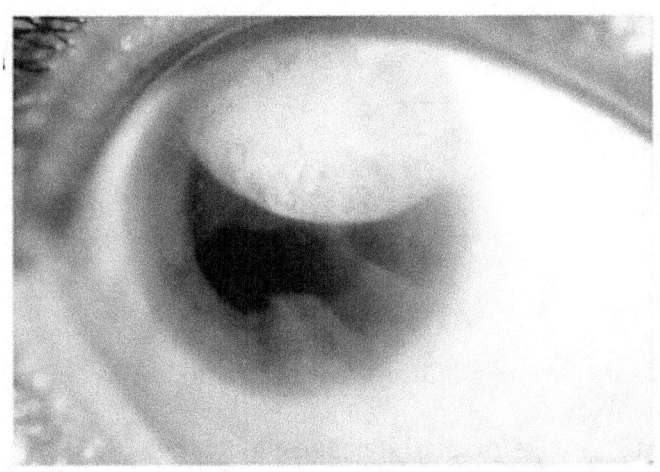

Sclerocornea[147] A portion of the clear cornea is covered by the white sclera; a birth defect. (Or, we could "imagine" as Darwin states[146], that the cornea has yet to evolve its' clarity). **Photo from**: Sclerocornea: http://upload.wikimedia.org/wikipedia/commons/c/c4/023-p-039-9426.jpg Neethirajan G, Krishnadas SR, Vijayalakshmi P, Shashikant S, Sundaresan P. BMC Med Genet. 2004 Apr 16;5:9. PAX6 gene variations associated with aniridia in south India.File:023-p-039-9426.jpg Uploaded by Filip em Created: November 20, 2007

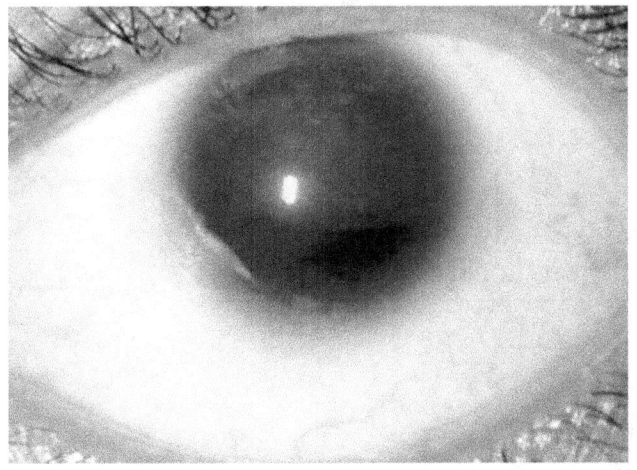

Photo from:
Aniridia:http://upload.wikimedia.org/ wikipedia/commons/c/c4/023-p-039-9426.jpg Neethirajan G, Krishnadas SR, Vijayalakshmi P, Shashikant S, Sundaresan P.
BMC Med Genet. 2004 Apr 16;5:9. PAX6 gene variations associated with aniridia in south India.
File:023-p-039-9426.jpg
Uploaded by Filip em
Created: November 20, 2007

Aniridia[148] The iris did not develop as a result of a birth defect. (Or, we could "imagine" as Darwin states[146], that the iris, the iris pupil dilator muscle and the iris pupil constrictor muscle has not yet evolved.)

285

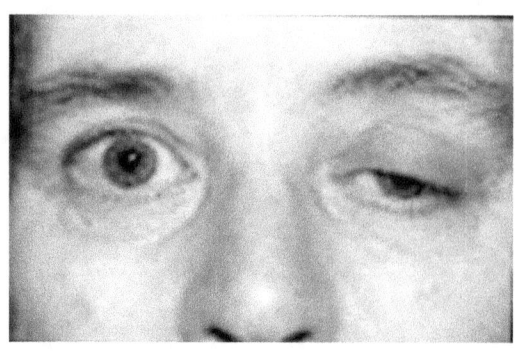

> The nerve that supplies the patient's left eyelid is non-functional.

Myesthenia Gravis[149] A condition which causes muscle weakness. Or, we could "imagine", as Darwin states[146], that the eyelid muscle has yet to evolve.
Photo from:http://en.wikipedia.org/wiki/Myasthenia_gravis#/media/File:Myasthenia.jpg

Description	**Nederlands:** Ogen van patient met Myasthenia Gravis
Date	21 November 2007 (original upload date)
	(Original text: *10-11-2007*)
Source	Transfered from nl.wikipedia
	(Original text: *eigen werk*)
Author	Original uploader was Cumulus at nl.wikipedia
	(Original text: *Cumulus*)
Permission (Reusing this file)	Released under the GNU Free Documentation License. (Original text: *Cumulus*)

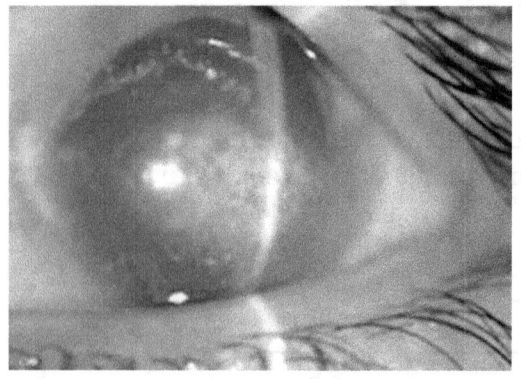

> The cornea is cloudy due to lack of adequate tear production.

Band Keratopathy[150] The cornea is opaque from chronic dry eye. (Or we could "imagine" as Darwin states[146], that the cornea has yet not evolved to a state of clarity, that is, the clarity of the cornea has not yet evolved.) How could the being survive? **Photo from:**http://upload.wikimedia.org/wikipedia/commons/a/a3/Band-keratopathy_left-eye.png *A photo of my left eye showing band keratopathy (a calcium buildup) on the cornea.* Author: Dr Jon Ruddle, Subject: Mr Paul Bone, Date: 2010-07-27, Copyright: (c) Jun Ruddle 2010 Licensed under the Creative Commons Attribution-Share Alike 3.0 Unported

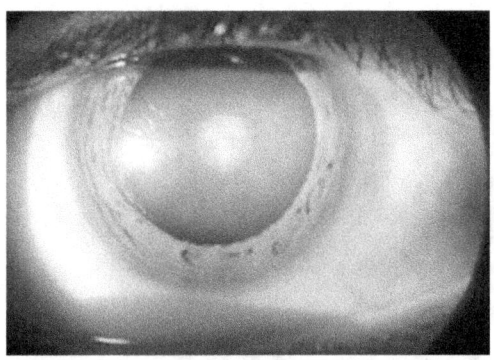

Cataract: The lens is cloudy and has an opaque center. Or if we "imagine" as Darwin states. the lens has not evolved to become "a thick layer of transparent tissue" as Darwin has "imagined" on page 158 of his book *The Origin of Species*[146].

Cataract[151] The lens inside the eye is cloudy and opaque, most often due to ageing, sometimes a congenital defect, sometimes secondary to trauma. (Or we could "imagine" as Darwin states[146], that the lens has not evolved to become "a thick layer of transparent tissue."

Photo from: Cataract http://upload.wikimedia.org/wikipedia/commons/b/ba/Cataract_in_human_ eye.png

Description: Cataract in Human Eye **Date:** 24 December 2005 **Source:** Own work **Author:** Rakesh Ahuja, MD **Permission (Reusing this file)** Multi-license This file is licensed under the Creative Commons Attribution-Share Alike 3.0 Unported license

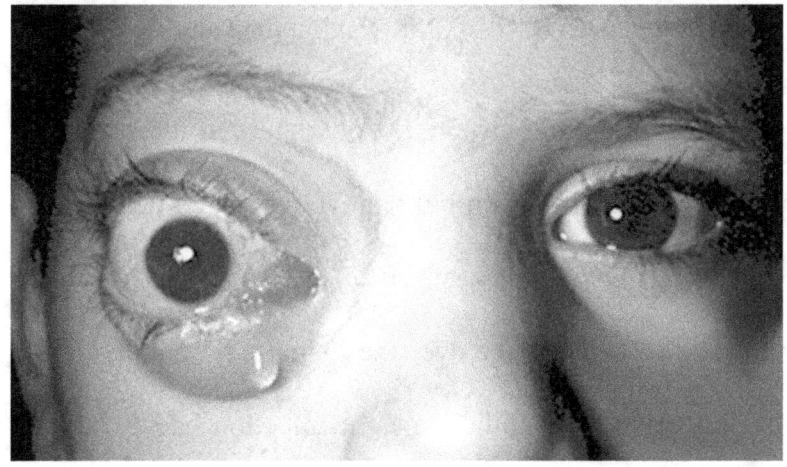

Orbital Tumor[152] The eye has a tumor behind it pushing the eye forward out of the eye socket. (Or, for example, if the eye were to have evolved, we could "imagine" as Darwin states[146] that the 8 bones of the eye socket have not yet evolved a deep enough eye socket at this stage of evolution so that the eye is not recessed into the eye socket).

Photo from: Page URL: Orbital Tumor

http://upload.wikimedia.org/wikipedia/commons/6/6d/406907P-PA-OCULAR.jpg

http://commons.wikimedia.org/wiki/File%3A406907P-PA-OCULAR.jpg

File URL: http://upload.wikimedia.org/wikipedia/commons/6/6d/406907P-PA-OCULAR.jpg

Attribution: By The Armed Forces Institute of Pathology [Public domain], via Wikimedia Commons

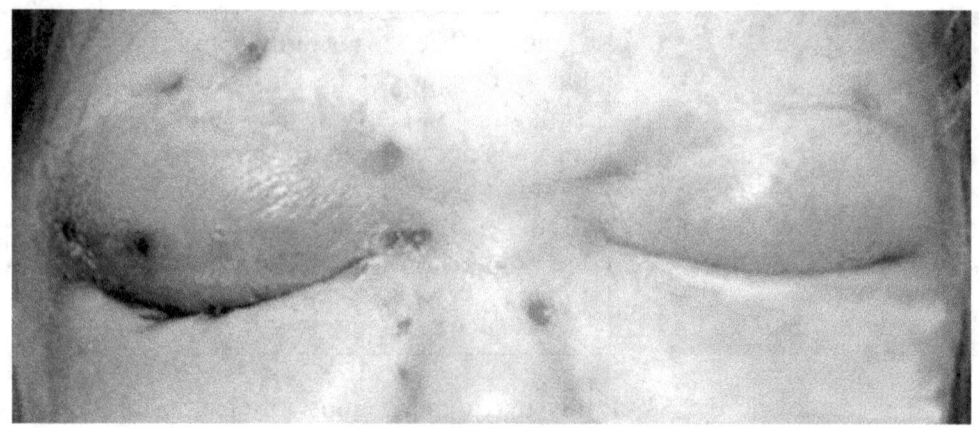

Orbital Cellulitis[153] The eye and all the adjacent tissues have an infection. (Or, for example, if the eye were to have evolved, we could "imagine" as Darwin states[146] that the muscle opening the lids has not yet fully and properly evolved.
Photo from: Orbital; Cellulitis
Page URL:http://commons.wikimedia.org/wiki/File%3AOrbital_cellulitis.jpg
File URL:http://upload.wikimedia.org/wikipedia/commons/8/8f/Orbital_cellulitis.jpg
Attribution: By Jonathan Trobe, M.D. - University of Michigan Kellogg Eye Center (The Eyes Have It) [CC BY 3.0 (http://creativecommons.org/licenses/by/3.0)], via Wikimedia Commons

Meditate on the reality necessary to evolve the immune response. Look again at the preceding photo. What happened to all the eyes that developed an infection but had *NOT* evolved the immune system necessary to fight the infection? None of them would have survived and the species would go extinct. The ability to fight infection cannot evolve because the eye (the being too) cannot survive while it is waiting for the immune system to evolve. The ability to fight infection had to be pre-planned and requires knowledge, foresight and anticipated intervention with a specific and *CORRECT* response. The immune response can only be explained by creation. Planned response to an adverse event requires foresight and knowledge, it had to be planned for because immunity is a specific response. Think about it, if the immune response wasn't present in the beginning nothing would have survived.

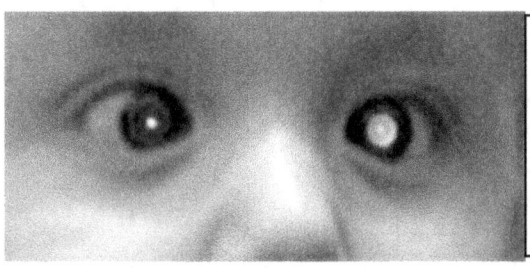

Here the retina has a tumor (cancer) growing in and on it (retinoblastoma). This cancer typically spreads to other parts of the body and if not detected early enough is usually fatal.

Retinoblasoma[154] A malignant tumor inside the eye of the retina. (Or, for example, if the eye were to have evolved, we could "imagine" as Darwin states[146] that the retina has not yet evolved to full and normal development and full and normal function.

Photo from:
http://en.wikipedia.org/wiki/Leukocoria#/media/File:Rb_whiteeye.PNG[154]
http://upload.wikimedia.org/wikipedia/commons/d/d8/Rb_whiteeye.PNG[154]
Description English: A child with a white eye reflection as a result of retinoblastoma.
Date: 28 October 2008 (original upload date)
Source: Transferred from en.wikipedia; transferred to Commons by User: Roberta F. using CommonsHelper. (Original text: *I created this work entirely by myself.*)
Author: J Morley-Smith (talk). Original uploader was Morleyj at en.wikipedia
Permission (Reusing this file) Released into the public domain (by the author).

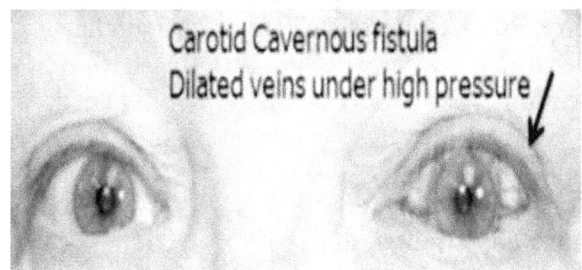

This show an abnormal connection between the carotid artery and the ophthalmic vein, a carotid cavernous sinus fistula or cc fistula. When this happens the artery is pumping blood into the vein. The eye and all the tissue surrounding the eye becomes engorged with blood.

Carotid-Cavernous Fistula[155] An abnormal connetion of an artery into a vein. (Or, for example, if the eye were to have evolved, we could "imagine" as Darwin states[146] that the arteries and veins have not yet evolved the normal routes of supply.

Photo from:
File URL:
http://upload.wikimedia.org/wikipedia/commons/b/bf/Corkscrew_blood_vessels_in_left_eye.jpg[138]
Page URL: http://en.wikipedia.org/wiki/Carotid-cavernous_fistula#/media/File:Corkscrew_blood_vessels_in_left_eye.jpg
Attribution: Sumeer Thinda, Mark R Melson and Rachel W Kuchtey -
http://www.biomedcentral.com/1471-2415/12/28
 CC BY 2.5
 File:Corkscrew blood vessels in left eye.jpg
 Uploaded by Kiatdd, Created: December 28, 2012

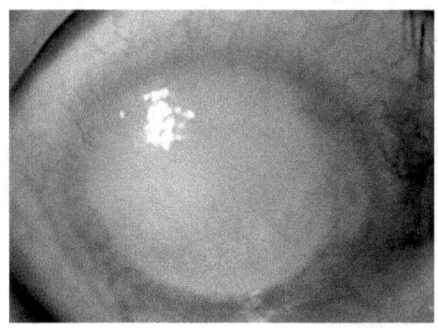

The cornea is cloudy from an infection, in this case an invasion by an amoebic organism.[156]

Acanthamoeba[156] The cornea is cloudy from an infection throughout the entire cornea. Or, you could imagine, as Darwin states[146], that the cornea has not yet *evolved* to a point of clarity. **Acanthamoeba photo from: Page URL:** http://en.wikipedia.org/wiki/Keratitis#/media/File:Parasite140120fig1_Acanthamoeba_ keratitis_Figure_1A.png. **File URL:** http://www.parasite journal.org/articles/parasite/full_html/ 2015/01/parasite140120/F1.html

Attribution: Jacob Lorenzo-Morales, Naveed A. Khan and Julia Walochnik - " (2015). "An update on *Acanthamoeba* keratitis: diagnosis, pathogenesis and treatment".

Parasite **22**: 10. DOI:10.1051/parasite/2015010. PMID 25687209. ISSN 1776-1042.

Figure 1A of published paper. Corneal melting and vascularization in a patient with Acanthamoeba keratitis Description: English: Figure 1A of published paper. Corneal melting and vascularization in a patient with Acanthamoeba keratitis Date: 18 February 2015 Source:" (2015). "An update on Acanthamoeba keratitis: diagnosis, pathogenesis and treatment". Parasite 22: 10. DOI:10.1051/ parasite/2015010. PMID 25687209. ISSN 1776-1042. Author: Jacob Lorenzo-Morales, Naveed A. Khan and Julia Walochnik

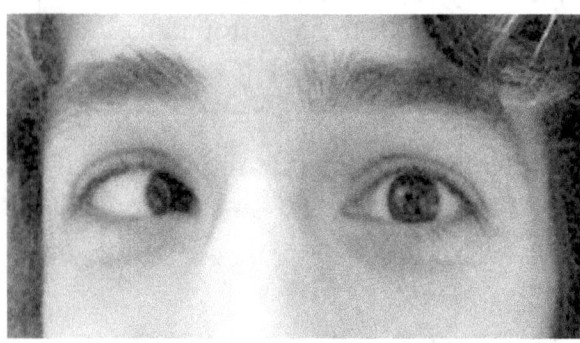

The patient's right Lateral rectus muscle is paralyzed. Or this is what would happen if the right lateral rectus muscle had not evolved. Permanent double vision results and the right eye cannot move to the right.

Esotropea[157] The right eye is abnormally crossed or tuned inward. Usually due to high hyperopea. sometimes secondary to elevated blood sugar, or stroke. Or, we imagine, as Darwin states[146] the right lateral rectus muscle has not evolved so the eye cannot turn out.

Photo from:
Esotropea:http://upload.wikimedia.org/wikipedia/commons/0/05/Andre_Filipe_Teixeira_Marques_E sotropia.jpg **Description: English:** Esotropia of right eye **Date:** 27 March 2010, 19:58:48 **Source:** Andre Filipe Teixeira Marques.jpg **Author:** Kakawere

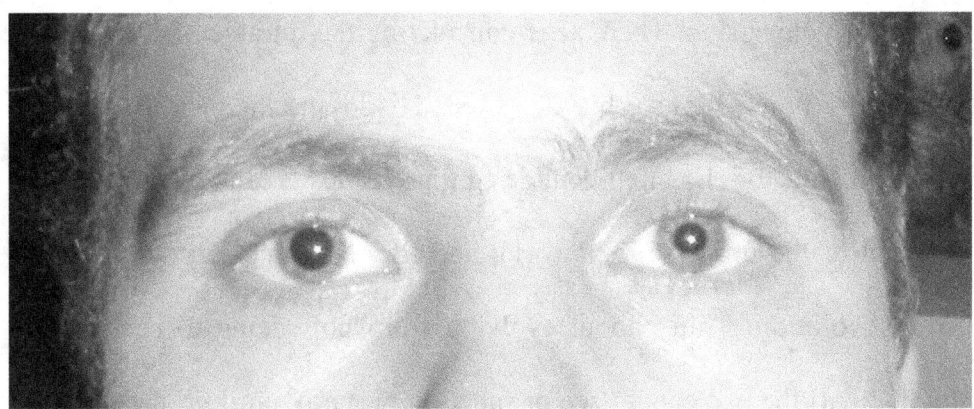

Right Dilated Pupil:[158] This is what the eye would look like if the iris constrictor muscle did not evolve. (A brain tumor can cause this, elevated blood sugar can cause this, a stroke or as in this case a dilating drop was instilled. This could also instead be a left constricted pupil. A constricted pupil can come from lower neck, shoulder, or upper thoracic vertebrae injury. A constricted pupil can also originate from a Superior Sympathetic Ganglion/Internal Carotid Artery problem, lung cancer, or stroke. Or, we could imagine, as Darwin states[146], that the nerve to the pupil constrictor muscle on the patient's right eye has not yet evolved. Or, we could imagine as Darwin states[139] that the iris dilator muscle on the patient's left eye has not yet evolved.

Description	**English:** Anisocoria (right eye instiled by tropicamide)
	Polski: Anizokoria (prawe oko zakroplne tropikamidem)
	Deutsch: medikamentös hervorgerufene de:Anisokorie durch Gabe von de:Tropicamid in das rechte Auge
Date	7 November 2006
Source	Own work
Author	Radomil talk

We take for granted the miracle of our eyes. When *everything is right* we don't think about what

a miracle *everything is right* is. Twenty-five different systems working together properly, seeing

properly; what a blessed miracle that is. Think about how your life would be if you could not see.

Imagine how your life would be - how you could function, how could you live, if you could not see.

Close your eyes for the next five minutes and think about it. Isn't it amazing! It only takes five minutes to appreciate the miracle of your vision. Then, after considering the miracle of your vision, thank God that you *can* see properly. Praise him for his Goodness, His loving kindness, His mercy and grace (unmerited favor)! Praise Him for His intelligent design in His design and Creation of the eye, and His Design and Creation of the universe, our world, and life. Now, close your eyes and think about Darwin's theory of evolution. Think about evolving from not seeing to seeing. The problem with the eye *evolving* is [that] the eye cannot see or function properly until *all* the parts are fully evolved and only if __*ALL*__ of them *evolved* **at the same time.** Fully-evolved simultaneous evolution of *ALL* the parts *ALL* at the same time can only happen by (and actually describes) *Creation!*

While the creationist might ask, "How could the eye evolve to see?" The evolutionist might ask, "Why would God make the eye so that we *can* see?" The answer may be that He *wants* us to see __*Him*__ and *know* __*He*__ exists. Again, the Bible says, "Since what may be known about God is *plain* to them, because God has made it *plain* to them. For since the creation of the world, God's invisible qualities—God's eternal power and divine nature—have been clearly __*SEEN*__, being understood from what has been made, so that men are without excuse[159]" (Romans 1:19–20, New International Version).

And, Colossians 1:16 may also give us an answer, "For it was in Him that all things were created, in Heaven and on earth, things __*seen*__ and things unseen, whether thrones, dominions, rulers, or authorities; all things were created and exist through Him and in and for Him[160].

It all comes down to this; the bottom line is: all things were made by Him, __*for Him!*__ He *wants* us to see Him, and know Him, and have a relationship with Him - because we were made by Him, __*for Him!*__

24. Conclusion

The primary purpose of this book is to answer Darwin's challenge and in doing so educate the reader about the structure and function of the eye for the reader to see and understand the ways in which the eye itself *proves* the eye could not have *evolved* and, Darwin's theory (as Darwin anticipated) "absolutely breaks down"[1].

However, a secondary but possibly more important result of the book has been realized. In answering Darwin's challenge a new theory, *Interdependent Evidence of Creation*, has been postulated and the evidence found in the eye *proves* the theory. Therefore, the conclusion is *Creation, not evolution*.

A third purpose hopefully realized is for those who are *not* presently Christian, to acknowledge God has revealed himself to you in the structure and function of the eye, and to help you understand the theory of Interdependent Evidence of Creation and how the eye proves the theory. Once you see the theory *Interdependent Evidence of Creation* is proven you will see Darwin's theory as Darwin himself anticipates "absolutely breaks down"[1]. I hope you will then believe in God. Thus, I hope to give you the ***proof*** you may be looking for by learning and understanding the workings of the eye. My goal is not to ask you to believe in God, just because I think you should, or someone else thinks you should. It is not my goal to persuade you to blindly believe in God, but through understanding and knowledge have ***an educated reason*** to believe in God and, thereby, be confident and comfortable believing in God. Believing in God because you are told *creation happened* is not enough. "There must be *proof*. I think God himself says that believing *just because* is not enough also. He has said (I am paraphrasing) to look at the evidence. He says, "The wind blows where it wills: and though you hear its sound, yet you neither know where it comes from or where it is going"[2] (John 3:8–12, The Amplified Bible). You can't see the wind; you know *not* where it comes from, yet you know it exists because you can see its effects. God has given us, the *proof* if we just look for it. **He Himself has said He has given us the proof!**[3] (Romans 1:19-20) By looking closely at the eye, (and all of His Creation) you can *see the proof.* You cannot come away thinking the eye evolved or it just happened on its own if you know the eye's many different structures and the *required* interdependent functions. These interdependent functions *require* totally completed functions. The eye will not work while waiting for functions and systems to "evolve". If you know the structures and the required totally completed interdependent functions, it is easy to see the eye *had* to be created.

Proof—it is important and ***necessary*** to look at the evidence to make an informed, educated decision about <u>any</u> theory to come to, or justify a *proof.* For example, when solving a crime what do they look for? PROOF! And for proof to be concluded they must have evidence; evidence that shows, and positively demonstrates the proof. They provide evidence to *prove* the crime. The same applies to <u>any</u> theory, to reach a conclusion you must have determined the proof. You can only determine the proof through evidence. Everyone should examine the ***evidence*** concerning Creation verses evolution. Don't just believe what someone has said, or something you have heard. Look at the evidence yourself and form a logical conclusion! Use the evidence to ***<u>prove</u>*** the theory!

Sometimes when attempting to prove a crime you may hear the term. "smoking gun". The term "smoking gun" was originally, and is still primarily, a reference to an object or fact which serves as conclusive evidence of a crime or similar act, just short of being caught *in flagrante delicto*[4] (caught in the very act as it is being perpetrated). The term smoking gun originally came from the idea of finding a smoking gun. A very recently fired gun on the person or suspect wanted for shooting someone, which in that situation would be nearly unshakable proof of having committed the crime[4]. (The phrase "smoking gun" originated in the Sherlock Holmes story, *The Adventure of the Gloria Scott*, 1893[4]) In addition its meaning has evolved in uses completely unrelated to criminal activity: For example, scientific evidence highly suggestive in favor of a particular hypothesis is sometimes called smoking gun evidence.[4]

Facts are facts! In ***any*** proposed theory[5] if the information presented is factually accurate, and scientifically correct, it is a fact, not just something "imagined" [6], "supposed" [6] or "believed" [6] as Darwin states. If the evidence ends in a conclusion which answers the theory, then the theory has been ***proven***. If after examining the eye, the conclusion, as Darwin himself states "***absolutely*** breaks down his theory[7]"; the conclusion at which we arrive *thereby* points to Creation and therefore - God.

However, not only does the evidence *thereby* point to God by "absolutely breaking down[7]" Darwin's theory, the eye *directly* demonstrates and ***proves*** the theory; Interdependent Evidence of Creation and thus, *directly* points to and affirms God.

Think about it; meditate on Darwin's statements! Darwin's theory states, "that probably all the organic beings which ever lived on this earth have descended from some one primordial form into which life was first breathed."[8] (Remember this would include the dinosaurs - and Darwin knew about them when he made this statement.)

Darwin also said he was led to the belief that, "animals and plants have descended from some one prototype.[8]" Think about the diversity of life and all the different "organic" life forms. We have germs; bacteria, viruses, funguses (fungi), molds, amoebae. We have plants; flowers, trees and grasses. Then we have insects; bees, butterflies, mosquitos, lady bugs. There are mulluscs; snails, slugs, oysters, clams, octopus, squid. Then there are fish; marlin, tuna, perch, bass, salmon. We have ragworms, earthworms, leeches. We have arachnids; spiders, crabs, scorpions. We have reptiles; lizards, turtles, snakes, alligators, dinosaurs. There are birds; hummingbirds, penguins, flamingos, ostriches, chickens, sparrows. We have mammals, bats, mice, elephants, whales, dolphins, dogs, cats, lions, lambs, chimps, and man. The diversity of life is VAST. All of these are organic beings; I just named a few of the millions. Darwin's theory states, "that probably all the organic beings which ever lived on this earth have descended from some one primordial form into which life was first breathed."[8] (Remember this would include the dinosaurs.) And he said, "animals and plants have descended from some one prototype"[8] Even if I try to *imagine*[9] it, as Darwin suggests[9], I cannot make sense of it. But MOST IMPORTANTLY, where is the evidence that *PROVES* his theory? If the evidence does noy ***PROVE*** THE THEORY IT IS FALSE !

Now, visualize these organic beings as you read their names, (plants are "organic" and since Darwin says they and animals "descended from some one prototype" we will refer to them as "beings"). Take a few moments and think about each one individually as you read. An octopus, tortoise (land turtle), humming bird, whale, bat, butterfly, scorpion, mouse, elephant, rose, earthworm, flea, salmon, bacteria, chimpanzee, dinosaur, ostrich, giant-redwood-tree, and man. Now re-read the list and think about how they can be related as you read. Now meditate on what Darwin's theory says, "<u>probably all the organic beings which ever lived on this earth have descended from some one primordial form into which life was first breathed.</u>"[10] How can they teach a theory as fact when Darwin himself introduces it with the word "***PROBABLY***"!

Darwin also said, and I am directly quoting him. "Analogy would lead me one step further, namely, to the ***belief*** that <u>all animals and plants have descended from some one prototype.</u>"[10] Again, how can they teach a theory as fact when Darwin himself introduces it with the word "***BELIEF***"[10] !

"***ALL*** ANIMALS AND PLANTS DESCENDED FROM SOME ***ONE PROTOTYPE***[10]" How someone can believe those statements and at the same time fault Christians for believing in God, I do not understand.

"Probably all the organic beings which ever lived on this earth have descended from some one primordial form into which life was first breathed."[10] Think that through a little further. Don't just take what Darwin says as fact. Think for yourself! First, where did the "some one primordial form" come from? Second, if there was just "one", and since according to Darwin, "<u>animals and plants have descended from some one prototype</u>[10]" on what did it feed or eat? There would not be ANY plants or animals on which it *could* eat. Was it carnivorous or herbivorous. In either case, neither plant or animals existed because it would have had to have been the "one primordial form[10]" so on what did it feed? And thirdly, how would it evolve the digestive system and at the same time survive while it was waiting on its digestive system to evolve. Fourthly, why would it reproduce since that would introduce competition for itself. And along those same lines, why would a plant evolve seeds? Why would a plant have a need to evolve seeds to reproduce, especially when it would find itself in competition for survival with its own offspring. Can a plant know it needs to reproduce? Now think about the last part of Darwin's statement, "all organic beings descended from some one primordial

form ***into which life was first breathed***"[10]. Fifthly, who "breathed into" the first breath?

All pictures used here, p. 290, were from files I purchased in *Art Explosion 525,000*

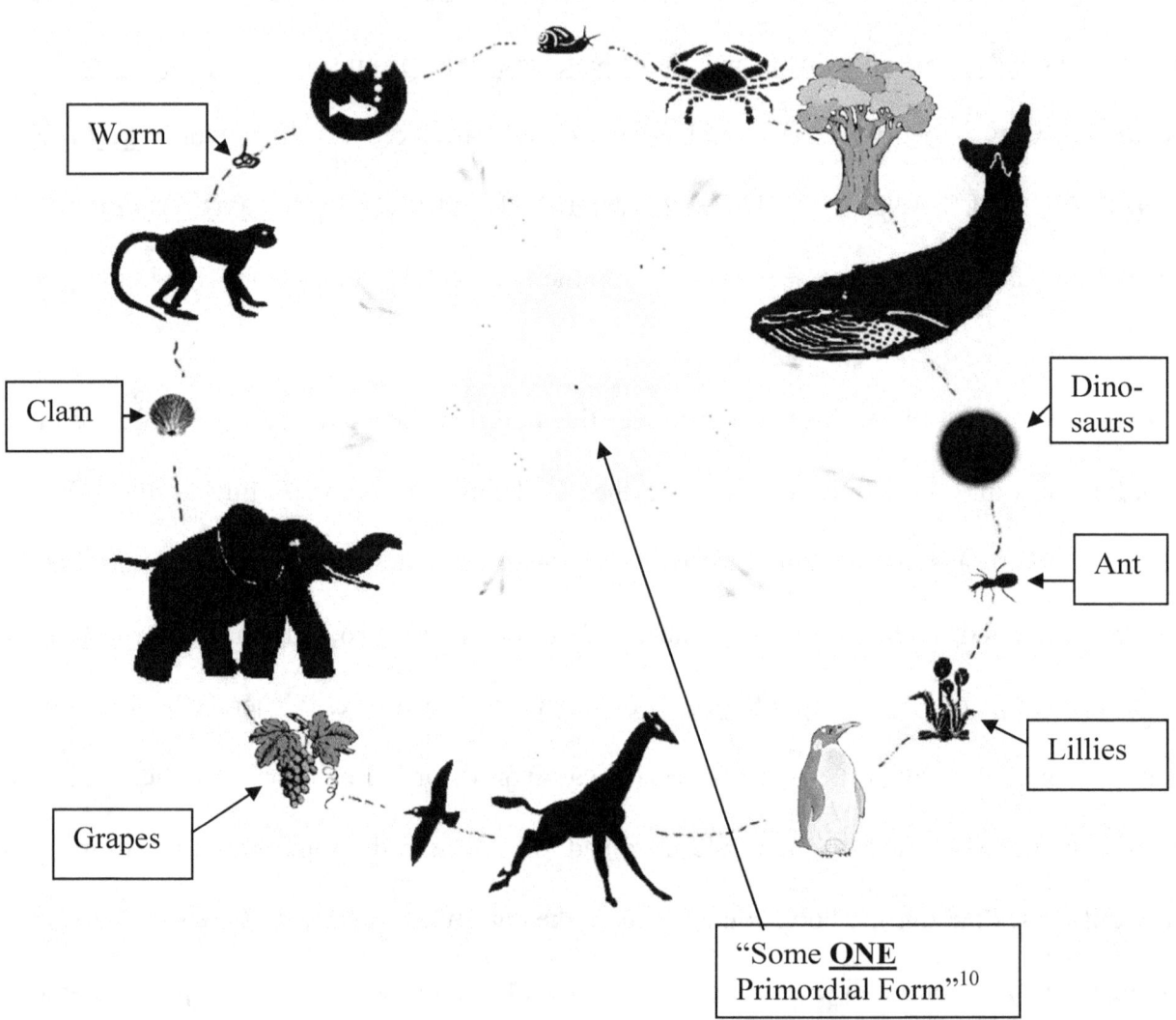

Worm

Clam

Grapes

Dino-saurs

Ant

Lillies

"Some **ONE** Primordial Form"[10]

Darwin's Circle of ~~Life~~ Lies © Rigney 2015

Now, think about the design and function of the eye. Darwin said, "If it could be demonstrated that any complex organ existed, which could not possibly have been formed by numerous, successive, slight modifications, *__my theory would absolutely break down__*"[11].

There are several examples of **"smoking gun evidence**[12]**"** when one looks at the Creation and Design of the eye that, *__without question__*, *absolutely breaks down Darwin's theory*[13]. In fact they are such strong evidence, they not only "absolutely break down"[13] Darwin's theory; they exhibit *__smoking gun evidence__*[12] that blows Darwin's theory[13] out of the primordial soup!

When we look at the design and function of the eye (which meets Darwin's criteria of a "complex organ[13]") his "theory absolutely breaks down[13]".

Items written in **bold** are **"smoking gun proofs"**.

1. The **trochlear bone loop** and insertion and attachment of **the superior oblique eye muscle through the loop and then back to the eye. (Also I.E.C.)**

2. The **routes of innervation of the iris dilator nerve and pupil constrictor nerve** to the **two opposing muscles of the iris. (Also I.E.C.)**

3. The interdependent aqueous producing **ciliary body inflow** - and the **trabecular meshwork Schlem's canal outflow** mechanism. (**Also I.E.C.**)

4. **The trabecular meshwork is a reverse pressure release valve.** Wisdom and *design* can only explain this. Think about it! WOW!

5. **Tear producing Lacrimal Gland;** *outer* upper part of eye **and tear drain (Puncta, Canalliculi, Lacrimal Fossa);** *inner* lower part into nose = **I.E.C.**

6. **Rod and Cone outer segments disintegrate to produce electrical stimulus**. (Darwin states, "several facts make me suspect that any sensitive nerve may be rendered sensitive to light, and likewise to those courser vibrations of the air which produce sound.[14]" THIS WE KNOW IS IMPOSSIBLE! He wasn't even close! This statement is in absolute error!

7. **The Optic Nerve carries impulses from 1.3 million nerve fibers of the retina (2.6 million nerve fibers for both eyes) yet the impulses do not short each other out or create conflicting or interfering episodes.**

8. The retina is split into 4 quadrants and exactly half of the 2.6 million nerve fibers cross over in the **Optic Chiasm. Each Quadrant is point for point identically, visually, spatially, equal in each eye. I.E.C.**

9. **The 3 pairs of external eye muscles all must have opposing eye muscles and all must be fully formed and fully function ALL at the same time = I.E.C.**

10. The **spacing, and alternating layers of the microfibril, fibril, lamella, sheets of the cornea. 250 alternating perpendicular layers.**

11. The **cornea gets its' oxygen by diffusion** from the air (Fick's Law)

12. The **curvature and the length of the cornea, lens and sclera are exact**.

13. Eyelid with muscles to open and close the lid. **Nerve feedback loop system of trigeminal nerve (#5) to sense dryness and oculomotor nerve (#3) to initiate the blink. Interdependent Evidence of Creation, I.E.C.**

14. **Retinal pigment epithelium takes away wastes** of outer segments of rods and cones.

15. **Dark, retinal pigment epithelium separates retina from white sclera.**

16. **Lens changes shape changing focus ability.**

17. **ALL blood vessels must be completely formed and All must be present and ALL must be present SIMULTANEOUSLY = I.E.C.**

18. **Visual cortex converts electrical impulses into vision.**

19. **The aqueous and vitreous are derived from the blood yet are clear.**

20. **Fourteen Bones with 35 *required* Foramen.**

There are *at least* 20 ***smoking-gun-proofs*** the eye was created. Stop and think about it; the **trochlear bone loop** *alone* is smoking gun evidence of Creation. ***Proof*** of Creation! The trochlea and Superior Oblique eye muscle all by itself "absolutely breaks down[15]" Darwin's theory. Interdependent Evidence of Creation!

The routes of innervation of the pupil dilating and constricting fibers *alone* is smoking gun evidence. **Proof!** And "absolutely breaks down[15]" Darwin's theory.

The fact **all twelve muscles MUST be present and MUST be fully functional ALL at the same time** otherwise the person will see double *alone* is smoking gun evidence. Proof! And "absolutely breaks down[15]" Darwin's theory.

The **lacrimal gland, lacrimal drain, and lacrimal fossa; the tear system** *alone* is smoking gun evidence. Proof! And "absolutely breaks down[15]" Darwin's theory.

The **ciliary-body (aqueous-fluid-producing) and trabecular meshwork (aqueous drain system)** *alone* is smoking gun evidence. Proof! And "absolutely breaks down[15]" Darwin's theory.

The **trabecular meshwork is a reverse pressure release valve.** That is, it closes under pressure ***preventing*** fluid loss and deflation of the eye. This *alone*, "absolutely breaks down[15]" Darwin's theory and is smoking gun evidence. Proof upon *Proof* upon <u>Proof</u> upon **Proof** upon <u>*Proof*</u> *upon* <u>***Proof***</u> upon PROOF upon *PROOF* upon <u>***PROOF !***</u>

In addition to the smoking gun proofs mentioned above, if you think about flagrante delicto[16]; or, *being caught in the very act of perpetration.* Sexual differentiation, sexual reproduction, pregnancy, birth, growth and puberty are examples of CREATION FLAGRANTE DELICTO![16] Think about that, meditate on that; sexual differentiation, sexual reproduction, pregnancy, birth, growth and puberty are examples of CREATION FLAGRANTE DELICTO! Creation caught in the very act! How would one <u>*evolve*</u> hormones necessary to go through puberty and necessary to sustain pregnancy?

Conclusion: The eye is a complex organ. Darwin HIMSELF states, "if it could be demonstrated that any complex organ existed, which could not have been formed by numerous, successive, slight modifications, my theory would absolutely break down."[17]

The design and/or function of the eye itself ***proves*** it could not have "been formed by numerous successive, slight modifications"[17]. Darwin's theory therefore, as he himself has stated, "absolutely breaks down[10]". The eye <u>***proves***</u> Darwin's theory of evolution is a false theory. Darwin's theory is "absolutely broken down."[17]

The eye itself demonstrates it was created and designed because of the eye's multiple systems and subsystems which are *all* dependent upon each other to function. Each one is necessary. Each one requires completed, totally developed function for the eye to see and for the eye to survive. This completed, totally developed function must occur simultaneously because they are all interdependent. We observe Interdependent Evidence of Creation! The evidence ***proves*** the eye could not have evolved. The eye proves interdependent Evidence of Creation. In fact, when you observe the evidence, it requires *more* faith to believe it could have just happened than is required to believe it was designed. Why, after looking at all the insurmountable evidence found just in the eye, would anyone conclude evolution?

I think people believe in *evolution* because they have been told it happened, and they blindly submit to what they have been told without really looking at the facts and the evidence. Plus, when they think of *creation*; they think; something or someone being able to make everything from nothing is, *too difficult to believe*. They think, and say to themselves, "logically, that can't be".

At the same time, I doubt if most people who believe in evolution know at the heart of Darwin's theory, the foundation upon which his theory is built is his "***belief***"[18] that (*page 380, The Origin of Species*) "***probably*** all organic beings which ever lived on this earth have descended from some one primordial form."[19] "**All organic beings**"[19] which "**ever lived**"[19], he ***believes*** -and thankfully he wrote it down. So, we do not have to assume, or infer, or guess; "What did he *really* think?" He wrote it down, and again it is on page 380 of his book, *The Origin of Species*. His theory includes his "***belief***"[19] all plants, all insects, all fish, all reptiles, all birds, all mammals, and this would also include all dinosaurs; and again, I am quoting him directly, "which **ever** lived on this earth, have descended from some one primordial form into which life was first breathed.[19]" And he also said, "animals and plants have descended from some one prototype."[19]

If most people who believe in evolution knew this is at the heart of his theory, they may see believing in *evolution* is not **_ANY_** different than believing in *creation*. I would hope again, they would think and say to themselves………… *"every organic being … which ever lived… descended from some one primordial form"[19]…… "*THAT IS TOO DIFFICULT TO BELIEVE!"

Likewise, when thinking about the evolution of the Universe and the "Big Bang" I hope you would say to yourself, "logically, that can't be!" "**EVERYTHING** from *NOTHING* by **_NOTHING_**" !

Which is more difficult to believe regarding the "Big Bang" *Everything…* from *nothing* – by…. **_nothing_**.

OR, God spoke, "let there be" and the Universe exploded into existence by His Power. Both require faith!

I think another reason people do *not* believe in *creation* is because they have not been given reasons *to* believe in it. The ironic thing about this statement is God says He has shown *everyone* creation - more than adequately,[20] in fact, He says He has shown everyone creation s*o thoroughly* that you have no excuse to *NOT* believe in creation.[20] It is a fearful thing to think one day you will stand before the Creator to give an account and He will say, "You have no excuse!"[20]

It is my hope and prayer I can help to show you a reason - in fact, more than 25 reasons - to believe in creation; that in **_looking at the evidence_** of the structure and function of the more than twenty-five interdependent systems and subsystems of the eye and in understanding the workings and interdependent workings of these systems and subsystems, you will see solid evidence, in fact,

"*smoking gun proofs*" of Creation. (1. How anyone can look at the trochlea and superior oblique muscle and think it evolved is beyond comprehension. (2. How anyone can look at the two routes of nerve supplies to the iris and think it evolved is beyond comprehension. (3. How anyone can look at the aqueous inflow outflow system and think it evolved is beyond comprehension. (4. How anyone can look at the tear gland and the tear drain locations and think they it evolved is beyond comprehension.) When you observe just these four smoking gun proofs, (and I could have listed at least twenty-one more) you will be able to look at the facts and make an *educated* decision to believe in creation and to accept Jesus as the Creator[21] and to ultimately accept Jesus as your savior from your sins.[22] I hope and pray after reading and learning about the eye you will *see God - in the eye*, and in looking at the eye you will think, "logically" this makes sense. Logically creation makes more sense than evolution - because, I can see **evidence** of a Creator. I see Interdependent Evidence of Creation.

Logically look at the evidence. For example, when looking at the trochlear bone loop and trochlear muscle system you should say to yourself, "there is no way a muscle could have evolved a way forward through the loop and then back again to the eye, and **at the same time** the trochlear bone loop evolved in the bone for the muscle to pass through it! Logically evolution just does not make sense for the trochlear loop and trochlear muscle system. Logically look at the evidence.

When looking at the pupil dilating and pupil constricting nerve routes you should say to yourself, "there is no way the pupil dilating nerve could have evolved a way down through the spinal cord, exit the spinal cord at the base of the neck, travel across the tip of the lung, then proceed down to the heart and the cardiac nerve plexus, then travel back up to the eye by way of the common carotid artery, to connect with other sympathetic systems in the superior cervical ganglion, then proceed on up with the now *internal* carotid artery, then to the eye by way of the ophthalmic artery and connect into the cliary ganglion in the back of the eye socket (and also the nasociliary division of the trigeminal nerve), then to the pupil dilator muscle. While at the same time the pupil constricting nerve proceeds from the brain stem through the oculomotor nerve, to the ciliary ganglion, then to the pupil constricting muscle of the pupil. I would hope you would say to yourself logically evolution just does not make sense for the pupil dilating nerve to run its' course and the pupil constricting nerve to run its' course. Think logically about the other ***smoking gun*** evidences I've written in ***bold*** in the preceding paragraphs. I would think you would say to yourself, logically evolution does not make sense. Logically, it can't be evolution! It must have been Created!

What I have observed in the evolutionary world and the astronomy world is the *experts* reasons for believing the world and the universe have been in existence for millions of years is they say the odds of life *evolving* is *highly probable*. The experts also say, given the fact we know there are billions and billions of stars and galaxies, the odds of us encountering another world such as ours is *highly probable*.

Well, if we are going to speak in *probabilities*, think about and meditate on the probabilities outlined in this book—probabilities such as the number of bones surrounding the eye and the number of nerves innervating the eye—their paths, their points of origin, how they form, how they function, and how they function in relation to each other. Then ponder_the eye's arterial blood supply, the sheer number of arteries, arterioles, and capillaries; their structure, function, path, and point of origin. Observe the many different courses they must take and the many different tissues they pass through and ultimately supply. Then look at the venous supply and again the vast numbers of veins and venules and their paths and the tissues they pass through. Then ponder the fact these soft tissue structures must penetrate the bones surrounding the eye. There must be the thirty-five foramen or holes for the many nerves, arteries, veins, tear glands, and tear drains to pass through! Then look at the number of systems and subsystems and the fact they are totally interdependent. Conclusion: There are at least 594 interdependent structures which create 25 different systems. These systems are dependent on systems, which are dependent upon systems, which are dependent upon systems, to the Nth degree! Ponder those probabilities! The eye and the body are a universe in and unto itself when you consider how loosely the term *universe* is used today. Then consider we have not even discussed the many different organ systems such as the kidneys, liver, intestines, lungs, heart, muscular, and skeletal systems which *must* be present to support the eye and keep the eye and the individual alive. If the individual does not live, the eye dies also. To this end there must also then be glandular functions: thyroid, pituitary, pancreas, adrenal, and mammary glands. Then we can go deeper into individual cellular structure and cellular function; then intercellular structures and cellular metabolism; and deeper yet into biochemistry, hormones, digestion, immunity, the fighting of infection, cellular repair and replacement, wound healing, waste excretion, sexual differentiation, sexual reproduction, pregnancy, birth, and provisions for bodily changes secondary to growth, just to name a few. Again, all are totally dependent upon each other. Intercellular component dependent upon other intercellular components, cell dependent upon other cells, tissue dependent upon other tissues, organ dependent upon other organs, system dependent upon other systems, function dependent upon other functions. These many, different cells, tissues, organs, systems - - billions, *upon billions,* **upon billions,** **_upon billions,_** UPON BILLIONS, *UPON BILLIONS,* ***UPON BILLIONS***, **_UPON BILLIONS_** of interdependent cells, tissues, organs, systems, and functions, which are totally interdependent upon each other for the individual to live. Remove any **_one_** system and the body cannot live.

The astronomer says there are billions of galaxies; therefore, the probabilities are high life could happen by chance here on earth or elsewhere in the universe. In reality we do not see any *evidence* it *has* happened elsewhere, and we have been looking elsewhere for a very long time. What is most disturbing is the astronomers *do* have *evidence* that at one time there was nothing, then all at once there was everything. And again, they know this based upon verified experimentation. The fact of the "Big Bang" is now *without question. That theory has been PROVEN!* Yet they still deny the Creator and think everything can come from nothing by the power of nothing. So… one of two things could have happened: Nothing became everything - by nothing. Or someone or something caused everything to come into being. Which makes more sense?

The laws governing how "the big bang" happened and the fact there are laws and theory which can explain it, all point to planned, purposed events. Planned, purposed events point to *a Creator*. Ironically, even *knowing* the "big bang" happened and having *multiple proofs* of it happening, many choose to believe, *it just happened by chance, and by nothing, for no reason.* Some believe nothing became everything by the power of nothing. Wow! Talk about having faith, and putting your faith in something. You would need *great* faith to think nothing became everything by means of nothing, In this case, you are talking about putting your faith literally in *nothing*! If you believe this, your faith is exceptionally great and yet you are putting your faith in nothing if you do not believe in GOD.

When contemplating probabilities, the creationists can say with just as much conviction, look at the billions and billions of interdependent cellular components, cells, tissues, organs, systems, and subsystems and interdependent functions–needed for the human body to function, then consider the probabilities of it happening by chance, all at once –because it has to happen all at once - otherwise, the body, with its billions of interdependent functions, cannot function if any one component was missing or waiting to evolve. The creationists, I think, can say with even greater conviction and ultimately even greater confidence it *had* to be created because we see *evidence* that it did happen we observe INTERDEPENDENT EVIDENCE OF CREATION – billions and billions of interdependent evidences.

CONCLUSION: The eye itself proves that it could not have been "formed by numerous, successive, slight modifications," as Darwin himself suposes.[23] That is, it could not have evolved. Thus (as Darwin himself stated) his "theory absolutely breaks down."[23] But also, since the conclusion is: Dawin's theory as Darwin anticipates," absolutely breaks down"[23] *the conclusion thereby is* that the eye is Created. Facts are facts! In ***any*** proposed theory if the information presented is factually accurate, and scientifically correct (that is, it is not just something I "imagine,"[24] "suppose,"[24] or "believe"[24] as Darwin states). If the evidence ends in a conclusion that answers the theory, then the theory has been ***proven***. Thus, if after examining the eye, the conclusion reached ***based upon the evidence observed***, as Darwin himself states, "***absolutely*** breaks down" his "theory;"[25] then the conclusion at which we then arrive *thereby*, points to Creation and therefore - God.

However, not only does the evidence *thereby* point to God by absolutely breaking down[25] Darwin's theory, the eye also ***directly*** demonstrates and ***proves*** the theory of Interdependent

Evidence of Creation and thus, **_directly_** points to and affirms God! The evidence <u>concludes</u> Creation therefore the evidence **_concludes_**: GOD!

ADDITIONAL IMPORTANT OBSERVATIONS AND CONSIDERATIONS:

I think Darwin anticipated that his "theory"[25] might later be disproved; at least he left his "theory"[25] as he himself called it, open for question, debate, and possible dispute. With his statement, "if it could be demonstrated that any complex organ existed, which could not have been formed by numerous, successive, slight modifications, my theory would absolutely break down[25]" he seems to welcome dispute, welcome questioning, and encourage the "breaking down" of his "theory."[25]

Furthermore, it is interesting that Darwin *himself* refers to his writings as "theory"[25] - and *He* is the *FATHER* of his "theory"[25]. Yet, most evolutionists don't regard evolution as "theory,"[25] most evolutionists, and most people in general believe evolution as fact even though Darwin himself –the father of the "theory"[25] calls it *"theory"*[25] numerous times. In fact, from page 145 to page 155 he wrote the word "theory" or the words "my theory" eight times[26]– eight times in 10 pages![26] It is *so* obvious that he said, "theory" or 'my theory" so many times; (in fact, too many times for me to attempt to count it throughout his whole book) that he wanted to make sure you *knew* it was a "theory".

I also see that Darwin, in his writings;[27] that is, <u>*he wrote it*</u>. He wrote it down so we don't have to infer or guess what he said or what he thought *because he wrote it down*. Darwin himself references the Creator, and when he references the Creator it is most often and numerous times in PRESENT

TENSE! Meaning he believes there *IS* a Creator not just that life *WAS* created. Furthermore, when he writes the word *Creator*, he writes it with a Capital "C", *I think*, out of respect and reverence for God the Creator.[27] And we see this *several* times throughout his book.[27] In fact, from page 156 to page 166 he wrote the word Creator, referencing the Creator or Creation with a capital "C" four times-four times in 10 pages).[27] He writes the word Creator so many times (and always with a capital "C") that I think he himself believed that at some point in time there was a *creation*. He certainly does acknowledge *the Creator* many, many times. And he also says, "into which life was first breathed[28]" which indicates his belief of a ***Creator***. Yet, even though he acknowledges "the Creator"[29] MANY, MANY TIMES, his followers; evolutionists, atheists etc., dismiss those statements he - ***without question, absolutely made*** - as if he never said them, while it is quite apparent - because he references "the Creator" (again - with a capital "C") *so many times*[29] that he believed there is a Creator. Why would he reference "the Creator" *in writing,* so many times if he did not feel life was Created. That would be stupid! *You can't dismiss the fact he said* and wrote *"Creator" and then come the conclusion he did not believe in creation or believe that his theory nullifies creation. It is obvious he himself believed there was "a Creator" and anyone who says otherwise is himself stating something Darwin did NOT say!* I doubt that many staunch evolutionists even know that he references a Creator (with a capital "C") in his book multiple, multiple times. It is clear he believed in a Creator, he wrote of "the Creator" and *always* with a capital "C!"[29] Know this! Meditate on this! If he wrote it, he specifically wanted the reader to know that. Do you think he thought as he was writing no one would read what he was writing? When you write, you are *intending* to convey information. He wanted his readers to know he believed in Creation, (with a capital C out of respect for the Creator). It is clear Darwin believed in **C**reation.[29] Remember he also said, "probably all the organic life forms which have ever lived on this earth have descended from some one primordial form, into which life was first breathed."[30] This would ***require*** a **C**reator!

Because of all these facts, and *especially* because Darwin wrote numerous times, "the **C**reator"[31] I do not feel I am speaking out of place or making assumptions on which I have no basis to stand, when I say *I think* if Darwin were alive today and knew the facts I am presenting concerning the smoking gun proofs of the eye, and the recent findings I write of in Chapter 23, he may change his thinking, remember, he was working with the knowledge base of the1800's.[32] Not to change his thinking that there is a Creator, (his writings demonstrate his belief in a Creator) but change his belief that, "probably all organic beings which have ever lived on this earth have descended from some one primordial form, into which life was first breathed,"[33] and instead believe that it's the DNA that dictates the species and that God Created each species utilizing the DNA as the common blueprint so that each species is created according to its' kind as is recorded in Genesis 1:12, 1:21,1:24, and 1:25.[34]

Most people choose not to believe in God because they think they must do something they don't want to do, or give up something they don't want to give up. They do not want to submit to some higher authority. They intentionally try to find reasons to doubt God's word the Holy Bible because they do not want to have to follow it or submit to it, or a higher power. Most people do not realize that the Bible was written by forty different authors over several hundreds of years and it is in absolute agreement with itself. They try to find ways to bend it out of order, and out of context, and are looking for a reason to not believe it and not to believe in God. As you can see, even Darwin believes there is a Creator. He believed it enough to put it in writing![35]

God wants you to see Him. He wants you to know Him. He has revealed Himself to you through His Creative Hand, through His Word, through Jesus, and through the Holy Bible. He is Just, and Fair, and Good. He is all powerful, and all knowing, and knows what is best. He is under all control in all circumstances (He is sovereign).

If you were to try to describe God in one word the best word to describe God in one word is Love. If you were to try to describe God in two words, the best two words to describe God in two words is Holy Love. You can trust Him. His nature is: more, higher, better, Just, Holy, Good, and Loving. He loves you and wants to have a relationship with you. He has reached out to all mankind through the provision of His Son Jesus. God has provided the means of restoration back into a relationship with Him if you accept it. Accept Jesus as your Savior.

God has shown us Himself through our eyes and in our eyes. He has also shown us *His nature* (His character, personality, temperament, and disposition) in our eyes. He doesn't do things half way or just enough to get by. His ways are higher than our ways. His ways are of *exceeding abundance*. Not only has he shown Himself in all of Creation and particularly in the Creation of the eye –and shown Himself to the point that no one will have an excuse not to believe in Him. At the same time, He also has shown us His character, personality, temperament, and disposition as we observe the eye. He has shown us His nature of *exceeding abundance* in our eyes. In John 10:10, Jesus says, "I came that they may have and enjoy life, and have it in abundance: to the full, till it overflows!"[36]

25. Beyond Evolution

How does evolution explain that we see so many examples of more than enough. Built in safe gaurds, automatic reflexes and preprogrammed responses. We have two eyes, two ears, two legs, two kidneys, five fingers, two arms, two legs; on, and on. It is obvious; we are observing evidence beyond evolution. Additionally we observe wound healing, infection defences, immune response and the ability to acquire immunity. Think about some of the reflexes that are built in which safeguard our wellbeing. Then there are responses such as the gag reflex, coughing, sneezing, and vomiting, Immediate jerking back when we touch something hot. Immediate jerking away when we sense something sharp when we are injured. We have chills with shivering when we get excessively cold. Then there is hunger, and thirst, and knowing when we need to stop eating or drinking. Sensing when we need to be warmer or colder. Then we have emotional safegauards and responses such as laughing, crying, joy, happiness, sadness, grief and love. How does evolution explain these things? Many aren't necessary for survival but they certainly make living, better. It is apparent we have gone beyond evolution when we observe these things. When we specifically look at the evidence of God when observing the eye we can see God's omniscience (all knowing power) in providing an exceeding abundance of what is needed so changes due to ageing, deterioration and degeneration do not cause the eye to cease to be able to function. We see God's wisdom and foresight further yet in creating the many automatic functions, safeguards against infection, provision for healing and wound repair, defense against infection, and automatic reflexes, to insure continued optimum function even after experiencing adverse events. As is His nature, that is, *exceeding abundance,* God also has provided ordered symmetry and beauty.

As you read each attribute of exceeding abundance meditate on the advantage and then question how evolution would explain it. Ask, "why is it this way? Why would it evolve in this manner?

1. We have two eyes so that in case one gets damaged we can still function.

2. The eyes are in the front of the head surrounded and protected by bone.

3. While the eyes are in the front of the head, the processing centers for vision are in the back of the head, the least likely area of the brain to get damaged in the event of a blow to the head. With this design, it is apparent Vision is to be preserved.

4. The autonomic (*automatic*) or non-thought *required* functions of the eye are in the deep inner recesses of the brain, the most protected areas and least prone to damage from stroke and degeneration.

5. The oil glands of the lid are placed (Interdependent Evidence of Creation) in the lid margin such that an oil layer floats on top of the water layer, preventing evaporation of the tears. (Oil floats on water).

6. There are 22 to 25 oil glands in each lid—that is, about 50 per eye.

7. There are two tear drainage ducts, one per lid, per eye. One on the top and one on the bottom of each lid Four tear drainage ducts total.

8. The tears drain into the *back* of the nose through the *bones of the nose* rather than running down the face, or out the *front* of the nose.

9. Eyelashes and eyebrows protect the eyes from dust and keep sweat and water from rolling into the eye when working in heat or rain *and* make the eye pretty and attractive.

10. Your eyes blink when needed, and you do not have to think about it. And, the blink rate changes as the conditions change. That is, your eyes don't just blink every seven seconds, they may blink more often or less often depending on dryness of the air, wind currents, temperature, etc.

11. The inside of the eyelids join the white part of the eye to prevent debris from getting lodged behind the eye.

12. The cornea is *clear* living tissue.

13. The cornea is living tissue yet with no arteries or veins.

14. You have to lose more than 50 percent of the endothelial cells of the cornea before the cornea becomes cloudy.

15. The curvature of the cornea is just the right curve—not too flat, not too steep— and is in the precise thickness and curvature. The precision of curvature is to about 15 microns or 1/10 of a hair. The precision of the thickness of the cornea is to about 100,000 nanometers or about 1/100 of a hair.

16. The aqueous fluid that fills the eye is derived from the blood yet it is clear fluid. It carries nutrients to the lens and the cornea.

17. The aqueous constantly flows in and out and the pressure is regulated precisely— not too high, not too low.

18. The eye does not loose pressure when it is pressed upon. There is a reverse pressure release valve! Wisdom and *design* can only explain this. Wow!

19. The lens of the eye is clear living tissue with no blood vessels. It changes shape to change focus, yet it remains clear as it changes shape and it does not distort the optics as it changes shape. Man has not been able to duplicate this phenomenon.

20. The vitreous is just the right consistency to protect the retina but at the same time does not transmit energy from blows. It is not too thin; it is not too thick—it is just right.

21. The length of the eye (the sclera) is precise to 1/3 of a millimeter or about 1/32 of an inch, or, .3mm which equals 300 microns, which is about 4 human hairs.

22. The six eye muscles are connected in the eye socket at exactly the right place and angle and have exactly the correct opposing eye muscle placement for the eye to aim together straight, move in all directions and to be able to rotate when the head is tilted to the left or right.

23. The trochlear eye muscle (superior oblique) comes from the back of the eye socket attached to the bone in the back of the eye socket; it comes forward and loops through a pulley-like bone, the trochlea, in the inside corner of the eye socket and then goes backward to the eye and attaches to the eye behind the equator of the eyeball. Meditate on that DESIGN! Evolution? NO! - I.E.C.!

24. The photoreceptors convert light to electrical impulses. The outer segments are continually replenished by the inner segments so that they don't wear out and we can see and continue seeing.

25. The retina contains million and millions (*3.52 billion* additional *functional cells* in both eyes) that are like internal relay switches that enable the eye to perform automatic functions such as locking onto moving targets, scanning, and instantaneous changing of gaze from one target to another without searching, overshooting, or undershooting.

26. The retina is split into exactly a left half and a right half, and exactly a top half and bottom half, such that what is seen in the right retina is exactly in the same place and corresponds exactly with the same exact location of the left eye.

27. The optic nerve carries the messages of the millions and millions of photoreceptors, and it is made up of millions of individual nerve fibers all carrying messages about your vision; the many different colors, shape, brightness, intensity of colors, movement of objects, distance of objects, nearness of objects, and on and on. Yet these messages do not get confused with each other or "short each other out."

28. The eyes are interconnected with the 12 cranial nerves such that if I hear something to my right, my eyes instantly move to the point of location of the source of the sound, thereby we experience visual location of objects making a sound. If something is attacking me or thrown at me, my eyes can determine from what direction and of what speed, and the eye interconnects with the muscles of the head, neck, and body to allow the body to move in the way needed to prevent injury.

29. The visual cortex receives millions upon millions of *sparks of electricity* (impulses) from the retina every second, and it somehow converts them into a picture that is in the brain exactly as it is in the real world.

God made your eyes. God made your eyes so you can see. He made our eyes so we could see all His wonderful creations. By seeing His creations, we know who He is and that He is real. He has revealed himself to us through His creations.

When one closely examines the eye, it is clear it was planned, it was designed, and the many parts required foresight and designed actions and reactions. *The eye was indeed engineered!* By studying the eye, we *know* God is real, and we can conclude He made everything. God has shown Himself to us through our eyes and all the things that He has made. Everything was made by Him and for Him, and by Him everything continues to exist.[1]

There are at least twenty-five major systems within the eye. *Each* of these systems is dependent on *each* of the other systems. They are *exponentially* and *absolutely interdependent*. Each system *must* be present and *fully* functional *all* at the same time. No *one* system can function without even *one* of the other twenty systems.

All twenty-five systems must be present; they *all* must be *fully functional, all* at once and *all* at the same time. The only way for; -*all* must be present, and *all* must be *fully functional*, and *all* at once, and *all* at the same time to happen, is by *Creation!* The evidence is beyond evolution; *we observe exceeding abundance*, which is the very nature of God. We observe *Interdependent Evidence of Creation; I.E.C.* when we observe the eye.

26. Intelligible Evidence Made Plain

God made your eyes so you can see. God made your eyes so you can see *Him* through His wonderful creations. Through the eye, ***everyone KNOWS*** He is God[1].

We can see God's wisdom, intelligence, foresight, planned prevention, exceeding abundance, beauty, wonder, and power in his creations, and in the working of the many miraculous things in the universe, in the world, in His Creations within the world, and in the structure and function of the eye. He has revealed Himself in everyone's heart, and he has revealed Himself through His Creation; Romans 1:19 & 20. And again, He says He has revealed Himself so effectively, everyone is without excuse if they do not acknowledge Him and accept Him.

"For that which is known about God is evident to them and made plain in their inner consciousness, because God Himself has shown it to them. For ever since the creation of the world His invisible nature and attributes, that is His eternal power and divinity, have been made intelligible and clearly discernable in and through the things that have been made. So men are without excuse (altogether without any defense or justification)" (Romans 1:19–20, Amplified Bible)[1].

God has revealed Himself to us in His creation of the things we see, and in His creation of the eye, *and* He has revealed Himself to us in His Word, the Bible.

There is much debate over the Bible; is it authentic, did people conspire to write it, is it accurate, does it have errors?

I have been a Christian for over forty years. I have been reading the Bible regularly, (I try to read it daily for the last 40+ years). After over forty years I have found by personal experience that I have never found God's Word, the Holy Bible, to be untrue, inaccurate, or unreliable. It is the *ONE THING* I can *ALWAYS* count on, I can rely on, and I can depend on. It is comforting for me to know that it is always as clear as black and white. God's Word, the Holy Bible is the ultimate truth. There are no gray areas, His Word is not "well maybe", it is yes, and no, and let it be so. To follow are a few interesting facts you may not be aware of or have not considered regarding the Bible.

1.) We have already discussed how science has confirmed there was a "BIG BANG", the moment everything came into existence, and how Einstein's theory [2,3] predicted what has now been confirmed by the Hubble telescope.[4, 5, 6, 7, 8, 9] And Einstein's theory which predicted the "Big Bang" has also been confirmed by the Red Shift,[10] and Bicep 2,[11, 12] (Background Imaging of Cosmic Extragalactic Polarization[10, 11, 12] (which also included researchers from Caltech, the Jet Propulsion Laboratory, Stanford University, and the University of Minnesota). Science has confirmed what the Bible says, when "God said, let there be light"[13] that is when the "Big Bang" happened. Or another way to say that is; science has confirmed there was an origin to the universe.[2, 3, 4, 5, 6, 7, 8, 9, 10, 11, 12] just as the Bible says; "In the beginning God created the heavens and the earth"[13]

2.) We have also discussed how God's word, the Bible, has reconciled what Creationists believe the age of the earth is, and what scientists, archeologists, paleontologists, and cosmologists believe the age of the earth is. Now this reconciliation by the Bible is now confirmed by man.[14] The age of the earth has now been reconciled by scientific confirmation.[15] We have scientific evidence that confirms what the Bible says. We have scientific evidence that confirms that the word of God is true and correct.

Here is some additional evidence of Bible truths you may not be aware of:

3.) The Bible manuscripts have been around for thousands of years. The Bible as we now know it was compiled several hundred years ago, if it was fake or false it would have fallen by the wayside long, long ago. It "still" remains, in fact, as time goes on it is showing itself to be more and more

accurate as archeologists, and geologist discover more and more artifacts which agree with the Bible.

4.) Science tells us there was a great flood; the Bible talks of the great flood.

5.) The Bible is full of prophecy that has come to pass.

6). The Bible lists names of places that have been confirmed.

7.) The Bible speaks in specifics regarding Kings of Israel and Kings of surrounding nations.

Many, many of those kingdoms have been confirmed.

8.) Science has confirmed what the Bible says in Genesis; the age of water on earth is older than

the sun![16] Part of the reason I am trying to emphasize this is that the media certainly is not going to

bring this up. Evolutionists, cosmologists, physicists aren't going to bring this up. Have you noticed;

they don't even talk about the "Big Bang" anymore (even though we KNOW FOR A FACT that it

occurred) because the "Big Bang" confirms what the Bible teaches. You won't hear about; the "Big

Bang", the earth being older than the sun, that the dinosaurs kept shrinking and shrinking, that the

DNA dictates the species, or the proof of the theory "Interdependent Evidence of Creation" because

they all confirm that there is God. Conversely, the media hammers and hammers on anything that

hints of evolution. Yet, most, if not all, of these so-called *hints of evolution* you could equally use the

term "adaptation" within the species instead of the term evolution. The media is biased against God.

However, God is as the Bible says, "the great I Am, meaning He has always been and He will

always be!

Over the years I have heard different people say, "The Bible is full of errors". I personally have

been reading it for over forty years and I have yet to find one place the Bible is in error. One place I have heard it most frequently said that the Bible is in error is; "Where did Cain get his wife?" (Cain was Adam and Eve's son). Consider this: According to Darwin, when he was studying the expression of traits transmitted over time from parent to offspring. he said:

"The elephant is reckoned to be the slowest breeder of all known animals, and I have taken some pains to estimate its probable minimum rate of natural increase: it will be under the mark to assume that it breeds when thirty years old, and goes on breeding till ninety years old, bringing forth three pair of young in this interval; if this be so, at the end of the fifth century there would be alive fifteen million elephants, descended from the first pair"[17].

He says each elephant will produce "three pair from age 30 to age 90", six offspring in a sixty-year timeframe. Darwin says in 500 years there are (*alive*) 15 million elephants[17]". WOW!

Humans can produce six offspring in six years - even less than six years. (My wife and I had four kids, <u>none</u> of which are twins, before our first child was five years old.) If humans could only produce six offspring in ten years - which is entirely "under the mark" [17] it would be *six times faster* than the elephants, which Darwin says produce six offspring in 60 years[17]. Using Darwin's numbers, humans *could* produce fifteen million humans six times faster than elephants; which would calculate out to fifteen million[17] humans in 83.33 years using Darwin's numbers and correlating that to humans. This seemed a little far-fetched to me so I recalculated with the following numbers.

If Adam and Eve produced six offspring in twenty years, (which would be a very conservative number, and very much "under the mark" [17]) and one half of them were females and one-half of them were males. Then each of those three females also produced six offspring in a 20-year period. In 100 years the population would be 728 people. Then if each of those three females produced six offspring each in a 20-year period, and half of them were males and half of them were female and those three females each produced six offspring and half of them were female; and that pattern continued through each generation. In 200 years, the population would be 118,098 if none of them had died in 200 years. In 300 years, the population would be 28,697,814, assuming there were no deaths and exactly half were males and half were females and that each female bore 6 children and no more than 6 children in 20 years, 3 being female and each female producing 6 children in 20 years. (According to the Bible before the flood, people lived hundreds of years, Methuselah living the longest at 782 years. According to the Bible Noah was 500 years old when he became the father of Shem, Ham, and Japheth[18]). If Noah did not have **_any_** kids until he was 500 years old it would not be unreasonable for Cain to not select a wife and have kids when he was- let's just say, 300 years old. Using this reproduction pattern, in 300 years the population of the earth would be 28,697,814. That is over 28.5 million. Keeping this in mind, it would be safe to say Cain probably married one of his sisters or possibly one of his cousins, or cousin's cousins, or nieces, or probably someone very far removed from his immediate family considering the population could be as high as 28.5 million by the time he was 300 years old. (Remember Noah had Shem, Ham, and Japeth when he was 500 years old).

9.) Even if Cain married just after the first hundred years of his life there could have been 728 people on earth at that time using this reproductive pattern. If half were females he would have had 364 women to choose from in just 100 years, and if you think about that, 364 women to choose from is a very large number to choose from. Remember, if instead we used Darwin's numbers for the elephant, and figuring that humans can reproduce at least six times faster per Darwin's figures[17], the population would be over 15 million[17], assuming half of them were female Cain would have had 7.5 million women to choose from for his wife in only 83.3 years! Darwin's numbers *would* be accurate!

10.) Another thing to think about that you often hear people say to refute the Bible is about Noah saving two of each animal on the arc during the flood. Many people think if Noah saved two of each animal on the arc[18] they couldn't replenish the earth, but again according to Darwin it would only take 500 years for two elephants to produce 15 million offspring and they only produce three offspring in 90 years, and they, according to Darwin, are "the slowest to reproduce."[19]

11.) Also on a separate note, I want to remind the reader that Genesis 1:20 says, "and God said, Let the waters bring forth abundantly and swarm with living creatures, and let birds fly over the earth in the open expanse of the heavens.[20] Then Genesis 1:21 says, "God created the great sea monsters and every living creature that moves."[20] some think this may be referring to God's creation of the dinosaurs.

You can rely on the Bible, it is true, it is without error, and it has the answers, guidance, and wisdom you need. The Bible is the instruction manual and guide for living and provides wise counsel for any question that may arise in day-to-day living. While there are more than 40 different authors in the Bible, the reason it is without error and in agreement with each of the different authors is that they were inspired by God exactly what to write, so the true author is God. The Bible says God breathed into each author His Word[21] (2 Timothy 3:16). Thank you, God, for giving us your Word!

27. Concerning Chance

The following Chapter; Chapter 27, is a difficult chapter. It is difficult for me to write. It is especially difficult if you, the person now reading this book are an atheist or devout evolutionist. However, I am praying that at this point; having observed all the evidence provided in this book, you now have the _**proof**_ you have been looking for in order to now believe in and now have accepted God. But know this; believing in God is not enough to be saved. The Bible says the devil and the demons KNOW there is God, and when they think about it they tremble and shudder in horror (James 2:19 page 1445 _The Amplified Bible_). You not only have to believe, but also accept God's free gift from sin and eternal life in hell, into salvation and eternal life in heaven through the atonement of Jesus Christ. Only by accepting Jesus Christ as Lord of your life can you be saved. Just as the blind begger had to listen and obey Jesus in order to accept his healing, you have to listen and obey Jesus and accept His free gift of salvation. Jesus lived and died for you and only _**He**_ can do that. _Anyone_ can die on a cross, but only Jesus _**lived**_ ALL OF HIS LIFE _**for you**_, without _**ever**_ sining. _**HE**_ is the only one who can take the punishment for your and my sin because HE is without sin – _**ALL**_ of his life, and even as He died on the cross He never sinned. But the following Chapter is not _Basic Christianity 101_, it is rather advanced Christianity and somewhat difficult for a new believer to accept and be at peace with. This is my attempt to answer some questions you may have as an atheist or evolutionist who hopefully has now become a beliving Christian. As a new believer (hopefully you are at this point) please remember this one point as you read the rest of this Chapter, "Concerning Chance," God IS love. He loves you and His love is genuine and perfect. His love is completely satisfyimg and completely fulfilling. Only His Love can satisfy all the desires of your heart. God is love, and God loves you completely, wholey, truly, and perfectly. But God is also HOLY!

Someone might ask, If God is love, why was I born blind?" Or they may ask, "If God is all powerful and in control of all things, why have I not been healed of my blindness?" Or as in the Bible when Jesus's disciples asked him about the man who was born blind at birth, "Rabi, who sinned, this man or his parents, that he should be born blind?" Jesus said, "It was not that this man or his parents sinned, but he was born blind in order that the workings of God should be manifested (displayed and illustrated) in Him.[1] (John 9:1–7, *The Amplified Bible.*

Jesus said the man was born blind so that Jesus could heal him, and the world would then know that Jesus **_is_** God.

Or, possibly in a different instance, *possibly*, God may know that a specific blind person while they are blind will continue to seek Him; whereas, if they were healed and could then see, they may instead turn away from God and get caught up in the cares of the world. So God in His wisdom and perfect judgment keeps the person blind so that the person will continue to seek Him.

There are so many *why* questions. *Why* do people get cancer? *Why* did she or he get killed in a car wreck? *Why* did that tornado kill all those people? *Why* did God allow that man to kill his six-year-old boy? *Why* did God allow that lady to kill her four children? *Why* did the hurricane,... earthquake,... tsunami,... war,... robbery,... fight, loss of job,... loss of finances,... loss of health,... etc. happen? If God is in control of all things... *why?* There are many different reasons for "Why?" only God knows the answer to Why? He has, though, given us some examples. He has made each of us according to His purpose and His good will and *however* He has chosen to make us. We are made by Him, for Him, as it pleases Him - to further **_His_** kingdom. For example, when He told Moses to go speak to the Pharaoh, Moses said, "O Lord, I am not eloquent, or a man of words, neither before or since You have spoken to Your servant: for I am slow of speech and have a heavy and awkward tongue." The Lord said to him, "Who has made man's mouth? Or who makes the dumb, or deaf, or the seeing, or the blind? Is it not I the Lord? Now therefore go, I will be with your mouth and will teach you what you shall say"[2] (Exodus 4:11–12 *The Amplified Bible*). God says **_He_** made Moses' tongue, **_He_** made the dumb, **_He_** made the deaf, **_He_** made the seeing and **_He_** made the blind. Then, the Lord used Moses to speak to Pharaoh even knowing of Moses's "heavy tongue" and awkward speech. Likewise, He can use us just as we are. He wants to use us, and He wants us to be useful to Him. He wants to have a relationship with us. We are all created according to His will and for His purpose. We were created by Him and for Him, "For it was in Him that all things were created, in heaven and on earth, things seen and things unseen, whether thrones, dominions, rulers, or authorities; all things were created and exist through Him (by His service, intervention) and in and for Him. And He existed before all things, and in Him all things consist, (cohere, and are held together)"[3] (Collosians1:16–17, *The Amplified Bible*). So, we were created by Him, for Him.

His word also says, "But who are you, a mere man, to criticize and contradict and answer back to God? Will what is formed say to Him that formed it, "Why have you made me thus?" Has the potter no right over the clay, to make out of the same mass (or lump) one vessel for beauty and distinction and honorable use, and another for menial or ignoble, or dishonorable use?"[4] Romans 9:20 & 21, (*The Amplified Bible*).

We must say to ourselves that God knows what is best under *all* circumstances and trust Him. Think about it, HE IS GOD! *He* is the Creator, and we are *His* created. He is the potter; we are the clay. He has revealed Himself to us through His creation.

Another answer to the *WHY* question may be found in Jeremiah 18:1–17 which says He allowed bad things to happen to His chosen people of Israel to cause them to, "return now each one from his evil way; reform your (accustomed) ways, and make your (individual) actions good and right.[5]"

Or, sometimes when bad things happen, it is because **we** got **ourselves** in a mess by direct disobedience. John 5:5-14 of the Amplified Bible, says, "Afterward when Jesus found him in the temple, He said to him, See, you are well! Stop sinning or something worse may happen to you."[6] Sometimes it is because of sin and disobedience in our lives when bad things happen, yet we wonder "Why?" and oftentimes we then blame God!

Sometimes someone else or the devil meant it for bad, but God meant it for good. For example, when many, many bad things happened to Joseph, one after another, after another, over much of the first portion of his life, he said, "As for you, (his brothers) you thought evil against me, but God meant it for good, to bring about that many people should be kept alive, as they are this day."[7] (Genesis 50:20, *The Amplified Bible*).

Sometimes bad things happen because God is pushing us, or using other people to push us in a direction he wants us to go, or is preventing us from going in a direction he doesn't want us to go. "He *makes* me to lie down in green pastures"[8] (Psalm 23, *The Amplified Bible*). That is, He did not say, "I suggest you lie down in green pastures."

Sometimes bad things happen as a direct attack from the devil as he wants us to sin against God (as illustrated in the book of Job), but even then God is in control of what is allowed.[9]

There are other reasons for *WHY*, but we may never know them and may not comprehend them if we did know them. Isaiah 55:9 says, "His ways are higher than our ways."[10] *Ultimately, God Knows what is best in all situations, and we should trust Him.*

Romans 3:23–24 says, "Since all have sinned and are falling short of the honor and glory which God bestows and receives. All are justified and made upright and in right standing with God, freely and gratuitously by His grace, (His unmerited favor and mercy, through the redemption which is provided) in Christ Jesus."[11] (*The Amplified Bible*)

God loves you. He has made Himself plain to you. He says, (and I paraphrase) "It is *so* plain that you therefore are without excuse."[12] That is; you cannot say, "I did not know God exists, no one ever told me." ***GOD*** has shown you!

Satan lies to you to attempt to make you believe in evolution and to keep you from seeing the evidence of God's creation and power. The devil wants you to follow him in sin to separate you from God. Your sins displease God and separate you from God. God is Holy and does not accept you because your sins (your disobedience to Him) make you unholy and unacceptable to God. Even though you continue to sin, God provided a *Way* to accept you. He wants a relationship with us so much that even though we sin, even though we rebel, He provided a Way to restore us into right standing with Him. This Way has been completed and is finished, but there is only *one Way*. God gave you His acceptance of you into His kingdom as a gift through Jesus' sinless life.

God said, (again I am paraphrasing) "I will allow Jesus to take your punishment for your sin." Jesus said, "I will take your punishment for your sins." Jesus *gave* his life on the cross to redeem you from God's punishment for your sins—even though Jesus never sinned. Actually, Jesus gave His life for you and me all the while He was alive too - because he resisted temptation for you and me every day, every time he was tempted and never sinned. No one else could have died for your sins or my sins because everyone else has sinned, and continues to sin, and therefore everyone except Jesus deserves God's punishment. Jesus is the only One who did not deserve to die because He never sinned and therefore He did not deserve *any* punishment. He is the only One who could give His sinless life as punishment and atonement for our sins. God accepts His death on the cross *(only because He was sinless)* as atonement for our sin[13]. But accepting Jesus' free gift of salvation is the only way to be forgiven. Thank you, God. Thank you, Jesus! (Paraphrased by the author, J. Jay Rigney, but based upon the book of Romans; Chapters 1– 8. *The Amplified Bible*)

Not only that, but while you are living you have a choice to live for Jesus or to live for yourself and to live for the devil. The devil wants to keep you separated from God and God's blessings. The devil wants to blind your eyes and deceive you so that you do not see God and His gift to you for the forgiveness of your sin, which allows you to be acceptable again unto God. Jesus is your Savior from your sin and eternal punishment in hell. He saves you from hell and eternal punishment, and, by Him, and *only* through Him, you may enter God's kingdom in heaven. When you die, you don't just go away and cease to exist, your soul will continue to live eternally, either in heaven or in hell, ***eternally***. Choose eternal life in heaven, not eternal life in hell. Ask Jesus right now to save you and be Lord of your life, and live for Him. He loves you and wants to have a personal relationship with you.

"Why should you, or why would you want to ask Jesus to save you and be <u>Lord</u> of your life?" you might ask yourself. There are many, many reasons; here are a few (that I like) of the many:

Jesus said, "Come unto me all of you who labor and are heavy-laden and overburdened, and I will cause you to rest (I will ease and relieve and refresh your souls). Take My yoke upon you and learn of Me, for I am gentle (meek) and humble (lowly) in heart, and you will find rest.) (relief and ease and refreshment and recreation and blessed quiet) for your souls. For My yoke is wholesome (useful, good- not harsh, hard, sharp, or pressing, but comfortable, gracious, and pleasant), and My burden is light and easy to be born[14] (Matthew 11:29–30).

Also, "The eyes of the Lord are toward the (uncompromisingly) righteous and His ears are open to their cry. The face of the Lord is against those who do evil, to cut off the remembrance of them from the earth. When the righteous cry for help, the Lord hears, and delivers them out of all their distress and troubles"[15] (Psalm 34:15–17).

And, "fools make a mock of sin and sin mocks the fool (who are its victims; a sin offering made by them only mocks them, bringing them disappointment and disfavor), but <u>among the upright there is the favor of God</u>"[16] (Proverbs 14:9).

And, "For the ways of man are directly before the eyes of the Lord, and He (who would have you live soberly, chastely, and godly) carefully weighs all man's goings"[17] (Proverbs 5:21).

Jesus said, "The thief (the devil) comes only in order to steal, and kill, and destroy, I (Jesus) came that they may have and enjoy life and have it in abundance, (to the full 'till it overflows"[18] (John 10:10).

He is GOD.[19] He is all powerful.[20] He is all knowing.[21] He is everywhere,[22]

He will give you his unmerited favor[23] (He will give you His Grace) if you acknowledge Him and let Him be LORD of your life.[23]

However, He has said in His word, just believing that there is a God is not sufficient. He reminds us that, "even demons believe there is a God - but when *they* think about it they shake with fear"[24] (Mark 3:29).

He also reminds us in His Word that just because we choose to accept Him as Lord of our lives that we will not have troubles, but He does say that He will help us and comfort us and be with us in our times of trouble. "Trouble and anguish have found and taken hold on me, yet, your commands are my delight"[25] (Psalm 119:143). And, "Blessed be the God and Father of our Lord Jesus Christ, the Father of sympathy, pity, and mercy; and the God Who is the Source of every comfort, consolation, and encouragement; Who comforts, consoles, and encourages us in every trouble, calamity, and affliction"[26] (2 Corinthians 1:3-4).

When we have troubles, we need to remember, He is God. He is in control of all things and, as He says in Romans 8: 28, "We are assured and know that God being a partner in their labor, all things work together and are fitting into a plan for good, to and for those who love God and are called according to His design and purpose"[27] (Romans 8:28). There will be a day of "judgment and destruction of the ungodly people"[28] (2nd Peter 3:7) "Being destined to receive punishment as the reward of their unrighteousness"[29] (2nd Peter 2:13). Then He will say to those at his left hand, "Be gone from Me, you cursed, into the eternal fire prepared for the devil and his angels"[30] (Matthew 25:41). Jesus himself said, "for the time is coming when all those who are in the tombs and graves shall hear His voice, and they shall come out- those who have practiced doing good will come out to the resurrection of new life, and those who have done evil will be raised for judgement, raised to meet their sentence of judgement and damnation."[31,32] (John 5:28-29, and John 5:29)

Therefore, "choose now, this day, whom you will serve"[33] (Joshua 24:14). Ask Jesus to save you from your sins[34] (Matthew 1:21). Accept His free gift of salvation, mercy, grace, and forgiveness of your sins into fellowship with Him again, and eternal life with God - not eternal life in Hell with the Devil.

God loves you. Jesus lived a sinless life for you. Jesus died for you. He wants to live within you and have fellowship with you. Accept his free gift of being saved from your sins. Accept Jesus and ask Him to save you and live within you and guide and direct you. Let Him be <u>Lord</u> of your life. "Love God with all your heart, all your soul and all your mind"[35] (Matthew 22:37).

Solomon was King of Israel. (His father was David; the King of Israel who slew Goliath.) Solomon was David's son. Solomon was specially anointed by God to have Wisdom. As a result, Solomon was the wisest man to ever live. Solomon; the most wise man to ever live, the smartest man to ever live- after contemplating, searching, and testing the reason for his, and mankind's existence said, "All has been heard: *the end of the matter is*: Fear God [revere and worship Him, knowing that He is] and keep His commandments, for this is the whole of man [the full, original purpose of his creation, the object of God's providence, the root of character, the foundation of all happiness, the adjustment to all inharmonious circumstances and conditions under the sun] and the whole [duty] for every man"[36] (Ecclesiastes 12:13).

God has said, "For that which is known about God is evident to them and made plain in their inner consciousness, because God Himself has shown it to them. For ever since the creation of the world His invisible nature and attributes, that is His eternal power and divinity, have been made intelligible and clearly discernable in and through the things that have been made. So men are without excuse."[37] (Romans 1:19–20) Would God say you are without excuse if He had not more than adequately proven Himself to you?

Even though God has said you are without excuse in not believing in Him. ***God Himself*** has made the way to restore our relationship to Him. Even though we sin, even though we reject God, even though we don't want to be subject to a higher authority, God so loves us that God has reached out to all of mankind; God Himself has provided His Son, Jesus, as a sacrifice to pay for our sins[38,39] with John 3:36). Jesus loves you so much that He willingly gave His life as payment in full for your sins, But ***YOU*** must receive it. Listen and obey God, Jesus and the Holy Spirit. Accept God's free gift, and ask Him to save you and be Lord of your life[40] (Romans 10:13).

References:

Title Page p.i

Table of Contents p.iii

About the Author p.v

Introduction p. vii
1. Charles Darwin, *The Origin of Species* (New York, New York: Barnes and Noble Books, 2004) page 158
2. Charles Darwin, *The Origin of Species* (New York, New York: Barnes and Noble Books, 2004) page 380
3. Charles Darwin, *The Origin of Species* (New York, New York: Barnes and Noble Books, 2004) page 7, 8. Charles Darwin, *The Origin of a Species* (New York, New York: Barnes and Noble Books, 2004) page 145, 146
4. Charles Darwin, *The Origin of Species* (New York, New York: Barnes and Noble Books, 2004) page 380
5. Charles Darwin, *The Origin of Species* (New York, New York: Barnes and Noble Books, 2004) page 159
6. R. C. Sproul Sr. Renewing your mind radio program
7. Charles Darwin, *The Origin of Species* (New York, New York: Barnes and Noble Books, 2004) page 158

Purpose p. xiii
1. Charles Darwin, *The Origin of Species* (New York, New York: Barnes and Noble Books, 2004) page 159
2. Charles Darwin, *The Origin of Species* (New York, New York: Barnes and Noble Books, 2004) page 158
3. Charles Darwin, *The Origin of Species* (New York, New York: Barnes and Noble Books, 2004) page 159

Chapter 1 External Eye Muscles p.1   (108 I.E.C.)

1. Henry Gray, F.R.S, *Anatomy of the Human Body Twentieth Edition* Lea & Febiger (Philadelphia and New York © 1918)

2. Henry Gray, F.R.S, *Anatomy of the Human Body Twentieth Edition* Lea & Febiger (Philadelphia and New York © 1918) page 817

3. Henry Gray, F.R.S, *Anatomy of the Human Body Twentieth Edition* Lea & Febiger (Philadelphia and New York © 1918) p.186

4. Henry Gray, F.R.S, *Anatomy of the Human Body Twentieth Edition* Lea & Febiger (Philadelphia and New York © 1918) p.1027

5. Henry Gray, F.R.S, *Anatomy of the Human Body Twentieth Edition* Lea & Febiger (Philadelphia and New York © 1918) p.1022

6 Henry Gray, F.R.S, *Anatomy of the Human Body Twentieth Edition* Lea & Febiger (Philadelphia and New York © 1918) p.1023

7. Charles Darwin, *The Origin of Species* (New York, New York: Barnes and Noble Books, 2004) page 159

8. Henry Gray, F.R.S, *Anatomy of the Human Body Twentieth Edition* Lea & Febiger (Philadelphia and New York © 1918) p.1022

9. Henry Gray, F.R.S, *Anatomy of the Human Body Twentieth Edition* Lea & Febiger (Philadelphia and New York © 1918) p.1023

10. Charles Darwin, *The Origin of Species* (New York, New York: Barnes and Noble Books, 2004) page 159

11. Henry Gray, F.R.S, *Anatomy of the Human Body Twentieth Edition* Lea & Febiger (Philadelphia and New York © 1918) p.885

12. Raymond E. Records, *Physiology of the Human Eye and Visual System* (Hagerstown Maryland: Harper & Row Publishers © 1979) pages 579-586

13. Henry Gray, F.R.S, *Anatomy of the Human Body Twentieth Edition* Lea & Febiger (Philadelphia and New York © 1918) p.885

14. Charles Darwin, *The Origin of Species* (New York, New York: Barnes and Noble Books, 2004) page 158

15. Charles Darwin, *The Origin of Species* (New York, New York: Barnes and Noble Books, 2004) page 159

16 Henry Gray, F.R.S, *Anatomy of the Human Body Twentieth Edition* Lea & Febiger (Philadelphia and New York © 1918) p.885

17. Charles Darwin, *The Origin of Species* (New York, New York: Barnes and Noble Books, 2004) page 159

Chapter 2 Cornea p.21   (36 I.E.C.)

1.Charles Darwin, *The Origin of Species* (New York, New York: Barnes and Noble Books, 2004) page 159

2. Francis Heed Adler, *Adler's Physiology of the Eye, Clinical Application* Sixth Edition Edited by Robert A. Moses, M.D.(Saint Louis, Missouri: The C.V. Moseby Company,© 1975) page 39

3. Raymond E. Records, *Physiology of the Human Eye and Visual System* (Hagerstown Maryland: Harper & Row Publishers © 1979) page 68-74

4. Henry Gray, F.R.S, *Anatomy of the Human Body Twentieth Edition* Lea & Febiger (Philadelphia and New York © 1918) Page 1008 with modifications and additions by Rigney

5. Gretchyn M. Bailey, NCLC, FAAO Editor in Chief, Content Channel Director; *Optometry Times®*, Sixth layer to human cornea discovered:Optometry Times®: Jul 01, 2013

6.Francis Heed Adler, *Adler's Physiology of the Eye, Clinical Application* Sixth Edition Edited by Robert A. Moses, M.D.(Saint Louis, Missouri: The C.V. Moseby Company,
© 1975) page 46

7. Raymond E. Records, *Physiology of the Human Eye and Visual System* (Hagerstown Maryland: Harper & Row Publishers © 1979) page 68

8. Eugene Wolff, *Anatomy of the Eye and Orbit* Seventh Edition (Philadelphia and Toronto: W.B. Saunders Company, 1981) page 34

9.Charles Darwin, *The Origin of Species* (New York, New York: Barnes and Noble Books, 2004) page 159

10. Francis Heed Adler, *Adler's Physiology of the Eye, Clinical Application* Sixth Edition Edited by Robert A. Moses, M.D. (Saint Louis, Missouri: The C.V. Moseby Company,
© 1975) page 46

11. Raymond E. Records, *Physiology of the Human Eye and Visual System* (Hagerstown Maryland: Harper & Row Publishers © 1979) page 81

12. Charles Darwin, *The Origin of Species* (New York, New York: Barnes and Noble Books, 2004) page 159

13. Henry Gray, F.R.S, *Anatomy of the Human Body* 28th Edition, Lea & Febiger (Philadelphia, Pennsylvania, © 1966) p.690

14. Raymond E. Records, *Physiology of the Human Eye and Visual System* (Hagerstown Maryland: Harper & Row Publishers © 1979) page 81

15. Eugene Wolff, *Anatomy of the Eye and Orbit* Seventh Edition (Philadelphia and Toronto: W.B. Saunders Company, 1981) page 441

16. Encyclopædia Britannica, Inc. ©2014
http://www.britannica.com/EBchecked/topic/376720/mesoderm

17. Raymond E. Records, *Physiology of the Human Eye and Visual System* (Hagerstown Maryland: Harper & Row Publishers © 1979) page 68

18. Francis Heed Adler, *Adler's Physiology of the Eye, Clinical Application* Sixth Edition Edited by Robert A. Moses, M.D. (Saint Louis, Missouri: The C.V. Moseby Company,
© 1975) page 39

19. Merrill Grayson M.D., *Diseases of the Cornea* (Saint Louis, Missouri: The C.V. Moseby Company,© 1979) page 203

20. Eugene Wolff, *Anatomy of the Eye and Orbit* Seventh Edition (Philadelphia and Toronto: W.B. Saunders Company, 1981) page 37

21. Raymond E. Records, *Physiology of the Human Eye and Visual System* (Hagerstown Maryland: Harper & Row Publishers © 1979) page 72

22. Eugene Wolff, *Anatomy of the Eye and Orbit* Seventh Edition (Philadelphia and Toronto: W.B. Saunders Company, 1981) page 41

23. Eugene Wolff, *Anatomy of the Eye and Orbit* Seventh Edition (Philadelphia and Toronto: W.B. Saunders Company, 1981) page 36-41

24. David F. Holmes, Christopher J. Gilpin, Clair Baldock, Ulrike Ziese, Abraham J. Koster, and Karl E. Kadler: *Corneal collagen fibril structure in three dimensions: Structural insights into fibril assembly, mechanical properties, and tissue organization* (Proceedings of the national academy of Sciences of the United States of America) vol. 98 no.13, www.pnas.org: David F. Holmes, 7307–7312, doi:10.1073/pnas.111150598

25. Eugene Wolff, *Anatomy of the Eye and Orbit* Seventh Edition (Philadelphia and Toronto: W.B. Saunders Company, 1981) page 36

26. Raymond E. Records, *Physiology of the Human Eye and Visual System* (Hagerstown Maryland: Harper & Row Publishers © 1979) page 70,72

27. Timothy J. Freegard M.D. F.R.C.S.FRCOphth. Royal Eye Infirmary, U.K. *The Physical Basis of the transparency of the Normal Cornea* (Eye) 1997 11,465-471
© 1997 Royal College of Ophthalmologists

28. **Description:** English: Congenital stromal dystrophy. Transmission electron microscopy of the corneal stroma showing normal collagen lamellae separated by abnormal randomly distributed collagen filaments in an electron-lucent extracellular matrix (Reproduced with permission from Bredrup et al.[101]). Klintworth Orphanet. *Journal of Rare Diseases* 2009 4:7 doi:10.1186/1750-1172-4-7. Русский: Врождённая дистрофия стромы роговицы.
Date: 2008 **Source:** Corneal dystrophies:
https://upload.wikimedia.org/wikipedia/commons/b/b9/Congenital_stromal_dystrophy_2.jpg
Website: http://ojrd.biomedcentral.com/articles/10.1186/1750-1172-4-7
Author: Klintworth GK.
Permission: © 2009 Klintworth; licensee BioMed Central Ltd. This is an Open Access article distributed under the terms of the Creative Commons Attribution License (http://creativecommons.org/licenses/by/2.0), which permits unrestricted use, distribution, and reproduction in any medium, provided the original work is properly cited.

29. This information can be verified on the internet, however the source I attempted to use would not give me permission to reference their site so I had to obtain it elsewhere.

30. This information can be verified on the internet, however the source I attempted to use would not give me permission to reference their site so I had to obtain it elsewhere.

31. This information can be verified on the internet, however the source I attempted to use would not give me permission to reference their site so I had to obtain it elsewhere.

32. Francis Heed Adler, *Adler's Physiology of the Eye, Clinical Application* Sixth Edition Edited by Robert A. Moses, M.D.(Saint Louis, Missouri: The C.V. Moseby Company, © 1975) page 54

33. Raymond E. Records, *Physiology of the Human Eye and Visual System* (Hagerstown Maryland: Harper & Row Publishers © 1979) page 80

34. Maurice D.M.:The Structire and Transparency of the Cornea. J Physiology 136:263-286,1957 as cited reference #41, Chapter 4 Cornea and Sclera, Raymond E. Records, *Physiology of the Human Eye and Visual System* (Hagerstown Maryland: Harper & Row Publishers © 1979) page 80

35. Charles Darwin, *The Origin of Species* (New York, New York: Barnes and Noble Books, 2004) page 158

36. Charles Darwin, *The Origin of Species* (New York, New York: Barnes and Noble Books, 2004) page 159

37. Charles Darwin, *The Origin of Species* (New York, New York: Barnes and Noble Books, 2004) page 158

38. http://marinesciencetoday.com/wp-content/uploads/2009/05/pacific-halibut.jpg

39.http://upload.wikimedia.org/wikipedia/commons/a/a5/Dorsal_side_of_Pacific_halibut_head.jpg

40. Raymond E. Records, *Physiology of the Human Eye and Visual System* (Hagerstown Maryland: Harper & Row Publishers © 1979) page 666-668

41. Charles Darwin, *The Origin of Species* (New York, New York: Barnes and Noble Books, 2004) page 159

42. **Description:** English: Congenital stromal dystrophy. Transmission electron microscopy of the corneal stroma showing normal collagen lamellae separated by abnormal randomly distributed collagen filaments in an electron-lucent extracellular matrix (Reproduced with permission from Bredrup et al.[101]). Klintworth Orphanet. *Journal of Rare Diseases* 2009 4:7 doi:10.1186/1750-1172-4-7. Русский: Врождённая дистрофия стромы роговицы. Date: 2008 **Source:** Corneal dystrophies:
https://upload.wikimedia.org/wikipedia/commons/b/b9/Congenital_stromal_dystrophy_2.jpg
Website: http://ojrd.biomedcentral.com/articles/10.1186/1750-1172-4-7
Author: Klintworth GK.
Permission: © 2009 Klintworth; licensee BioMed Central Ltd. This is an Open Access article distributed under the terms of the Creative Commons Attribution License (http://creativecommons.org/licenses/by/2.0), which permits unrestricted use, distribution, and reproduction in any medium, provided the original work is properly cited.

43. Eugene Wolff, *Anatomy of the Eye and Orbit* Seventh Edition (Philadelphia and Toronto: W.B. Saunders Company, 1981) page 40

44. Eugene Wolff, *Anatomy of the Eye and Orbit* Seventh Edition (Philadelphia and Toronto: W.B. Saunders Company, 1981) page 43

45. Eugene Wolff, *Anatomy of the Eye and Orbit* Seventh Edition (Philadelphia and Toronto: W.B. Saunders Company, 1981) page 42

46. Written by The Editors of Encyclopædia Britannica, *Last Updated 8-7-2014* (All contents of the Services are © Encyclopædia Britannica, Inc. or its licensors. All rights reserved. Encyclopædia Britannica is copyrighted 1994-2014 by Encyclopædia Britannica, Inc.)

47. J. Grote, R. Zander: *Corneal Oxygen Supply Conditions: Oxygen Transport to Tissue — II* (Advances in Experimental Medicine and Biology): Volume 75, 1976, pp 449-455

48. http://www.usc.edu/dept/biomed/bme403/Section_1/fick.html

49. Chhabra, Mahendra*; Prausnitz, John M.*; Radke, Clayton J. (Modeling Corneal Metabolism and Oxygen Transport During Contact Lens Wear): Optometry & Vision Science: *May 2009 - Volume 86 - Issue 5 - pp 454-466*

50. Sho C. Takatori, Percy Lazon de la Jara, Brien Holden, Klaus Ehrmann, Arthur Ho and Clayton J. Radke *In Vivo Oxygen Uptake into the Human Cornea* (Investigative Ophthalmology and Visual Science) Published online before print July 26, 2012, doi:10.1167/iovs.12-10059 Invest. Ophthalmol. Vis. Sci.September 19, 2012 vol. 53 no. 10 6331-6337

51. Fatt I, Bieber MT. The steady-state distribution of oxygen and carbon dioxide in the in vivo cornea. I. The open eye in air and the closed eye. Exp Eye Res. 1968 Jan;7(1):103-12.

52. Efron N, Carney LG. *Oxygen levels beneath the closed eyelid* (Investigative Ophthalmology & Vision Science) 1979 Jan;18(1):93-5.

53. Charles Darwin, *The Origin of Species* (New York, New York: Barnes and Noble Books, 2004) page 158

54. Francis Heed Adler, *Adler's Physiology of the Eye, Clinical Application* Sixth Edition Edited by Robert A. Moses, M.D.(Saint Louis, Missouri: The C.V. Moseby Company,© 1975) page 38

55. Francis Heed Adler, *Adler's Physiology of the Eye, Clinical Application* Sixth Edition Edited by Robert A. Moses, M.D.(Saint Louis, Missouri: The C.V. Moseby Company,© 1975) page 518

56. Francis Heed Adler, *Adler's Physiology of the Eye, Clinical Application* Sixth Edition Edited by Robert A. Moses, M.D. (Saint Louis, Missouri: The C.V. Moseby Company,© 1975) page 38

57. Charles Darwin, *The Origin of Species* (New York, New York: Barnes and Noble Books, 2004) page 158

58. Charles Darwin, *The Origin of Species* (New York, New York: Barnes and Noble Books, 2004) page 159

59. Charles Darwin, *The Origin of Species* (New York, New York: Barnes and Noble Books, 2004) page 158

Chapter 3 Iris p.61   (62 I.E.C.)

1. Henry Gray, F.R.S, *Anatomy of the Human Body Twentieth Edition* Lea & Febiger (Philadelphia and New York © 1918) Page 1012 p.1013

2. Charles Darwin, *The Origin of Species* (New York, New York: Barnes and Noble Books, 2004) page 159

3. Henry Gray, F.R.S, *Anatomy of the Human Body Twentieth Edition* Lea & Febiger (Philadelphia and New York © 1918) Page 968- p.972

4. Henry Gray, F.R.S, *Anatomy of the Human Body* 28th Edition, Lea & Febiger (Philadelphia, Pennsylvania, © 1966) p.1016

5. Henry Gray, F.R.S, *Anatomy of the Human Body* 28th Edition, Lea & Febiger (Philadelphia, Pennsylvania, © 1966) p.1012 & 1016

6. Netter, Frank H., M.D. *Ciba Collection of Medical Illustrations Vol 1*, Viba Pharmaceutical Co. (U.S.A.) © 1972) p.92

7. Henry Gray, F.R.S, *Anatomy of the Human Body* 28th Edition, Lea & Febiger (Philadelphia, Pennsylvania, © 1966) p.917

8. Henry Gray, F.R.S, *Anatomy of the Human Body Twentieth Edition* Lea & Febiger (Philadelphia and New York © 1918) Page 972

9. Henry Gray, F.R.S, *Anatomy of the Human Body Twentieth Edition* Lea & Febiger (Philadelphia and New York © 1918) Pages 776, 1021, 541, 887 combined by Rigney

10. Charles Darwin, *The Origin of Species* (New York, New York: Barnes and Noble Books, 2004) page 159

11. Henry Gray, F.R.S, *Anatomy of the Human Body* 28th Edition, Lea & Febiger (Philadelphia, Pennsylvania, © 1966) p.1012 & p.1016

12. Charles Darwin, *The Origin of Species* (New York, New York: Barnes and Noble Books, 2004) page 159

13. Henry Gray, F.R.S, *Anatomy of the Human Body Twentieth Edition* Lea & Febiger (Philadelphia and New York © 1918) Pages 566, 569, 527, 766 combined by Rigney

14. Henry Gray, F.R.S, *Anatomy of the Human Body Twentieth Edition* Lea & Febiger (Philadelphia and New York © 1918) Page 1013

15. Charles Darwin, *The Origin of Species* (New York, New York: Barnes and Noble Books, 2004) page 159

Chapter 4 Aqueous p.73   (32 I.E.C.)

1. Henry Gray, F.R.S, *Anatomy of the Human Body Twentieth Edition* Lea & Febiger (Philadelphia and New York © 1918) Page 1006

2. Henry Gray, F.R.S, *Anatomy of the Human Body Twentieth Edition* Lea & Febiger (Philadelphia and New York © 1918) Page 1021

3. Henry Gray, F.R.S, *Anatomy of the Human Body Twentieth Edition* Lea & Febiger (Philadelphia and New York © 1918) Page 1012

Chapter 5 Aqueous Flow p.77   (I.E.C. listed in Chapter 4)

1. Charles Darwin, *The Origin of Species* (New York, New York: Barnes and Noble Books, 2004) page 159

2. Henry Gray, F.R.S, *Anatomy of the Human Body* 20th Edition, Lea & Febiger (Philadelphia & New York, © 1918) Page 1012[2] with additions by Rigney

Chapter 6 Trabecular Meshwork p.79
(I.E.C. listed in Chapter 4)

1. Henry Gray, F.R.S, *Anatomy of the Human Body* 28th Edition, Lea & Febiger (Philadelphia, Pennsylvania, © 1966) page 1063

2. Francois, J. (1948) Bull. Soc. Belg ophthal.,#, 55

3. Leber, T. after Maggiore, L. 1917; -24) Ann. Ottalm., 40, 317;52, 625

4. Rochon-Duvigneaud, A. (1903) Encycl. Franc. d'Ophtal., 1, 369

5. Henry Gray, F.R.S, *Anatomy of the Human Body* 20th Edition, Lea & Febiger (Philadelphia & New York, © 1918) Page 1019 with additions by Rigney

6. Henry Gray, F.R.S, *Anatomy of the Human Body* 20th Edition, Lea & Febiger (Philadelphia & New York, © 1918) Page 1011 with additions by Rigney

7. Henry Gray, F.R.S, *Anatomy of the Human Body* 20th Edition, Lea & Febiger (Philadelphia & New York, © 1918) Page 1007 with additions by Rigney

8. Charles Darwin, *The Origin of Species* (New York, New York: Barnes and Noble Books, 2004) page 159

Chapter 7 Eyelids p.87   (11 I.E.C.)

1. Henry Gray, F.R.S, *Anatomy of the Human Body* 20th Edition, Lea & Febiger (Philadelphia & New York © 1918) p.379

2. J. G. Chusid, Correlative Neuroanatomy &Functional Neurology 17th edition, Lange (Los Altos, California, © 1979) p.99

3. J. G. Chusid, Correlative Neuroanatomy &Functional Neurology 17th edition, Lange (Los Altos, California, © 1979) p.94

4. Francis Heed Adler, *Adler's Physiology of the Eye, Clinical Application* Sixth Edition Edited by Robert A. Moses, M.D, (Saint Louis, Missouri: The C.V. Moseby Company,
© 1975) page 15

5 Henry Gray, F.R.S, *Anatomy of the Human Body* 20th Edition, Lea & Febiger (Philadelphia & New York © 1918) p.1027

6. Charles Darwin, *The Origin of Species* (New York, New York: Barnes and Noble Books, 2004) page 158, p159

7. Henry Gray, F.R.S, *Anatomy of the Human Body* 20th Edition, Lea & Febiger (Philadelphia & New York © 1918) p.1021

8. Francis Heed Adler, *Adler's Physiology of the Eye, Clinical Application* Sixth Edition Edited by Robert A. Moses, M.D. (Saint Louis, Missouri: The C.V. Moseby Company,
© 1975) page 19

9. Mark B. Abelson, MD, CM, FRCSC, FARVO, Darlene Dartt, PhD, and James McLaughlin, PhD, Andover, Mass., Mucins: Foundation of A Good Tear Film,
Review of Ophthalmology November 7, 2011

10. Henry Gray, F.R.S, *Anatomy of the Human Body* 20th Edition, Lea & Febiger (Philadelphia & New York © 1918) p.1027

Chapter 8 Tears p.93   (13 I.E.C.)

1. Francis Heed Adler, *Adler's Physiology of the Eye, Clinical Application* Sixth Edition Edited by Robert A. Moses, M.D. (Saint Louis, Missouri: The C.V. Moseby Company, © 1975) page 18, 21, 22
2 Henry Gray, F.R.S, *Anatomy of the Human Body* 20th Edition, Lea & Febiger (Philadelphia & New York © 1918) p.1027 & p.1029
3 Henry Gray, F.R.S, *Anatomy of the Human Body* 20th Edition, Lea & Febiger (Philadelphia & New York © 1918) p.885
4. Francis Heed Adler, *Adler's Physiology of the Eye, Clinical Application* Sixth Edition Edited by Robert A. Moses, M.D.(Saint Louis, Missouri: The C.V. Moseby Company, © 1975) page 18, 21, 22
5. Henry Gray, F.R.S, *Anatomy of the Human Body* 20th Edition, Lea & Febiger (Philadelphia & New York © 1918) p.887
6. Charles Darwin, *The Origin of Species* (New York, New York: Barnes and Noble Books, 2004) page 159
7. Charles Darwin, *The Origin of Species* (New York, New York: Barnes and Noble Books, 2004) page 158
8. Charles Darwin, *The Origin of Species* (New York, New York: Barnes and Noble Books, 2004) page 159

Chapter 9 Tear Flow p.97   (17 I.E.C.)

1. Henry Gray, F.R.S, *Anatomy of the Human Body* 28th Edition, Lea & Febiger (Philadelphia, Pennsylvania, © 1966) page 1070
2. Henry Gray, F.R.S, *Anatomy of the Human Body* 20th Edition, Lea & Febiger (Philadelphia & New York, © 1918) Page 186
3 Henry Gray, F.R.S, *Anatomy of the Human Body* 20th Edition, Lea & Febiger (Philadelphia & New York, © 1918) Page 1026 & p.1027
4. Charles Darwin, *The Origin of Species* (New York, New York: Barnes and Noble Books, 2004) page 159
5. Charles Darwin, *The Origin of Species* (New York, New York: Barnes and Noble Books, 2004) page 158
6. Charles Darwin, *The Origin of Species* (New York, New York: Barnes and Noble Books, 2004) page 159

Chapter 10 Lens p.101   (I.E.C. Listed in Chapter 11)

1. Henry Gray, F.R.S, *Anatomy of the Human Body* 20th Edition, Lea & Febiger (Philadelphia &New York, © 1918) Page 1006

Chapter 11 Focusing Control p.103   (9 I.E.C.)

1. Henry Gray, F.R.S, *Anatomy of the Human Body* 20th Edition, Lea & Febiger (Philadelphia &New York, © 1918) Page 1006

Chapter 12 Bones of the Eye p.107   (7 I.E.C.)

1 Henry Gray, F.R.S, *Anatomy of the Human Body* 20th Edition, Lea & Febiger (Philadelphia & New York, © 1918) Page 186

2 Henry Gray, F.R.S, *Anatomy of the Human Body* 20th Edition, Lea & Febiger (Philadelphia & New York, © 1918) Page 160

3. Henry Gray, F.R.S, *Anatomy of the Human Body* 20th Edition, Lea & Febiger (Philadelphia & New York, © 1918) Page 187

4. Henry Gray, F.R.S, *Anatomy of the Human Body* 20th Edition, Lea & Febiger (Philadelphia & New York, © 1918) Page 1022

5. Henry Gray, F.R.S, *Anatomy of the Human Body* 20th Edition, Lea & Febiger (Philadelphia & New York, © 1918) Page 1027

Chapter 13 Foramen p.113   (39 I.E.C.)

1. Charles Darwin, *The Origin of Species* (New York, New York: Barnes and Noble Books, 2004) page 159

2. Henry Gray, F.R.S, *Anatomy of the Human Body* 20th Edition, Lea & Febiger (Philadelphia & New York, © 1918) Page 186

3. Charles Darwin, *The Origin of Species* (New York, New York: Barnes and Noble Books, 2004) page 159

4. Henry Gray, F.R.S, *Anatomy of the Human Body* 20th Edition, Lea & Febiger (Philadelphia & New York, © 1918) Page 817

5. Henry Gray, F.R.S, *Anatomy of the Human Body* 20th Edition, Lea & Febiger (Philadelphia & New York, © 1918) Page 186

6. Henry Gray, F.R.S, *Anatomy of the Human Body* 20th Edition, Lea & Febiger (Philadelphia & New York, © 1918) Page 191

7. Henry Gray, F.R.S, *Anatomy of the Human Body* 20th Edition, Lea & Febiger (Philadelphia & New York, © 1918) Page 1026, p.1027, p.1029

8. Charles Darwin, *The Origin of Species* (New York, New York: Barnes and Noble Books, 2004) page 159

9. Henry Gray, F.R.S, *Anatomy of the Human Body* 20th Edition, Lea & Febiger (Philadelphia & New York, © 1918) Page 186, p.1026, p. 1027 & p.1029

Chapter 14 Vitreous p.123   (3 I.E.C.)

1. Henry Gray, F.R.S, *Anatomy of the Human Body* 20th Edition, Lea & Febiger (Philadelphia & New York, © 1918) Page 1006

2 Henry Gray, F.R.S, *Anatomy of the Human Body* 20th Edition, Lea & Febiger (Philadelphia & New York, © 1918) Page 1006

Chapter 15 Sclera p.125   (I.E.C. Listed in Vitreous)

1. Henry Gray, F.R.S, *Anatomy of the Human Body* 28th Edition, Lea & Febiger (Philadelphia, Pennsylvania, © 1966) p.1051

2. Henry Gray, F.R.S, *Anatomy of the Human Body* 20th Edition, Lea & Febiger (Philadelphia & New York, © 1918) Page 1006

3.Henry Gray, F.R.S, *Anatomy of the Human Body* 20th Edition, Lea & Febiger (Philadelphia & New York, © 1918) Page 1012

4 Henry Gray, F.R.S, *Anatomy of the Human Body* 20th Edition, Lea & Febiger (Philadelphia & New York, © 1918) Page 1021

5. Henry Gray, F.R.S, *Anatomy of the Human Body* 20th Edition, Lea & Febiger (Philadelphia & New York, © 1918) Page 1009

6. Charles Darwin, *The Origin of Species* (New York, New York: Barnes and Noble Books, 2004) page 159

7. Henry Gray, F.R.S, *Anatomy of the Human Body* 20th Edition, Lea & Febiger (Philadelphia & New York, © 1918) Page 1006

8. Francis Heed Adler, *Adler's Physiology of the Eye, Clinical Application* Sixth Edition Edited by Robert A. Moses, M.D. (Saint Louis, Missouri: The C.V. Moseby Company, © 1975) page 518

Chapter 16 Retina p.131   (111 I.E.C.)

1. Henry Gray, F.R.S, *Anatomy of the Human Body* 28th Edition, Lea & Febiger (Philadelphia, Pennsylvania, © 1966) p.1060

2. Raymond E. Records, *Physiology of the Human Eye and Visual System* (Hagerstown Maryland: Harper & Row Publishers © 1979) p.312

3. Charles Darwin, *The Origin of Species* (New York, New York: Barnes and Noble Books, 2004) page 159

4. Charles Darwin, *The Origin of Species* (New York, New York: Barnes and Noble Books, 2004) page 158

5. Henry Gray, F.R.S, *Anatomy of the Human Body* 20th Edition, Lea & Febiger (Philadelphia & New York, © 1918) Page 1016

6. R. Greeff (1900) Handbuch der gesamten Augenheilkunde, 2nd ed, vol.1., Graefe and Saemisch, Leipzig.

7. By OpenStax College [CC BY 3.0 (http://creativecommons.org/licenses/by/3.0)], via Wikimedi Commons.File:1414 Rods and Cones.jpg (File:1414 Rods and Cones - ru.svg) via Wikimedi Commons.File:1414 Rods and Cones.jpg (File:1414 Rods and Cones - ru.svg)[7] (They do not endorse me, or my use of their work)

8. R. Greeff (1900) Handbuch der gesamten Augenheilkunde, 2nd ed, vol.1., Graefe and Saemisch, Leipzig; additions by J. Jay Rigney, *Darwin's Challenge Answered* p.134

9. Henry Gray, F.R.S, *Anatomy of the Human Body* 20th Edition, Lea & Febiger (Philadelphia & New York, © 1918) Page 1016

10. Raymond E. Records, *Physiology of the Human Eye and Visual System* (Hagerstown Maryland: Harper & Row Publishers © 1979) p.312-315

11. Raymond E. Records, *Physiology of the Human Eye and Visual System* (Hagerstown Maryland: Harper & Row Publishers © 1979) p.307

12. Raymond E. Records, *Physiology of the Human Eye and Visual System* (Hagerstown Maryland: Harper & Row Publishers © 1979) p.311

13. Henry Gray, F.R.S, *Anatomy of the Human Body* 20th Edition, Lea & Febiger (Philadelphia & New York, © 1918) Page 1016

14. Raymond E. Records, *Physiology of the Human Eye and Visual System* (Hagerstown Maryland: Harper & Row Publishers © 1979) p.305

15. Raymond E. Records, *Physiology of the Human Eye and Visual System* (Hagerstown Maryland: Harper & Row Publishers © 1979) p.323

16. Raymond E. Records, *Physiology of the Human Eye and Visual System* (Hagerstown Maryland: Harper & Row Publishers © 1979) p.303, 536

17. Charles Darwin, *The Origin of Species* (New York, New York: Barnes and Noble Books, 2004) page 159

18. By OpenStax College [CC BY 3.0 (http://creativecommons.org/licenses/by/3.0)], via Wikimedi Commons. File:1414 Rods and Cones.jpg (File:1414 Rods and Cones - ru.svg) (http://creativecommons.org/licenses/by/3.0)], via Wikimedi Commons. File:1414 Rods and Cones.jpg (File:1414 Rods and Cones - ru.svg)

19. The use of the term additional "functional retinal cells" and "function and function automatically" was made by Rigney and not Wolff and not Records. J. Jay Rigney *Darwin's Challenge Answered* © 2016) p.138; Rigney; not Wolff, not Records

20. Eugene Wolff, *Anatomy of the Eye and Orbit* Seventh Edition (Philadelphia and Toronto: W.B. Saunders Company, 1981) pages 118-128

21. Dowling and Boycott, 1966. Cajal (1896) and Polyak (1941). Eugene Wolff, *Anatomy of the Eye and Orbit* Seventh Edition (Philadelphia and Toronto: W.B. Saunders Company, 1981) pages 118-128

22. 506 million is based upon the fact that there are 253 million photoreceptors (rods and cones) Rigney took that number, 253 million, along with the fact that there appears approximately 17 rods and cones and 30 additional "functional" cells in Henry Gray, F.R.S, *Anatomy of the Human Body* 20th Edition, Lea & Febiger (Philadelphia & New York, © 1918) illustration on page 1016. (Reference number 13 Chapter 16 on page 136 of this book *Darwin's Challenge Answered* © 2016 by Rigney). That is a ratio of approximately 2:1; therefore, there appears to be approximately twice as many functional cells of the retina as there are rods and cones.

23. Eugene Wolff, *Anatomy of the Eye and Orbit* Seventh Edition (Philadelphia and Toronto: W.B. Saunders Company, 1981) page 121

24. The use of the term "functional retinal cells" and "function and function automatically" was made by Rigney, not Wolff, and not Records. J. Jay Rigney *Darwin's Challenge Answered* © 2016, p.138-141; Rigney, not Wolff; Rigney, not Records

25. Eugene Wolff, *Anatomy of the Eye and Orbit* Seventh Edition (Philadelphia and Toronto: W.B. Saunders Company, 1981) p.121

26. Eugene Wolff, *Anatomy of the Eye and Orbit* Seventh Edition (Philadelphia and Toronto: W.B. Saunders Company, 1981) p.325

27. Eugene Wolff, *Anatomy of the Eye and Orbit* Seventh Edition (Philadelphia and Toronto: W.B. Saunders Company, 1981) p.341

28. Eugene Wolff, *Anatomy of the Eye and Orbit* Seventh Edition (Philadelphia and Toronto: W.B. Saunders Company, 1981) p.102

29. Eugene Wolff, *Anatomy of the Eye and Orbit* Seventh Edition (Philadelphia and Toronto: W.B. Saunders Company, 1981) p. 341

30. J. Jay Rigney *Darwin's Challenge Answered* © 2016) p.137-147; Rigney and not Wolff; Rigney, and not Records.

31. Eugene Wolff, *Anatomy of the Eye and Orbit* Seventh Edition (Philadelphia and Toronto: W.B. Saunders Company, 1981) p.119-128

32. J. Jay Rigney, *Darwin's Challenge Answered* ©2016 p.137-143; Rigney not Wolff; Rigney not Records.

33. Henry Gray, F.R.S, *Anatomy of the Human Body* 28th Edition, Lea & Febiger (Philadelphia, Pennsylvania, © 1966) p.812

34. Henry Gray, F.R.S, *Anatomy of the Human Body* 28th Edition, Lea & Febiger (Philadelphia, Pennsylvania, © 1966) p.806 -846

35. J.G. Chusid, *Correlative Neuroanatomy & Functional Neurology* 17th Edition (Los Altos, California 1979) Page 88 & 90 and Pages 84-110

36. J. Jay Rigney, *Darwin's Challenge Answered* © 2016 p.145, The use of the term "survival by design not survival of the fittest." was made by Rigney and not Wolff, not Chusid, not Records, not Darwin.

37. Henry Gray, F.R.S, *Anatomy of the Human Body* 20th Edition, Lea & Febiger (Philadelphia & New York, © 1918) Page 1016

38 Henry Gray, F.R.S, *Anatomy of the Human Body* 20th Edition, Lea & Febiger (Philadelphia & New York, © 1918) Page 781

39. Art explosion 525,000 purchased by J. Jay Rigney, O.D.

40. Art explosion 525,000 purchased by J. Jay Rigney, O.D.

41. Charles Darwin, *The Origin of Species* (New York, New York: Barnes and Noble Books, 2004) page 158

42. Charles Darwin, *The Origin of Species* (New York, New York: Barnes and Noble Books, 2004) page 380

43. Charles Darwin, *The Origin of Species* (New York, New York: Barnes and Noble Books, 2004) page 380

44. Charles Darwin, *The Origin of Species* (New York, New York: Barnes and Noble Books, 2004) page 156

45. Charles Darwin, *The Origin of Species* (New York, New York: Barnes and Noble Books, 2004) page 159

46. Charles Darwin, *The Origin of Species* (New York, New York: Barnes and Noble Books, 2004) page 380

47. Charles Darwin, *The Origin of Species* (New York, New York: Barnes and Noble Books, 2004) page 156

48. Charles Darwin, *The Origin of Species* (New York, New York: Barnes and Noble Books, 2004) page 159

49. Charles Darwin, *The Origin of Species* (New York, New York: Barnes and Noble Books, 2004) page 158

50. Charles Darwin, *The Origin of Species* (New York, New York: Barnes and Noble Books, 2004) page 380

51. Charles Darwin, *The Origin of Species* (New York, New York: Barnes and Noble Books, 2004) page 158

52. Charles Darwin, *The Origin of Species* (New York, New York: Barnes and Noble Books, 2004) page 156-172

53. Charles Darwin, *The Origin of Species* (New York, New York: Barnes and Noble Books, 2004) page 380

54. Charles Darwin, *The Origin of Species* (New York, New York: Barnes and Noble Books, 2004) page 158

55. Charles Darwin, *The Origin of Species* (New York, New York: Barnes and Noble Books, 2004) page 380

56. Charles Darwin, *The Origin of Species* (New York, New York: Barnes and Noble Books, 2004) page 158

57. Charles Darwin, *The Origin of Species* (New York, New York: Barnes and Noble Books, 2004) page 159

58. Charles Darwin, *The Origin of Species* (New York, New York: Barnes and Noble Books, 2004) page 158

Chapter 17 Optic Nerve p.155   (6 I.E.C.)

1. Eugene Wolff, *Anatomy of the Eye and Orbit* Seventh Edition (Philadelphia and Toronto: W.B. Saunders Company, 1981) page 341

2.Kupfer, C., Chumbley, L. and Donner, J. (1967), Journal Anataomy, 101, 393

3. Eugene Wolff, *Anatomy of the Eye and Orbit* Seventh Edition (Philadelphia and Toronto: W.B. Saunders Company, 1981) page 325

4. Charles Darwin, *The Origin of Species* (New York, New York: Barnes and Noble Books, 2004) page 157

5. Charles Darwin, *The Origin of Species* (New York, New York: Barnes and Noble Books, 2004) page 159

6. Henry Gray, F.R.S, *Anatomy of the Human Body* 28th Edition, Lea & Febiger (Philadelphia, Pennsylvania, © 1966) p.870, 871

7. Eugene Wolff, *Anatomy of the Eye and Orbit* Seventh Edition (Philadelphia and Toronto: W.B. Saunders Company, 1981) page 380

8.Les Voies Optiques:
http://www.campusdanatomie.org/sites/default/files/users/admin/voiesoptiques_sans_legende.pdf
https://creativecommons.org/licenses/by-sa/3.0

9. Charles Darwin, *The Origin of Species* (New York, New York: Barnes and Noble Books, 2004) page 159

Chapter 18 Visual Cortex p.163   (24 I.E.C.)

1.Eugene Wolff, *Anatomy of the Eye and Orbit* Seventh Edition (Philadelphia and Toronto: W.B. Saunders Company, 1981) page 341

2. Charles Darwin, *The Origin of Species* (New York, New York: Barnes and Noble Books, 2004) page 159

3. Eugene Wolff, *Anatomy of the Eye and Orbit* Seventh Edition (Philadelphia and Toronto: W.B. Saunders Company, 1981) page 325

4.Eugene Wolff, *Anatomy of the Eye and Orbit* Seventh Edition (Philadelphia and Toronto: W.B. Saunders Company, 1981) page 341

5. Polyak, S. L. (1941) *The Retina,* Chicago (1957) *The Vertebrate Visual System*, Chicago

6. Oppel, O. (1963) *ALlbrect v. Graefes Archive of Ophthalmology*, 166, 19

7.Eugene Wolff, *Anatomy of the Eye and Orbit* Seventh Edition (Philadelphia and Toronto: W.B. Saunders Company, 1981) page 375

8. Charles Darwin, *The Origin of Species* (New York, New York: Barnes and Noble Books, 2004) page 158

9. Charles Darwin, *On the Origin of Species* (An Electronic Classics Series Publication, New York, New York: Barnes and Noble Books, 2004) page 429

10.David Hambling, February 2007, *Questioning perceptual blindness: I see no ships European* explorers found indigenous peoples unable to see their tallships – or did they? http:www.forteantimes.com/strangedays/science/20/questioning_perceptual_blindness.html

11. Charles Darwin, *The Origin of Species* (New York, New York: Barnes and Noble Books, 2004) page 157

12. Charles Darwin, *On the Origin of Species* (An Electronic Classics Series Publication, New York, New York: Barnes and Noble Books, 2004) page 429

13.David Hambling, February 2007, *Questioning perceptual blindness: I see no ships European* explorers found indigenous peoples unable to see their tallships – or did they? http:www.forteantimes.com/strangedays/science/20/questioning_perceptual_blindness.html

14. Charles Darwin, *On the Origin of Species* (An Electronic Classics Series Publication, New York, New York: Barnes and Noble Books, 2004) page 429

15.Eugene Wolff, *Anatomy of the Eye and Orbit* Seventh Edition (Philadelphia and Toronto: W.B. Saunders Company, 1981) page 375

16. The Zondervan Corporation, The Amplified Bible, Expanded Edition, Romans 1:19 (La Habra, California: The Zondervan Corporation and the Lochman Corporation, 1965) page 1298. Scripture taken from THE AMPLIFIED BIBLE, Old Testament copyright© 1965, 1987 by the Zondervan Corporation. The Amplified New Testament copyright ©1958, 1987 by the Lochman Foundation. Used by permission.

Chapter 19 Nerves of the Eye p.171 (60 I.E.C.)

1. Henry Gray, F.R.S, *Anatomy of the Human Body* 20th Edition, Lea & Febiger (Philadelphia & New York, © 1918) p.817

2. J.G. Chusid, *Correlative Neuroanatomy & Functional Neurology* 17th Edition (Los Altos, California 1979) Page 84-110

3. The Zondervan Corporation, The Amplified Bible, Expanded Edition, Proverbs 3:5-7 p.703 (La Habra, California: The Zondervan Corporation and the Lochman Corporation, 1965) page 702. Scripture taken from THE AMPLIFIED BIBLE, Old Testament copyright© 1965, 1987 by the Zondervan Corporation. The Amplified New Testament copyright ©1958, 1987 by the Lochman Foundation. Used by permission.

4. The Zondervan Corporation, The Amplified Bible, Expanded Edition, John 10:10 p.1227 (La Habra, California: The Zondervan Corporation and the Lochman Corporation, 1965) Scripture taken from THE AMPLIFIED BIBLE, Old Testament copyright© 1965, 1987 by the Zondervan Corporation. The Amplified New Testament copyright ©1958, 1987 by the Lochman Foundation. Used by permission.

5. Eugene Wolff, *Anatomy of the Eye and Orbit* Seventh Edition (Philadelphia and Toronto: W.B. Saunders Company, 1981) page 275-324

6. J.G. Chusid, *Correlative Neuroanatomy & Functional Neurology* 17th Edition (Los Altos, California 1979) Page 84-110

7. Eugene Wolff, *Anatomy of the Eye and Orbit* Seventh Edition (Philadelphia and Toronto: W.B. Saunders Company, 1981) page 316

8. Henry Gray, F.R.S, *Anatomy of the Human Body* 20th Edition, Lea & Febiger (Philadelphia & New York, © 1918) p.887

9. J.G. Chusid, *Correlative Neuroanatomy & Functional Neurology* 17th Edition (Los Altos, California 1979) Page 98, 99

10. Eugene Wolff, *Anatomy of the Eye and Orbit* Seventh Edition (Philadelphia and Toronto: W.B. Saunders Company, 1981) page 319

11. J.G. Chusid, *Correlative Neuroanatomy & Functional Neurology* 17th Edition (Los Altos, California 1979) Page 101, Pages 101-102

12. J.G. Chusid, *Correlative Neuroanatomy & Functional Neurology* 17th Edition (Los Altos, California 1979) Page 103

13. J.G. Chusid, *Correlative Neuroanatomy & Functional Neurology* 17th Edition (Los Altos, California 1979) Page 97

14. J.G. Chusid, *Correlative Neuroanatomy & Functional Neurology* 17th Edition (Los Altos, California 1979) Page 98

15. Eugene Wolff, *Anatomy of the Eye and Orbit* Seventh Edition (Philadelphia and Toronto: W.B. Saunders Company, 1981) page 316, & p.321

16. Eugene Wolff, *Anatomy of the Eye and Orbit* Seventh Edition (Philadelphia and Toronto: W.B. Saunders Company, 1981) page 316

17. J.G. Chusid, *Correlative Neuroanatomy & Functional Neurology* 17th Edition (Los Altos, California 1979) Page 98,103,105 &106

18. Frank H. Netter, M.D., *Ciba Collection of Medical Illustrations Volume 1* Eleventh Printing (Summit, New Jersey: Case-Hoyt Corp., Rochester, New York 1972) Volume 1 p.84

19. J.G. Chusid, *Correlative Neuroanatomy & Functional Neurology* 17th Edition (Los Altos, California 1979) Page 106,107

20. Frank H. Netter, M.D., *Ciba Collection of Medical Illustrations Volume 1* Eleventh Printing (Summit, New Jersey: Case-Hoyt Corp., Rochester, New York 1972) Volume 1 p.42 &43

21. Frank H. Netter, M.D., *Ciba Collection of Medical Illustrations Volume 1* Eleventh Printing (Summit, New Jersey: Case-Hoyt Corp., Rochester, New York 1972) Volume 1 p.84

22. J.G. Chusid, *Correlative Neuroanatomy & Functional Neurology* 17th Edition (Los Altos, California 1979) Page 108

23. Frank H. Netter, M.D., *Ciba Collection of Medical Illustrations Volume 1* Eleventh Printing (Summit, New Jersey: Case-Hoyt Corp., Rochester, New York 1972) Volume 1 p.84

24. Henry Gray, F.R.S, *Anatomy of the Human Body* 20th Edition, Lea & Febiger (Philadelphia & New York, © 1918) Pages 881, 1006, 1022, 890, 898, 1022, 927, 379, 1034, 1048, 909, 527, 1165, 1034, and 379 Combined and arranged by Rigney

25. Henry Gray, F.R.S, *Anatomy of the Human Body* 20th Edition, Lea & Febiger (Philadelphia & New York, © 1918) Page 817

26. Henry Gray, F.R.S, *Anatomy of the Human Body* 20th Edition, Lea & Febiger (Philadelphia & New York, © 1918) Page 1027

27. Henry Gray, F.R.S, *Anatomy of the Human Body* 20th Edition, Lea & Febiger (Philadelphia & New York, © 1918) Page 885

28. Henry Gray, F.R.S, *Anatomy of the Human Body* 20th Edition, Lea & Febiger (Philadelphia & New York, © 1918) Page 887

29. Charles Darwin, *The Origin of Species* (New York, New York: Barnes and Noble Books, 2004) page 159

30. Henry Gray, F.R.S, *Anatomy of the Human Body* 20th Edition, Lea & Febiger (Philadelphia & New York, © 1918) Page 900

31. Charles Darwin, *The Origin of Species* (New York, New York: Barnes and Noble Books, 2004) page 159

Chapter 20 Blood Supply to the Eye p.185   (56 I.E.C.)

1.Nova *Cut to the Heart:* (Amazing Heart Facts)
http://www.pbs.org/wgbh/nova/heart/heartfacts.html © 1997WGBH

2. http://learn.fi.edu/learn/heart/vessels/vessels.html *The Human Heart* Blood Vessels © 1996-2014 The Franklin Institute, All Rights Reserved. webteam@www.fi.edu

3. Mark L. Entman, M.D. Last Updated 9-4-2014 (Encyclopædia Britannica) Human cardiovascular system ©2014 Encyclopædia Britannica, Inc. http://www.britannica.com/EBchecked/topic/95628/human-cardiovascular-system/33570/The-arteries

4. Eugene Wolff, *Anatomy of the Eye and Orbit* Seventh Edition (Philadelphia and Toronto: W.B. Saunders Company, 1981) page 406,407

5. Henry Gray, F.R.S, *Anatomy of the Human Body* 20th Edition, Lea & Febiger (Philadelphia & New York, © 1918) Pages 566, 885 combined with modifications by Rigney

6. Henry Gray, F.R.S, *Anatomy of the Human Body* 20th Edition, Lea & Febiger (Philadelphia & New York, © 1918) Pages 572, 574

7. From Henry Gray, F.R.S, *Anatomy of the Human Body* 20th Edition, Lea & Febiger (Philadelphia & New York © 1918) Pages 572

8. From Henry Gray, F.R.S, *Anatomy of the Human Body* 20th Edition, Lea & Febiger (Philadelphia & New York © 1918) Pages 1009

9. From Henry Gray, F.R.S, *Anatomy of the Human Body* 20th Edition, Lea & Febiger (Philadelphia & New York © 1918) Pages 1010

10. From Henry Gray, F.R.S, *Anatomy of the Human Body* 20th Edition, Lea & Febiger (Philadelphia & New York © 1918) Pages 1012

11. From Henry Gray, F.R.S, *Anatomy of the Human Body* 20th Edition, Lea & Febiger (Philadelphia & New York © 1918) Pages 644

12. From Henry Gray, F.R.S, *Anatomy of the Human Body* 20th Edition, Lea & Febiger (Philadelphia & New York © 1918) Pages 659

13. Charles Darwin, *The Origin of Species* (New York, New York: Barnes and Noble Books, 2004) page 159

14. Dr. R.C. Sproul, Renewing Your Mind Radio Program. Ligonier Ministries http://www.ligonier.org/

15. Dr. Stephen C. Meyer, *The Case for the Creator by Lee Strobel* Illustra Medea, http://www.thecaseforacreator.com/scientists.php

Chapter 21 Additional Scientific Evidence of Creation (p. 197)

1. Charles Darwin, *Voyage of the Beagle* (London, England: Pinguin Books, 1989) page x

2. Charles Darwin, *The Origin of Species* (New York, New York: Barnes and Noble Books, 2004) page x

3. (http://cosmology.carnegiescience.edu/timeline/1929)

4. (http://www.space.com/15665-edwin-powell-hubble.html).

5. *The Case for the Creator by Lee Strobel* Illustra Medea, http://www.thecaseforacreator.com/scientists.php

6. Alan Lightman, Posted 09.09.97 (NOVA) *Relativity and the Cosmos* http://www.pbs.org/wgbh/nova/physics/relativity-and-the-cosmos.html

7. *The Case for the Creator by Lee Strobel* Illustra Medea, http://www.thecaseforacreator.com/scientists.php

8. Edwin R. Hubble Expansion of the Universe http://www.pbs.org/wgbh/nova/physics/relativity-and-the-cosmos.html

9. http://cosmictimes.gsfc.nasa.gov/online_edition/1929cosmic/expanding.html)

10. (http://www.latimes.com/science/la-sci-cosmic-inflation-20140318story.html#axzz2wkwisjlg)

11. http://www.scientificamerican.com/article/gravity-waves-cmb-b-mode polarization/

12. Charles Darwin, *The Origin of Species* (New York, New York: Barnes and Noble Books, 2004) page 380

13. Miller, Stanley L.; Harold C. Urey (July 1959). "Organic Compound Synthesis on the Primitive Earth". *Science* **130** (3370): 245–51. Bibcode:1959Sci...130..245M. doi:10.1126/science.130.3370.245. PMID 13668555.

14. *The Case for the Creator by Lee Strobel* Illustra Medea, http://www.thecaseforacreator.com/scientists.php

15. Charles Darwin, *The Origin of Species* (New York, New York: Barnes and Noble Books, 2004) page 380

16. The Zondervan Corporation, The Amplified Bible, Expanded Edition, Matthew Chapters 24 &25 (La Habra, California: The Zondervan Corporation and the Lochman Corporation, 1965) page 1110 - 1113. Scripture taken from THE AMPLIFIED BIBLE, Old Testament copyright© 1965, 1987 by the Zondervan Corporation. The Amplified New Testament copyright ©1958, 1987 by the Lochman Foundation. Used by permission.

17. The Zondervan Corporation, The Amplified Bible, Expanded Edition, Exodus 3:14 (La Habra, California: The Zondervan Corporation and the Lochman Corporation, 1965) page 71. Scripture taken from THE AMPLIFIED BIBLE, Old Testament copyright

© 1965, 1987 by the Zondervan Corporation. The Amplified New Testament copyright ©1958, 1987 by the Lochman Foundation. Used by permission.

18. The Zondervan Corporation, The Amplified Bible, Expanded Edition, Romans 1:19 (La Habra, California: The Zondervan Corporation and the Lochman Corporation, 1965) page 1298. Scripture taken from THE AMPLIFIED BIBLE, Old Testament copyright

© 1965, 1987 by the Zondervan Corporation. The Amplified New Testament copyright ©1958, 1987 by the Lochman Foundation. Used by permission.

19. Creation Ministries International Ltd., Creation.com http://creation.com/how-old-is-the-earth

20. The Zondervan Corporation, The Amplified Bible, Expanded Edition, 2 Timothy 3:16 (La Habra, California: The Zondervan Corporation and the Lochman Corporation, 1965) page 1416. Scripture taken from THE AMPLIFIED BIBLE, Old Testament copyright

© 1965, 1987 by the Zondervan Corporation. The Amplified New Testament copyright ©1958, 1987 by the Lochman Foundation. Used by permission.

21. The Zondervan Corporation, The Amplified Bible, Expanded Edition, Genesis 1:1-19 page 1&2. (La Habra, California: The Zondervan Corporation and the Lochman Corporation, 1965) Scripture taken from THE AMPLIFIED BIBLE, Old Testament copyright© 1965, 1987 by the Zondervan Corporation. The Amplified New Testament copyright ©1958, 1987 by the Lochman Foundation. Used by permission.

22. The Zondervan Corporation, The Amplified Bible, Expanded Edition, Genesis 1:1 page 1. (La Habra, California: The Zondervan Corporation and the Lochman Corporation, 1965) Scripture taken from THE AMPLIFIED BIBLE, Old Testament copyright© 1965, 1987 by the Zondervan Corporation. The Amplified New Testament copyright ©1958, 1987 by the Lochman Foundation. Used by permission.

23. The Zondervan Corporation, The Amplified Bible, Expanded Edition, Genesis 1:2 page 1&2. (La Habra, California: The Zondervan Corporation and the Lochman Corporation, 1965) Scripture taken from THE AMPLIFIED BIBLE, Old Testament copyright© 1965, 1987 by the Zondervan Corporation. The Amplified New Testament copyright ©1958, 1987 by the Lochman Foundation. Used by permission.

24. The Zondervan Corporation, The Amplified Bible, Expanded Edition, Genesis 1:1-19 page 1&2. (La Habra, California: The Zondervan Corporation and the Lochman Corporation, 1965) Scripture taken from THE AMPLIFIED BIBLE, Old Testament copyright© 1965, 1987 by the Zondervan Corporation. The Amplified New Testament copyright ©1958, 1987 by the Lochman Foundation. Used by permission.

25. September 26, 2014 Vol.345 no.6204, pp.1590-1593 DOI: 10.1126/science.1258055 in the Journal *"Science"*

26. The Zondervan Corporation, The Amplified Bible, Expanded Edition, Genesis 1:1-19 page 1&2. (La Habra, California: The Zondervan Corporation and the Lochman Corporation, 1965) Scripture taken from THE AMPLIFIED BIBLE, Old Testament copyright© 1965, 1987 by the Zondervan

Corporation. The Amplified New Testament copyright ©1958, 1987 by the Lochman Foundation. Used by permission.

27. September 26, 2014 Vol.345 no.6204, pp.1590-1593 DOI: 10.1126/science.1258055 in the Journal *"Science"*

28. The Zondervan Corporation, The Amplified Bible, Expanded Edition, Genesis 1:1-19 page 1&2. (La Habra, California: The Zondervan Corporation and the Lochman Corporation, 1965) Scripture taken from THE AMPLIFIED BIBLE, Old Testament copyright© 1965, 1987 by the Zondervan Corporation. The Amplified New Testament copyright ©1958, 1987 by the Lochman Foundation. Used by permission.

Chapter 22 Vision! (p. 215)

1. Charles Darwin, *The Origin of Species* (New York, New York: Barnes and Noble Books, 2004) page 157

2. Eugene Wolff, *Anatomy of the Eye and Orbit* Seventh Edition (Philadelphia and Toronto: W.B. Saunders Company, 1981) page 375

3. Henry Gray, F.R.S, *Anatomy of the Human Body* 20th Edition, Lea & Febiger (Philadelphia & New York, © 1918) Pages 766, 1006 Additions by Rigney

4. Charles Darwin, *The Origin of Species* (New York, New York: Barnes and Noble Books, 2004) page 156-159

5. Raymond E. Records, *Physiology of the Human Eye and Visual System* (Hagerstown Maryland: Harper & Row Publishers © 1979) pages 665-668

6. Charles Darwin, *The Origin of Species* (New York, New York: Barnes and Noble Books, 2004) page 156-159

7. Charles Darwin, *The Origin of Species* (New York, New York: Barnes and Noble Books, 2004) pages 159

8. Raymond E. Records, *Physiology of the Human Eye and Visual System* (Hagerstown Maryland: Harper & Row Publishers © 1979) pages 665-668

9. Wade N.J.: *On interocular transfer of the movement aftereffect in individuals with and without normal binocular vision.* Perception 5:113-118, 1976

10. Homann A. Creutzfeldt, O.D.: *Squint and the development of binocularity in humans.* Nature 254:613-614, 1975

11. Banks M.S., Aslin R.N. Letson R.D.: *Sensitive period for the development of human binocular vision.* Science 190:675-677, 1975

12. Raymond E. Records, *Physiology of the Human Eye and Visual System* (Hagerstown Maryland: Harper & Row Publishers © 1979) pages 665-668

13. Veagan, Taylor D. *Critical period for deprivation amblyopea in children.* Transactions of the Ophthalmological Societies of the U.K. 1979; 99:432-439

14. Francis Heed Adler, *Adler's Physiology of the Eye, Clinical Application* Sixth Edition Edited by Robert A. Moses, M.D.(Saint Louis, Missouri: The C.V. Moseby Company, © 1975) page 39

15. Raymond E. Records, *Physiology of the Human Eye and Visual System* (Hagerstown Maryland: Harper & Row Publishers © 1979) pages 666

16. Charles Darwin, *The Origin of Species* (New York, New York: Barnes and Noble Books, 2004) page 156-158

17. Blanca Ruiz de Zarate, Jamie Tejedor; *Current Concepts in the Management of Amblyopea.* Clinical Ophthalmolgy, 2007, Dec; 1(4): 403-414

18. Pediatric Eye Disease Investigator Group: *A randomized trial of atropine vs. patching for treatment of moderate amblyopea in children.* Archives of Ophthalmology 2002 Mar; 120(3):268-78

19. Longmuir S, Pfeiffer W, Scott W, Olson R, *Effect of occlusion amblyopia after prescribed full-time occlusion on long-term visual acuity outcomes.*.Journal of Pediatric Ophthalmology and Strabismus, 2013 Mar-April; 50(2):94-101

20. The Zondervan Corporation, The Amplified Bible, Expanded Edition, Mark 8:22-25 (La Habra, California: The Zondervan Corporation and the Lochman Corporation, 1965) page 1135-1136. Scripture taken from THE AMPLIFIED BIBLE, Old Testament copyright © 1965, 1987 by the Zondervan Corporation. The Amplified New Testament copyright ©1958, 1987 by the Lochman Foundation. Used by permission.

21. The Zondervan Corporation, The Amplified Bible, Expanded Edition, Romans 1:19 (La Habra, California: The Zondervan Corporation and the Lochman Corporation, 1965) page 1298. Scripture taken from THE AMPLIFIED BIBLE, Old Testament copyright

© 1965, 1987 by the Zondervan Corporation. The Amplified New Testament copyright ©1958, 1987 by the Lochman Foundation. Used by permission.

22. The Zondervan Corporation, The Amplified Bible, Expanded Edition, John 9:1-33 (La Habra, California: The Zondervan Corporation and the Lochman Corporation, 1965) page 1225-1226. Scripture taken from THE AMPLIFIED BIBLE, Old Testament copyright© 1965, 1987 by the Zondervan Corporation. The Amplified New Testament copyright ©1958, 1987 by the Lochman Foundation. Used by permission.

23. The Zondervan Corporation, The Amplified Bible, Expanded Edition, John 9:7&8 (La Habra, California: The Zondervan Corporation and the Lochman Corporation, 1965) page 1225. Scripture taken from THE AMPLIFIED BIBLE, Old Testament copyright

© 1965, 1987 by the Zondervan Corporation. The Amplified New Testament copyright ©1958, 1987 by the Lochman Foundation. Used by permission.

24. The Zondervan Corporation, The Amplified Bible, Expanded Edition, Romans 1:19 (La Habra, California: The Zondervan Corporation and the Lochman Corporation, 1965) page 1298. Scripture taken from THE AMPLIFIED BIBLE, Old Testament copyright

© 1965, 1987 by the Zondervan Corporation. The Amplified New Testament copyright ©1958, 1987 by the Lochman Foundation. Used by permission.

25. The Zondervan Corporation, The Amplified Bible, Expanded Edition, John 9:7 (La Habra, California: The Zondervan Corporation and the Lochman Corporation, 1965) page 1225. Scripture taken from THE AMPLIFIED BIBLE, Old Testament copyright© 1965, 1987 by the Zondervan Corporation. The Amplified New Testament copyright ©1958, 1987 by the Lochman Foundation. Used by permission.

26. The Zondervan Corporation, The Amplified Bible, Expanded Edition, Romans 6:23 p.1306 & Romans 10:13 p. 1312 (La Habra, California: The Zondervan Corporation and the Lochman Corporation, 1965) page 1298. Scripture taken from THE AMPLIFIED BIBLE, Old Testament copyright© 1965, 1987 by the Zondervan Corporation. The Amplified New Testament copyright ©1958, 1987 by the Lochman Foundation. Used by permission

27. The Zondervan Corporation, The Amplified Bible, Expanded Edition, John 3:16 (La Habra, California: The Zondervan Corporation and the Lochman Corporation, 1965) page 1211. Scripture taken from THE AMPLIFIED BIBLE, Old Testament copyright © 1965, 1987 by the Zondervan Corporation. The Amplified New Testament copyright ©1958, 1987 by the Lochman Foundation. Used by permission.

28. The Zondervan Corporation, The Amplified Bible, Expanded Edition, Isaiah 35:5 42:7 (La Habra, California: The Zondervan Corporation and the Lochman Corporation, 1965) page 1298. Scripture taken from THE AMPLIFIED BIBLE, Old Testament copyright© 1965, 1987 by the Zondervan Corporation. The Amplified New Testament copyright ©1958, 1987 by the Lochman Foundation. Used by permission.

29. The Zondervan Corporation, The Amplified Bible, Expanded Edition, Isaiah 9:6 p.763, Isaiah 42:1-7 p.798 (La Habra, California: The Zondervan Corporation and the Lochman Corporation, 1965) page 1298. Scripture taken from THE AMPLIFIED BIBLE, Old Testament copyright© 1965, 1987 by the Zondervan Corporation. The Amplified New Testament copyright ©1958, 1987 by the Lochman Foundation. Used by permission.

30. The Zondervan Corporation, The Amplified Bible, Expanded Edition, John 11:14-45 p.1229-1230 (La Habra, California: The Zondervan Corporation and the Lochman Corporation, 1965) page 1298. Scripture taken from THE AMPLIFIED BIBLE, Old Testament copyright© 1965, 1987 by the Zondervan Corporation. The Amplified New Testament copyright ©1958, 1987 by the Lochman Foundation. Used by permission.

31. The Zondervan Corporation, The Amplified Bible, Expanded Edition, Romans 1:19 & 20 (La Habra, California: The Zondervan Corporation and the Lochman Corporation, 1965) page 1298. Scripture taken from THE AMPLIFIED BIBLE, Old Testament copyright© 1965, 1987 by the Zondervan Corporation. The Amplified New Testament copyright ©1958, 1987 by the Lochman Foundation. Used by permission.

Chapter 23 The DNA Dictates the Species (p. 229)

1. The Zondervan Corporation, The Amplified Bible, Expanded Edition, Romans 1:19 (La Habra, California: The Zondervan Corporation and the Lochman Corporation, 1965) page 1298. Scripture taken from THE AMPLIFIED BIBLE, Old Testament copyright © 1965, 1987 by the Zondervan Corporation. The Amplified New Testament copyright ©1958, 1987 by the Lochman Foundation. Used by permission.

2. **Pond Scene:** Art Explosion 525,000

3. Sunrays and Sky:
   Description English: Sunray through clouds
   Deutsch: Sonnenstrahlen
   Date August 2007
   Source Own work
   Author: user:AngMoKio
   http://upload.wikimedia.org/wikipedia/commons/0/01/Sunray_clouds_amk.jpgClouds and sun:
   http://www.enjoylifebalance.com/wpcontent/uploads/2013/02/
   Sun-rays-through-clouds.jpg
   4. Geese Flying: Migrating Geese:
   http://commons.wikimedia.org/wiki/File:PSM_V84_D218_Flocking_habit_of_migratorybirds_fi
   g6.jpg#mediaviewer/File:PSM_V84_D218_Flocking_habit_of_migratory_birds_fig6.jpghttp://u
   pload.wikimedia.org/wikipedia/commons/0/0b/PSM_V84_D218_Flocking_habit_of_migratory_
   birds_fig6.jpg
   H. K. Job - C. C. Trowbridge: *On the origin of the flocking habit of migratory birds*. The Popular
   science monthly, Volume 84, p214. New York, Popular Science Pub. Co., March 1914. Online:
   archive.org.
5. **Whitetail Buck:** deer:http://upload.wikimedia.org/wikipedia/commons/b/b7/White-
   tailed_deer.jpg
6. **Wolf:**Wolf:http://upload.wikimedia.org/wikipedia/commons/5/53/Scandinavian_grey_wolf_
   Canis_lupus.jpg
7. **Fish Jumping:** https://openclipart.org/detail/524/jumping%20fish
   https://openclipart.org/image/300px/svg_to_png/524/johnny-automatic-jumping-fish.png
8. **Butterfly:** Butterfly:commons.wikimedia.org/wiki/File:Monarch_Butterfly_-
   _Danaus_plexippus_(5890526585).jpg
   http://upload.wikimedia.org/wikipedia/commons/0/0e/Monarch_Butterfly_-
   _Danaus_plexippus_%285890526585%29.jpg
9. **Frog:** Frog:http://upload.wikimedia.org/wikipedia/commons/4/49/Green_treefrog.jpg
10. **Dragonfly:**
    dragonfly:http://upload.wikimedia.org/wikipedia/commons/e/e4/Dragonfly_ran-
    384.jpgdragonfly:http://upload.wikimedia.org/wikipedia/commons/e/e4/Dragonfly_ran-384.jpg
11.**Turtle:**
    turtle:http://en.wikipedia.org/wiki/Turtle#/media/File:Florida_Box_Turtle_Digon3_re-edited.jpg
    http://upload.wikimedia.org/wikipedia/commons/f/f4/Florida_Box_Turtle_Digon3_re-edited.jpg
12. **Dandelions:**
    dandelion:http://upload.wikimedia.org/wikipedia/commons/8/8d/Dandelion_closeup_at_Marymo
    or_Park.jpg
13. The Zondervan Corporation, The Amplified Bible, Expanded Edition, Romans 1:19 (La Habra,
California: The Zondervan Corporation and the Lochman Corporation, 1965) page $$#1298.
Scripture taken from THE AMPLIFIED BIBLE, Old Testament copyright© 1965, 1987 by the

Zondervan Corporation. The Amplified New Testament copyright ©1958, 1987 by the Lochman Foundation. Used by permission.

14. Charles Darwin, *The Origin of Species* (New York, New York: Barnes and Noble Books, 2004) page 380

15. Charles Darwin, *The Origin of Species* (New York, New York: Barnes and Noble Books, 2004) {paleontology references} pages 232-250

16. Charles Darwin, *The Origin of Species* (New York, New York: Barnes and Noble Books, 2004) page 380

17. Charles Darwin, *The Origin of Species* (New York, New York: Barnes and Noble Books, 2004) pages 159

18. **Minnow:** http://en.wikipedia.org/wiki/Rainbow_shiner
    http://upload.wikimedia.org/wikipedia/commons/a/a7/Rainbow_Shiner.jpg

19. **Trout:** http://en.wikipedia.org/wiki/Trout
    http://en.wikipedia.org/wiki/Trout#mediaviewer/File:Rainbow_Trout.jpg

20. **Salmon:** http://en.wikipedia.org/wiki/Chinook_salmon
    http://upload.wikimedia.org/wikipedia/commons/d/d8/Chinook_Salmon_Adult_Male.jpg

21. **Shark:** http://www.google.com/url?q=http://en.wikipedia.org/wiki
    Great_white_shark&sa=U&ei=CRryVJ3TJZKWyATK94FY&ved=0CCgQ9QEwCQ&usg=AFQ
    jCNHlVxTwBMcA1_YtTe2ryrUXjecygA
    http://upload.wikimedia.org/wikipedia/commons/8/85/Great_white_shark_size_comparison.svg

22. **Orca:** http://no.wikipedia.org/wiki/Spekkhogger
    http://upload.wikimedia.org/wikipedia/commons/d/d4/Orca_size-2.svg

23. **Humpback Whale:** http://en.wikipedia.org/wiki/Humpback_whale
    http://en.wikipedia.org/wiki/Humpback_whale#mediaviewer/File:Humpback_whale_size.svg

24. **Humingbird:** http://en.wikipedia.org/wiki/Hummingbird
    http://upload.wikimedia.org/wikipedia/commons/1/16/Archilochus-alexandri-002-edit.jpg

25. **Finch:** http://en.wikipedia.org/wiki/Finch
    http://en.wikipedia.org/wiki/Finch#mediaviewer/File:Coccothraustes_coccothraustes_1_(Marek_
    Szczepanek).jpg

26. **Bat:** http://en.wikipedia.org/wiki/Bat
    http://upload.wikimedia.org/wikipedia/commons/7/77/Big-eared-townsend-fledermaus.jpg

27. **Condor:** http://en.wikipedia.org/wiki/Condor
    http://en.wikipedia.org/wiki/Condor#mediaviewer/File:California-condor.jpg

28. **Penguin**: http://en.wikipedia.org/wiki/Penguin
    http://en.wikipedia.org/wiki/Emperor_penguin
    http://en.wikipedia.org/wiki/Emperor_penguin#mediaviewer/File:Aptenodytes_forsteri_-
    Snow_Hill_Island,_Antarctica_-adults_and_juvenile-8.jpg

29. **Ostrich:** Art Explosion 525,00

30. Answers in Genesis, https://cdn-assets.answersingenesis.org
    https://cdn-assets.answersingenesis.org/img/articles/ee/v2/ape-to-man-evolution.jpg

31. Charles Darwin, *The Origin of Species* (New York, New York: Barnes and Noble Books, 2004) page 159

32. Charles Darwin, *The Origin of Species* (New York, New York: Barnes and Noble Books, 2004) page 380

33. Charles Darwin, *The Origin of Species* (New York, New York: Barnes and Noble Books, 2004) page 158

34. Charles Darwin, *The Origin of Species* (New York, New York: Barnes and Noble Books, 2004) page 380

35. http://darwiniana.org/jaws.htm

36. Robert J Asher, Thomas Lehmann (Dental eruption in afrotherian mammals) BMC Biology, March 2008, 6:14,11. doi:10.1186/1741-7007-6-14

37. http://palaeo.gly.bris.ac.uk/palaeofiles/elephants/ and http://www.tfcg.org/pdf/article_afrotheria.pdf

38. **Elephant photo** reference: http://en.wikipedia.org/wiki/African_bush_elephant#mediaviewer/File:Loxodonta_africana_-_old_bull_(Ngorongoro,_2009).jpg

39. **Hyrax photo** reference: http://en.wikipedia.org/wiki/Hyrax http://upload.wikimedia.org/wikipedia/commons/0/0c/Yellow-spotted_Rock_Hyrax.jpg

40. **Manatee photo** reference: http://en.wikipedia.org/wiki/Manatee http://upload.wikimedia.org/wikipedia/commons/1/1e/Manatee at Sea World Orlando_Mar_10.JPG

41. Dumbacher, John P,. Rathburn, Galen B., Osborn, timothy O., Griffin, Michael, Eiseb, Seth J. *Journal of Mammalogy,* {95(3):443-454,2014}June 26, 2014

42. http://www.thewestsidestory.net/2014/06/28/13562/sengi-mouse-species-elephant-genes-discovered-africa/

43. Charles Darwin, *The Origin of Species* (New York, New York: Barnes and Noble Books, 2004) page 159

44. Charles Darwin, *The Origin of Species* (New York, New York: Barnes and Noble Books, 2004) page 59

45. Charles Darwin, *The Origin of Species* (New York, New York: Barnes and Noble Books, 2004) page 73

46. Dumbacher, John P,. Rathburn, Galen B., Osborn, timothy O., Griffin, Michael, Eiseb, Seth J. *Journal of Mammalogy,* {95(3):443-454,2014}June 26, 2014

47. Charles Darwin, *The Origin of Species* (New York, New York: Barnes and Noble Books, 2004) page 159

48. **Sengi Mouse** reference: http://en.wikipedia.org/wiki/Elephant_shrew http://en.wikipedia.org/wiki/Elephant_shrew#mediaviewer/File:Rhynchocyon_petersione.JPG

49. **Male African Elephant** reference: elephant picture references: http://en.wikipedia.org/wiki/African_bush_elephant http://en.wikipedia.org/wiki/African_bush_elephant#mediaviewer/File:Loxodontaafricana_-_old_bull_(Ngorongoro,_2009).jpg

50. Dumbacher, John P,. Rathburn, Galen B., Osborn, timothy O., Griffin, Michael, Eiseb, Seth J. *Journal of Mammalogy,* {95(3):443-454,2014}June 26, 2014

51. Charles Darwin, *The Origin of Species* (New York, New York: Barnes and Noble Books, 2004) page 159

52. Dumbacher, John P,. Rathburn, Galen B., Osborn, timothy O., Griffin, Michael, Eiseb, Seth J. *Journal of Mammalogy,* {95(3):443-454,2014}June 26, 2014

53. Charles Darwin, *The Origin of Species* (New York, New York: Barnes and Noble Books, 2004) page 380

54. **Sengi Mouse** reference: http://en.wikipedia.org/wiki/Elephant_shrew http://en.wikipedia.org/wiki/Elephant_shrew#mediaviewer/File:Rhynchocyon_ petersione.JPG

55. **Male African Elephant** reference: elephant picture references: http://en.wikipedia.org/wiki/African_bush_elephant http://en.wikipedia.org/wiki/African_bush_elephant#mediaviewer/File:Loxodontaafricana_- _old_bull_(Ngorongoro,_2009).jpg

56. Answers in Genesis, https://cdn-assets.answersingenesis.org https://cdn-assets.answersingenesis.org/img/articles

57. http://www.reuters.com/article/2014/07/31/us-science-dinosaurs-i dUSKBN0G02KJ20140731

58.Yee, Michael S. Y., Cau, Andrea, Naish D., Dyke, Gareth J. *Science* 1 August 2014, Vol. 345 no. 6196 pp. 562-566, DOI: 10.1126/science.1252243

59. http://palaeo.gly.bris.ac.uk/palaeofiles/elephants/ and http://www.tfcg.org/pdf/article_afrotheria.pdf

60. Dumbacher, John P,. Rathburn, Galen B., Osborn, timothy O., Griffin, Michael, Eiseb, Seth J. *Journal of Mammalogy,* {95(3):443-454,2014}June 26, 2014

61. http://www.reuters.com/article/2014/07/31/us-science-dinosaurs-idUSKBN0G02KJ20140731

62. Yee, Michael S. Y., Cau, Andrea, Naish D., Dyke, Gareth J. *Science* 1 August 2014, Vol. 345 no. 6196 pp. 562-566, DOI: 10.1126/science.1252243

63. Charles Darwin, *The Origin of Species* (New York, New York: Barnes and Noble Books, 2004) page 380

64. Yee, Michael S. Y., Cau, Andrea, Naish D., Dyke, Gareth J. *Science* 1 August 2014, Vol. 345 no. 6196 pp. 562-566, DOI: 10.1126/science.1252243

65. Charles Darwin, *The Origin of Species* (New York, New York: Barnes and Noble Books, 2004) page 380

66. Charles Darwin, *The Origin of Species* (New York, New York: Barnes and Noble Books, 2004) {paleontology references} pages 232-250

67. Charles Darwin, *The Origin of Species* (New York, New York: Barnes and Noble Books, 2004) page 380

68.-85. Dinosaur and other "Organic Life" pictures (all the other pictures) used on my *Supposed evolutionary tree of life before and after the meteor strike* (Evolutionary Tree Illustration, Page 246) were acquired through *Google Images* under the Creative Commons license. The pictures were changed from their original use and arranged into a new format entirely of <u>my</u> own making. I do not in any way suggest the licensor endorses me or my use.

68. - 85. Source of use are from:

68. Bottom-Right 6 branches; these six photos used (except the ptaradactyl) were from:

File:Hell Creek dinosaurs and pterosaurs by durbed.jpg

Uploaded by I. J. Reid

Created: December 31, 2012

http://upload.wikimedia.org/wikipedia/commons/b/b6/Hell_Creek_dinosaurs_and_pterosaurs_by _durbed.jpg

CC BY-SA 3.0 view terms

69. Pterodactyl:

http://en.wikipedia.org/wiki/Pterodactylus

http://upload.wikimedia.org/wikipedia/commons/a/ae/Pterodactylus_holotype_flymmartyniuk.png

CC BY-SA 3.0 view terms

70. Apatosaurus:

http://dinosaurs.wikia.com/wiki/Apatosaurus

http://img2.wikia.nocookie.net/__cb20140716003332/dinosaurs/images/e/e9/Shutterstock_44190 772.jpg

CC BY-SA 3.0 view terms

71. Triceratops:

http://en.wikipedia.org/wiki/Triceratops

http://en.wikipedia.org/wiki/Triceratops#mediaviewer/File:Hell_Creek_dinosaursand_pterosaurs _by_durbed.jpg

CC BY-SA 3.0 view terms

72. Stegosaurus:

http://en.wikipedia.org/wiki/Stegosaurus

http://en.wikipedia.org/wiki/Stegosaurus#mediaviewer/File:Stegosaurus_BW.jpg

http://upload.wikimedia.org/wikipedia/commons/7/70/Stegosaurus_BW.jpg

CC BY-SA 3.0 view terms

73. Turkey:

http://en.wikipedia.org/wiki/Domesticated_turkey

http://upload.wikimedia.org/wikipedia/commons/d/d0/Male_north_american_turkey_supersatura ted.jpg

CC BY-SA 3.0 view terms

74. Peacock:

http://en.wikipedia.org/wiki/Indian_peafowl

http://upload.wikimedia.org/wikipedia/commons/3/39/Peacock_by_Nihal_Jabin.jpg

CC BY-SA 3.0 view terms

75. Chicken:
http://en.wikipedia.org/wiki/Poltava_(chicken)
http://upload.wikimedia.org/wikipedia/commons/f/f4/Poltava_chicken_breed_male2.jpg
CC BY-SA 3.0 view terms

76. Dove:
http://en.wikipedia.org/wiki/Columbidae
http://upload.wikimedia.org/wikipedia/commons/8/89/India_Dove.jpg
CC BY-SA 3.0 view terms

77. Sparrow:
http://en.wikipedia.org/wiki/Eurasian_tree_sparrow
http://upload.wikimedia.org/wikipedia/commons/7/73/Tree_Sparrow_Japan_Flip.Jpg
CC BY-SA 3.0 view terms

78. Bluejay:
http://commons.wikimedia.org/wiki/File:Cyanocitta_cristata_blue_jay.jpg
http://upload.wikimedia.org/wikipedia/commons/4/47/Cyanocitta_cristata_blue_jay.jpg
CC BY-SA 3.0 view terms

79. Robin:
http://en.wikipedia.org/wiki/American_robin
http://upload.wikimedia.org/wikipedia/commons/b/b8/Turdus-migratorius-002.jpg
CC BY-SA 3.0 view terms

80. Ostrich:
Art explosion 525,000

81. Horse:
http://en.wikipedia.org/wiki/Equine_anatomy
http://upload.wikimedia.org/wikipedia/commons/c/cc/RCMP_Farm_Hannoverian2.jpg
CC BY-SA 3.0 view terms

82. Bear:
http://en.wikipedia.org/wiki/Kodiak_bear
http://upload.wikimedia.org/wikipedia/commons/8/81/Kodiak_bear_in_germany.jpg
CC BY-SA 3.0 view terms

83. Lion:
http://en.wikipedia.org/wiki/Lion
http://upload.wikimedia.org/wikipedia/commons/7/73/Lion_waiting_in_Namibia.Jpg
CC BY-SA 3.0 view terms

84. Turtle:
http://en.wikipedia.org/wiki/Turtle
http://upload.wikimedia.org/wikipedia/commons/f/f4/Florida_Box_Turtle_Digon3_re-edited.jpg
CC BY-SA 3.0 view terms

85. Man:

Answers in Genesis, https://cdn-assets.answersingenesis.org

https://cdn-assets.answersingenesis.org/img/articles

86. Charles Darwin, *The Origin of Species* (New York, New York: Barnes and Noble Books, 2004) page 158

87. Charles Darwin, *The Origin of Species* (New York, New York: Barnes and Noble Books, 2004) page 159

88. - 101. Vehicle Pictures used on <u>my</u> **Vehicular Evolutionary Tree Illustration**, (Page 221) were acquired through *Google Images* under the Creative Commons license. The pictures were changed from their original use and arranged into a new format entirely of <u>my</u> own making. I do not in any way suggest the licensor endorses me or my use.

88. Unicycle:

https://openclipart.org/image/300px/svg_to_png/76657/unicycle.png

89. Bicycle: https://openclipart.org/image/300px/svg_to_png/166988/1326274846.png

90. Tricycle:

https://openclipart.org/detail/212409/Old%20Tricycle

https://openclipart.org/image/300px/svg_to_png/212409/tricycle-1921-vector.png

91. Motorcycle:

https://openclipart.org/detail/180082/Harley%20Motorcycle

https://openclipart.org/image/300px/svg_to_png/180082/harleydavidson.png

92. Buggy:

https://openclipart.org/image/300px/svg_to_png/194516/buggy.png

93. Model T: https://openclipart.org/image/300px/svg_to_png/46909/modelt1906car.png

94. Corvette:

<u>https://openclipart.org/image/300px/svg_to_png/168425/53-</u>Corvette.png

95. Pickup:

<u>https://openclipart.org/image/300px/svg_to_png/211221/pickup-</u>truck.png

96. Boxtruck:

https://openclipart.org/image/300px/svg_to_png/193675/Truck-Woofer.png

97. Dumptruck:

https://openclipart.org/image/300px/svg_to_png/183902/dumptruck.png

98. Big Rig:

https://openclipart.org/image/300px/svg_to_png/47365/big-truck-01.png

99. Race car:

https://openclipart.org/download/173850/race-car.svg

100. Bulldozer:

https://openclipart.org/download/27062/egore911-bulldozer.svg

101. Formula 1 car:

<u>https://openclipart.org/image/300px/svg_to_png/8554/Gerald-G-</u>Formula-One-Car.png

102. Henry Gray, F.R.S, *Anatomy of the Human Body* 20th Edition, Lea & Febiger (Philadelphia & New York, © 1918) Page 186

103. Henry Gray, F.R.S, *Anatomy of the Human Body* 20th Edition, Lea & Febiger (Philadelphia & New York, © 1918) Page 1027

104. Henry Gray, F.R.S, *Anatomy of the Human Body* 20th Edition, Lea & Febiger (Philadelphia & New York, © 1918) Page 1022

105. Henry Gray, F.R.S, *Anatomy of the Human Body* 20th Edition, Lea & Febiger (Philadelphia & New York, © 1918) Page 1022

106. Henry Gray, F.R.S, *Anatomy of the Human Body* 20th Edition, Lea & Febiger (Philadelphia & New York, © 1918) Page 885

107. Henry Gray, F.R.S, *Anatomy of the Human Body* 20th Edition, Lea & Febiger (Philadelphia & New York, © 1918) Pages 1026,1027

108. Henry Gray, F.R.S, *Anatomy of the Human Body* 20th Edition, Lea & Febiger (Philadelphia & New York, © 1918) Page 817

109. Henry Gray, F.R.S, *Anatomy of the Human Body* 20th Edition, Lea & Febiger (Philadelphia & New York, © 1918) Page 885

110. Henry Gray, F.R.S, *Anatomy of the Human Body* 20th Edition, Lea & Febiger (Philadelphia & New York, © 1918) Page 379

111. Henry Gray, F.R.S, *Anatomy of the Human Body* 20th Edition, Lea & Febiger (Philadelphia & New York, © 1918) Pages 566 & 569

112. Henry Gray, F.R.S, *Anatomy of the Human Body* 20th Edition, Lea & Febiger (Philadelphia & New York, © 1918) Page 644

113. Henry Gray, F.R.S, *Anatomy of the Human Body* 20th Edition, Lea & Febiger (Philadelphia & New York, © 1918) Page 570

114. Henry Gray, F.R.S, *Anatomy of the Human Body* 20th Edition, Lea & Febiger (Philadelphia & New York, © 1918) Page 1009

115 Henry Gray, F.R.S, *Anatomy of the Human Body* 20th Edition, Lea & Febiger (Philadelphia & New York, © 1918) Page 1010

116. Henry Gray, F.R.S, *Anatomy of the Human Body* 20th Edition, Lea & Febiger (Philadelphia & New York, © 1918) Page 972

117. Henry Gray, F.R.S, *Anatomy of the Human Body* 20th Edition, Lea & Febiger (Philadelphia & New York, © 1918) Pages 566, 569, 527, 766

118. Henry Gray, F.R.S, *Anatomy of the Human Body* 20th Edition, Lea & Febiger (Philadelphia & New York, © 1918) Page 1006

119. Henry Gray, F.R.S, *Anatomy of the Human Body* 20th Edition, Lea & Febiger (Philadelphia & New York, © 1918) Page 1021

120. Henry Gray, F.R.S, *Anatomy of the Human Body* 20th Edition, Lea & Febiger (Philadelphia & New York, © 1918) Page 1012

121. OpenStax College [CC BY 3.0 (http://creativecommons.org/licenses/by/3.0)], via Wikimedi Commons. File:1414 Rods and Cones.jpg (File:1414 Rods and Cones - ru.svg)

122. R. Greeff (1900) Handbuch der gesamten Augenheilkunde, 2nd ed, vol.1., Graefe and Saemisch, Leipzig.

123.Henry Gray, F.R.S, *Anatomy of the Human Body* 20th Edition, Lea & Febiger (Philadelphia & New York, © 1918) Page 1016

124. Henry Gray, F.R.S, *Anatomy of the Human Body* 20th Edition, Lea & Febiger (Philadelphia & New York, © 1918) Page 1006

125. Les Voies Optiques:
http://www.campusdanatomie.org/sites/default/files/users/admin/voiesoptiques_sans_legende.pdf
http://www.campusdanatomie.org/sites/default/files/users/admin/voiesoptiques_sans_legende.pdf
https://creativecommons.org/licenses/by-sa/3.0

126. Henry Gray, F.R.S, *Anatomy of the Human Body* 20th Edition, Lea & Febiger (Philadelphia & New York, © 1918) Page 1026 &1027

127. Henry Gray, F.R.S, *Anatomy of the Human Body* 20th Edition, Lea & Febiger (Philadelphia & New York, © 1918) Page 1021

128. Henry Gray, F.R.S, *Anatomy of the Human Body* 20th Edition, Lea & Febiger (Philadelphia & New York, © 1918) Page 1026 & 1027

129. Henry Gray, F.R.S, *Anatomy of the Human Body* 20th Edition, Lea & Febiger (Philadelphia & New York, © 1918) Page 1021

130. Henry Gray, F.R.S, *Anatomy of the Human Body* 20th Edition, Lea & Febiger (Philadelphia & New York, © 1918) Page 1022

131. Les Voies Optiques:
http://www.campusdanatomie.org/sites/default/files/users/admin/voiesoptiques_sans_legende.pdf
http://www.campusdanatomie.org/sites/default/files/users/admin/voiesoptiques_sans_legende.pdf
https://creativecommons.org/licenses/by-sa/3.0

132. R.C. Sproul Sr., *Renewing Your Mind* radio program

133. Charles Darwin, *The Origin of Species* (New York, New York: Barnes and Noble Books, 2004) page 380

134. The Zondervan Corporation, The Holy Bible, New International Version, Matthew 13:49 (La Habra, California: The Zondervan Corporation and the Lochman Corporation, 1965) page 1094. Scripture taken from THE Holy Bible, New International Version® copyright© 1973,1978, 1984 by International Bible Society. Used by permission.

135. Charles Darwin, *The Origin of Species* (New York, New York: Barnes and Noble Books, 2004) page 158

136. Charles Darwin, *The Origin of Species* (New York, New York: Barnes and Noble Books, 2004) page 380

137. Charles Darwin, *The Origin of Species* (New York, New York: Barnes and Noble Books, 2004) page 158

138. Charles Darwin, *The Origin of Species* (New York, New York: Barnes and Noble Books, 2004) page 380

139. Charles Darwin, *The Origin of Species* (New York, New York: Barnes and Noble Books, 2004) page 159

140. Charles Darwin, *The Origin of Species* (New York, New York: Barnes and Noble Books, 2004) page 380

141. Charles Darwin, *The Origin of Species* (New York, New York: Barnes and Noble Books, 2004) page 158

142. Charles Darwin, *The Origin of Species* (New York, New York: Barnes and Noble Books, 2004) page 159

143. Charles Darwin, *The Origin of Species* (New York, New York: Barnes and Noble Books, 2004) page 158

144. Charles Darwin, *The Origin of Species* (New York, New York: Barnes and Noble Books, 2004) page 145

145. Charles Darwin, *The Origin of Species* (New York, New York: Barnes and Noble Books, 2004) page 159

146. Charles Darwin, *The Origin of Species* (New York, New York: Barnes and Noble Books, 2004) page 158

147. Sclerocornea: http://upload.wikimedia.org/wikipedia/commons/c/c4/023-p-039-9426.jpg
Neethirajan G, Krishnadas SR, Vijayalakshmi P, Shashikant S, Sundaresan P.
BMC Med Genet. 2004 Apr 16;5:9.
PAX6 gene variations associated with aniridia in south India.
File:023-p-039-9426.jpg
Uploaded by Filip em, Created: November 20, 2007

148. Aniridia:http://upload.wikimedia.org/wikipedia/commons/c/c4/023-p-039-9426.jpg
Neethirajan G, Krishnadas SR, Vijayalakshmi P, Shashikant S, Sundaresan P.
BMC Med Genet. 2004 Apr 16;5:9.
PAX6 gene variations associated with aniridia in south India.
File:023-p-039-9426.jpg   Uploaded by Filip em Created: November 20, 2007

149. Myesthenia Gravis
http://en.wikipedia.org/wiki/Myasthenia_gravis#/media/File:Myasthenia.jpg
Description: Nederlands: Ogen van patient met Myasthenia Gravis
Date: 21 November 2007 (original upload date)(Original text: *10-11-2007*)
Source: Transfered from nl.wikipedia (Original text: *eigen werk*)
Author: Original uploader was Cumulus at nl.wikipedia (Original text: *Cumulus*)
Permission: (Reusing this file) Released under the GNU Free Documentation License. (Original text: *Cumulus*)

150. Band Keratopathy http://upload.wikimedia.org/wikipedia/commons/a/a3/Band-keratopathy_left-eye.png *A photo of my left eye showing band keratopathy (a calcium buildup) on the cornea.* Author: Dr Jon Ruddle, Subject: Mr Paul Bone,
Date: 2010-07-27, Copyright: (c) Jun Ruddle 2010 Licensed under the Creative Commons Attribution-Share Alike 3.0 Unported

151. Cataract
http://upload.wikimedia.org/wikipedia/commons/b/ba/Cataract_in_human_eye.png
Description: Cataract in Human Eye

Date: 24 December 2005, Source: Own work, Author: Rakesh Ahuja, MD
Permission (Reusing this file) Multi-license This file is licensed under the Creative Commons
Attribution-Share Alike 3.0 Unported license.

152. Orbital tumor: Photo from: Page URL:Orbital Tumor:
http://upload.wikimedia.org/wikipedia/commons/6/6d/406907P-PA-OCULAR.jpg
http://commons.wikimedia.org/wiki/File%3A406907P-PA-OCULAR.jpg
File URL: http://upload.wikimedia.org/wikipedia/commons/6/6d/406907P-PA-OCULAR.jpg
Attribution: By The Armed Forces Institute of Pathology [Public domain], via Wikimedia Commons

153. Orbital; Cellulitis:
Photo from: Orbital; Cellulitis:
Page URL:http://commons.wikimedia.org/wiki/File%3AOrbital_cellulitis.jpg
File URL:http://upload.wikimedia.org/wikipedia/commons/8/8f/Orbital_cellulitis.jpg
Attribution: By Jonathan Trobe, M.D. - University of Michigan Kellogg Eye Center (The Eyes Have
It) [CC BY 3.0 (http://creativecommons.org/licenses/by/3.0)], via Wikimedia Commons

154. Retinoblasoma:
http://en.wikipedia.org/wiki/Leukocoria#/media/File:Rb_whiteeye.PNG
http://upload.wikimedia.org/wikipedia/commons/d/d8/Rb_whiteeye.PNG
Description English: A child with a white eye reflection as a result of retinoblastoma.
Date: 28 October 2008 (original upload date)
Source: Transferred from en.wikipedia; transferred to Commons by User:Roberta F. using
CommonsHelper. (Original text: *I created this work entirely by myself.*)
Author:J Morley-Smith (talk). Original uploader was Morleyj at en.wikipedia Permission (Reusing
this file) Released into the public domain (by the author).

155. Carotid-Cavernous Fistula:
File:http://upload.wikimedia.org/wikipedia/commons/b/bf/Corkscrew_blood_vessels_in_left_eye.jpg
Page URL: http://en.wikipedia.org/wiki/Carotid-
cavernous_fistula#/media/File:Corkscrew_blood_vessels_in_left_eye.jpg
Attribution: Sumeer Thinda, Mark R Melson and Rachel W Kuchtey -
http://www.biomedcentral.com/1471-2415/12/28
CC BY 2.5
File:Corkscrew blood vessels in left eye.jpg
Uploaded by Kiatdd, Created: December 28, 2012

156. Acanthamoeba:
Page URL: http://en.wikipedia.org/wiki/Keratitis#/media/File:Parasite140120fig1_Acanthamoeba_
keratitis_Figure_1A.png
File URL: http://www.parasite
journal.org/articles/parasite/full_html/2015/01/parasite140120/F1.html
Attribution: Jacob Lorenzo-Morales, Naveed A. Khan and Julia Walochnik - " (2015). "An update
on *Acanthamoeba* keratitis: diagnosis, pathogenesis and treatment".
Parasite 22: 10. DOI:10.1051/parasite/2015010. PMID 25687209. ISSN 1776-1042.

Figure 1A of published paper. Corneal melting and vascularization in a patient with *Acanthamoeba* keratitisDescription: English: Figure 1A of published paper. Corneal melting and vascularization in a patient with *Acanthamoeba* keratitisDate: 18 February 2015Source:" (2015). "An update on *Acanthamoeba* keratitis: diagnosis, pathogenesis and treatment". *Parasite* 22: 10. DOI:10.1051/parasite/2015010. PMID 25687209. ISSN 1776-1042. Author:Jacob Lorenzo-Morales, Naveed A. Khan and Julia Walochnik141.

157. Esotropea:

Esotropea:http://upload.wikimedia.org/wikipedia/commons/0/05/Andre_Filipe_Teixeira_Marques_E sotropia.jpg Description: English: Esotropia of right eye Date: 27 March 2010, 19:58:48 Source: Andre Filipe Teixeira Marques.jpg Author:Kakawere

158. Right dilated pupil:

Description, English: Anisocoria (right eye instiled by tropicamide) Polski: Anizokoria (prawe oko zakroplne tropikamidem)Deutsch: medikamentös hervorgerufene de:Anisokorie durch Gabe von de:Tropicamid in das rechte Auge. Date, 7 November 2006. Source, Own work, Author, Radomil talk. This file is licensed under the Creative Commons Attribution-Share Alike 3.0 Unported license.
159. The Zondervan Corporation, The Holy Bible, New International Version, Romans 1:19 & 20 (La Habra, California: The Zondervan Corporation and the Lochman Corporation, 1965) page 1298. Scripture taken from THE Holy Bible, New International Version® copyright© 1973,1978, 1984 by International Bible Society. Used by permission.
160. The Zondervan Corporation, The Amplified Bible, Expanded Edition, Collosians 1:16 (La Habra, California: The Zondervan Corporation and the Lochman Corporation, 1965) page 1390. Scripture taken from THE AMPLIFIED BIBLE, Old Testament copyright© 1965, 1987 by the Zondervan Corporation. The Amplified New Testament copyright ©1958, 1987 by the Lochman Foundation. Used by permission.

Chapter 24 Conclusion p.293   (594 TOTAL I.E.C.)

1. Charles Darwin, *The Origin of Species* (New York, New York: Barnes and Noble Books, 2004) page 159
2. The Zondervan Corporation, The Amplified Bible, Expanded Edition, John 3:8-12 (La Habra, California: The Zondervan Corporation and the Lochman Corporation, 1965) page 1210. Scripture taken from THE AMPLIFIED BIBLE, Old Testament copyright
© 1965, 1987 by the Zondervan Corporation. The Amplified New Testament copyright ©1958, 1987 by the Lochman Foundation. Used by permission.
3. The Zondervan Corporation, The Amplified Bible, Expanded Edition, Romans 1:19 (La Habra, California: The Zondervan Corporation and the Lochman Corporation, 1965) page 1298. Scripture taken from THE AMPLIFIED BIBLE, Old Testament copyright
© 1965, 1987 by the Zondervan Corporation. The Amplified New Testament copyright ©1958, 1987 by the Lochman Foundation. Used by permission.
4. Smoking Gun: From: Wikipedia® is a registered trademark of the Wikimedia Foundation, Inc., a non-profit organization. http://en.wikipedia.org/wiki/Smoking_gun

5. Charles Darwin, *The Origin of Species* (New York, New York: Barnes and Noble Books, 2004) page 159

6. Charles Darwin, *The Origin of Species* (New York, New York: Barnes and Noble Books, 2004) page 158

7. Charles Darwin, *The Origin of Species* (New York, New York: Barnes and Noble Books, 2004) page 159

8. Charles Darwin, *The Origin of Species* (New York, New York: Barnes and Noble Books, 2004) page 380

9. Charles Darwin, *The Origin of Species* (New York, New York: Barnes and Noble Books, 2004) page 158

10. Charles Darwin, *The Origin of Species* (New York, New York: Barnes and Noble Books, 2004) page 380

11. Charles Darwin, *The Origin of Species* (New York, New York: Barnes and Noble Books, 2004) page 159

12. Smoking Gun: From: Wikipedia® is a registered trademark of the Wikimedia Foundation, Inc., a non-profit organization. http://en.wikipedia.org/wiki/Smoking_gun

13. Charles Darwin, *The Origin of a Species* (New York, New York: Barnes and Noble Books, 2004) page 159

14. Charles Darwin, *The Origin of Species* (New York, New York: Barnes and Noble Books, 2004) page 157

15. Charles Darwin, *The Origin of Species* (New York, New York: Barnes and Noble Books, 2004) page 159

16. Flagrante Delicto: From: Wikipedia® is a registered trademark of the Wikimedia Foundation, Inc., a non-profit organization. http://en.wikipedia.org/wiki/Smoking_gun

17. Charles Darwin, *The Origin of Species* (New York, New York: Barnes and Noble Books, 2004) page 159

18. Charles Darwin, *The Origin of Species* (New York, New York: Barnes and Noble Books, 2004) page 158

19. Charles Darwin, *The Origin of Species* (New York, New York: Barnes and Noble Books, 2004) page 380

20. The Zondervan Corporation, The Amplified Bible, Expanded Edition, Romans 1:19-20 (La Habra, California: The Zondervan Corporation and the Lochman Corporation, 1965) page 1298. Scripture taken from THE AMPLIFIED BIBLE, Old Testament copyright© 1965, 1987 by the Zondervan Corporation. The Amplified New Testament copyright ©1958, 1987 by the Lochman Foundation. Used by permission.

21. The Zondervan Corporation, The Amplified Bible, Expanded Edition, John 1:1-5 (La Habra, California: The Zondervan Corporation and the Lochman Corporation, 1965) page 1207. Scripture taken from THE AMPLIFIED BIBLE, Old Testament copyright © 1965, 1987 by the Zondervan Corporation. The Amplified New Testament copyright ©1958, 1987 by the Lochman Foundation. Used by permission.

22. The Zondervan Corporation, The Amplified Bible, Expanded Edition, Romans 4:7-25

(La Habra, California: The Zondervan Corporation and the Lochman Corporation, 1965) page 1207. Scripture taken from THE AMPLIFIED BIBLE, Old Testament copyright
© 1965, 1987 by the Zondervan Corporation. The Amplified New Testament copyright ©1958, 1987 by the Lochman Foundation. Used by permission.

23. Charles Darwin, *The Origin of Species* (New York, New York: Barnes and Noble Books, 2004) page 159

24. Charles Darwin, *The Origin of Species* (New York, New York: Barnes and Noble Books, 2004) page 158

25. Charles Darwin, *The Origin of Species* (New York, New York: Barnes and Noble Books, 2004) page 159

26. Charles Darwin, *The Origin of Species* (New York, New York: Barnes and Noble Books, 2004) page 145-155

27. Charles Darwin, *The Origin of Species* (New York, New York: Barnes and Noble Books, 2004) pages 156-166

28. Charles Darwin, *The Origin of Species* (New York, New York: Barnes and Noble Books, 2004) page 380

29. Charles Darwin, *The Origin of Species* (New York, New York: Barnes and Noble Books, 2004) pages 156-166

30. Charles Darwin, *The Origin of Species* (New York, New York: Barnes and Noble Books, 2004) page 380

31. Charles Darwin, *The Origin of Species* (New York, New York: Barnes and Noble Books, 2004) pages 156-166

32. Charles Darwin, *The Origin of Species* (New York, New York: Barnes and Noble Books, 2004) pages VIII- X

33. Charles Darwin, *The Origin of Species* (New York, New York: Barnes and Noble Books, 2004) page 380

34. The Zondervan Corporation, The Amplified Bible, Expanded Edition, Genesis 1:12-25 (La Habra, California: The Zondervan Corporation and the Lochman Corporation, 1965) page 2&3. Scripture taken from THE AMPLIFIED BIBLE, Old Testament copyright© 1965, 1987 by the Zondervan Corporation. The Amplified New Testament copyright ©1958, 1987 by the Lochman Foundation. Used by permission.

35. Charles Darwin, *The Origin of Species* (New York, New York: Barnes and Noble Books, 2004) pages 156-166

36. The Zondervan Corporation, The Amplified Bible, Expanded Edition, John 10:10 (La Habra, California: The Zondervan Corporation and the Lochman Corporation, 1965) page 1227. Scripture taken from THE AMPLIFIED BIBLE, Old Testament copyright© 1965, 1987 by the Zondervan Corporation. The Amplified New Testament copyright ©1958, 1987 by the Lochman Foundation. Used by permission.

Chapter 25 Beyond Evolution (p. 315)

1. The Zondervan Corporation, The Amplified Bible, Expanded Edition, Collosians 1:16 (La Habra, California: The Zondervan Corporation and the Lochman Corporation, 1965) page 1390. Scripture taken from THE AMPLIFIED BIBLE, Old Testament copyright © 1965, 1987 by the Zondervan Corporation. The Amplified New Testament copyright ©1958, 1987 by the Lochman Foundation. Used by permission.

Chapter 26 Intelligible Evidence Made Plain (p. 321)

1. The Zondervan Corporation, The Amplified Bible, Expanded Edition, Romans 1:19-20 (La Habra, California: The Zondervan Corporation and the Lochman Corporation, 1965) page 1298. Scripture taken from THE AMPLIFIED BIBLE, Old Testament copyright © 1965, 1987 by the Zondervan Corporation. The Amplified New Testament copyright ©1958, 1987 by the Lochman Foundation. Used by permission.

2. *The Case for the Creator by Lee Strobel* Illustra Medea, http://www.thecaseforacreator.com/scientists.php

3. Alan Lightman, Posted 09.09.97 (NOVA) *Relativity and the Cosmos* http://www.pbs.org/wgbh/nova/physics/relativity-and-the-cosmos.html2. Creation Ministries International Ltd.,Creation.com http://creation.com/how-old-is-the-earth

4. (http://cosmology.carnegiescience.edu/timeline/1929)

5. (http://www.space.com/15665-edwin-powell-hubble.html).

6. *The Case for the Creator by Lee Strobel* Illustra Medea, http://www.thecaseforacreator.com/scientists.php

7. Alan Lightman, Posted 09.09.97 (NOVA) *Relativity and the Cosmos* http://www.pbs.org/wgbh/nova/physics/relativity-and-the-cosmos.html

8. *The Case for the Creator by Lee Strobel* Illustra Medea, http://www.thecaseforacreator.com/scientists.php

9. Edwin R. Hubble Expansion of the Universe http://www.pbs.org/wgbh/nova/physics/relativity-and-the-cosmos.html

10. http://cosmictimes.gsfc.nasa.gov/online_edition/1929cosmic/expanding.html)

11. (http://www.latimes.com/science/la-sci-cosmic-inflation-20140318story.html#axzz2wkwisjlg)

12. http://www.scientificamerican.com/article/gravity-waves-cmb-b-mode_polarization/

13. The Zondervan Corporation, The Amplified Bible, Expanded Edition, Genesis 1:1-19 page 1&2. (La Habra, California: The Zondervan Corporation and the Lochman Corporation, 1965) Scripture taken from THE AMPLIFIED BIBLE, Old Testament copyright© 1965, 1987 by the Zondervan Corporation. The Amplified New Testament copyright ©1958, 1987 by the Lochman Foundation. Used by permission.

14. September 26, 2014 Vol.345 no.6204, pp.1590-1593 DOI: 10.1126/science.1258055 in the Journal *"Science"*

15. Rigney, J. Jay, O.D. *Darwin's Challenge Answered © 2016, p193-204*

16. September 26, 2014 Vol.345 no.6204, pp.1590-1593 DOI: 10.1126/science.1258055 in the Journal *"Science"*

17. Charles Darwin, *The Origin of Species* (New York, New York: Barnes and Noble Books, 2004) page 62

18. The Zondervan Corporation, The Amplified Bible, Expanded Edition, Genesis 6:19 page 8. (La Habra, California: The Zondervan Corporation and the Lochman Corporation, 1965) Scripture taken from THE AMPLIFIED BIBLE, Old Testament copyright© 1965, 1987 by the Zondervan Corporation. The Amplified New Testament copyright ©1958, 1987 by the Lochman Foundation. Used by permission.

19. Charles Darwin, *The Origin of Species* (New York, New York: Barnes and Noble Books, 2004) page 62

20. The Zondervan Corporation, The Amplified Bible, Expanded Edition, Genesis 1:20, 21 page 2. (La Habra, California: The Zondervan Corporation and the Lochman Corporation, 1965) Scripture taken from THE AMPLIFIED BIBLE, Old Testament copyright© 1965, 1987 by the Zondervan Corporation. The Amplified New Testament copyright ©1958, 1987 by the Lochman Foundation. Used by permission.

21. The Zondervan Corporation, The Amplified Bible, Expanded Edition, 2 Timothy 3:16 page 1416. (La Habra, California: The Zondervan Corporation and the Lochman Corporation, 1965) Scripture taken from THE AMPLIFIED BIBLE, Old Testament copyright© 1965, 1987 by the Zondervan Corporation. The Amplified New Testament copyright ©1958, 1987 by the Lochman Foundation. Used by permission.

Chapter 27 Concerning Chance (p. 327)

1. The Zondervan Corporation, The Amplified Bible, Expanded Edition, John 9:1-7 page 1225. (La Habra, California: The Zondervan Corporation and the Lochman Corporation, 1965) Scripture taken from THE AMPLIFIED BIBLE, Old Testament copyright © 1965, 1987 by the Zondervan Corporation. The Amplified New Testament copyright ©1958, 1987 by the Lochman Foundation. Used by permission.

2. The Zondervan Corporation, The Amplified Bible, Expanded Edition, Exodus 4:11-12 page 72. (La Habra, California: The Zondervan Corporation and the Lochman Corporation, 1965) Scripture taken from THE AMPLIFIED BIBLE, Old Testament copyright© 1965, 1987 by the Zondervan Corporation. The Amplified New Testament copyright ©1958, 1987 by the Lochman Foundation. Used by permission.

3. The Zondervan Corporation, The Amplified Bible, Expanded Edition, Collosions 1:16&17 page 1390. (La Habra, California: The Zondervan Corporation and the Lochman Corporation, 1965) Scripture taken from THE AMPLIFIED BIBLE, Old Testament copyright© 1965, 1987 by the Zondervan Corporation. The Amplified New Testament copyright ©1958, 1987 by the Lochman Foundation. Used by permission.

4. The Zondervan Corporation, The Amplified Bible, Expanded Edition, Romans 9:20&21 page 1311. (La Habra, California: The Zondervan Corporation and the Lochman Corporation, 1965) Scripture taken from THE AMPLIFIED BIBLE, Old Testament copyright© 1965, 1987 by the Zondervan Corporation. The Amplified New Testament copyright ©1958, 1987 by the Lochman Foundation. Used by permission.

5. The Zondervan Corporation, The Amplified Bible, Expanded Edition, Jeremiah 18:1-17 page 850-851. (La Habra, California: The Zondervan Corporation and the Lochman Corporation, 1965) Scripture taken from THE AMPLIFIED BIBLE, Old Testament copyright© 1965, 1987 by the Zondervan Corporation. The Amplified New Testament copyright ©1958, 1987 by the Lochman Foundation. Used by permission.

6. The Zondervan Corporation, The Amplified Bible, Expanded Edition, John 5:5-14 page 1215. (La Habra, California: The Zondervan Corporation and the Lochman Corporation, 1965) Scripture taken from THE AMPLIFIED BIBLE, Old Testament copyright© 1965, 1987 by the Zondervan Corporation. The Amplified New Testament copyright ©1958, 1987 by the Lochman Foundation. Used by permission.

7. The Zondervan Corporation, The Amplified Bible, Expanded Edition, Genesis 50:20 page 67. (La Habra, California: The Zondervan Corporation and the Lochman Corporation, 1965) Scripture taken from THE AMPLIFIED BIBLE, Old Testament copyright© 1965, 1987 by the Zondervan Corporation. The Amplified New Testament copyright ©1958, 1987 by the Lochman Foundation. Used by permission.

8. The Zondervan Corporation, The Amplified Bible, Expanded Edition, Psalm 23:1-6 page 621. (La Habra, California: The Zondervan Corporation and the Lochman Corporation, 1965) Scripture taken from THE AMPLIFIED BIBLE, Old Testament copyright© 1965, 1987 by the Zondervan Corporation. The Amplified New Testament copyright ©1958, 1987 by the Lochman Foundation. Used by permission.

9. The Zondervan Corporation, The Amplified Bible, Expanded Edition, Job 1:1 through Job 42:17 page 571-607. (La Habra, California: The Zondervan Corporation and the Lochman Corporation, 1965) Scripture taken from THE AMPLIFIED BIBLE, Old Testament copyright© 1965, 1987 by the Zondervan Corporation. The Amplified New Testament copyright ©1958, 1987 by the Lochman Foundation. Used by permission.

10. The Zondervan Corporation, The Amplified Bible, Expanded Edition, Isaiah 55:9 p.813 (La Habra, California: The Zondervan Corporation and the Lochman Corporation, 1965) Scripture taken from THE AMPLIFIED BIBLE, Old Testament copyright © 1965, 1987 by the Zondervan Corporation. The Amplified New Testament copyright ©1958, 1987 by the Lochman Foundation. Used by permission.

11. The Zondervan Corporation, The Amplified Bible, Expanded Edition, Romans 3:23-24 page 1302. (La Habra, California: The Zondervan Corporation and the Lochman Corporation, 1965) Scripture taken from THE AMPLIFIED BIBLE, Old Testament copyright© 1965, 1987 by the Zondervan Corporation. The Amplified New Testament copyright ©1958, 1987 by the Lochman Foundation. Used by permission.

12. The Zondervan Corporation, The Amplified Bible, Expanded Edition, Romans 1:19-20 page 1298. (La Habra, California: The Zondervan Corporation and the Lochman Corporation, 1965) Scripture taken from THE AMPLIFIED BIBLE, Old Testament copyright© 1965, 1987 by the Zondervan Corporation. The Amplified New Testament copyright ©1958, 1987 by the Lochman Foundation. Used by permission.

13. The Zondervan Corporation, The Amplified Bible, Expanded Edition, Romans1:1 through Romans 8:39 page 1297-1310. (La Habra, California: The Zondervan Corporation and the Lochman Corporation, 1965) Scripture taken from THE AMPLIFIED BIBLE, Old Testament copyright© 1965, 1987 by the Zondervan Corporation. The Amplified New Testament copyright ©1958, 1987 by the Lochman Foundation. Used by permission.

14. The Zondervan Corporation, The Amplified Bible, Expanded Edition, Matthew 11:29-30 page 1089. (La Habra, California: The Zondervan Corporation and the Lochman Corporation, 1965) Scripture taken from THE AMPLIFIED BIBLE, Old Testament copyright© 1965, 1987 by the Zondervan Corporation. The Amplified New Testament copyright ©1958, 1987 by the Lochman Foundation. Used by permission.

15. The Zondervan Corporation, The Amplified Bible, Expanded Edition, Psalm 34:15-17 page 626. (La Habra, California: The Zondervan Corporation and the Lochman Corporation, 1965) Scripture taken from THE AMPLIFIED BIBLE, Old Testament copyright© 1965, 1987 by the Zondervan Corporation. The Amplified New Testament copyright ©1958, 1987 by the Lochman Foundation. Used by permission.

16. The Zondervan Corporation, The Amplified Bible, Expanded Edition, Proverbs 14:9 page 714. (La Habra, California: The Zondervan Corporation and the Lochman Corporation, 1965) Scripture taken from THE AMPLIFIED BIBLE, Old Testament copyright© 1965, 1987 by the Zondervan Corporation. The Amplified New Testament copyright ©1958, 1987 by the Lochman Foundation. Used by permission.

17. The Zondervan Corporation, The Amplified Bible, Expanded Edition, Proverbs 5:21 page 704. (La Habra, California: The Zondervan Corporation and the Lochman Corporation, 1965) Scripture taken from THE AMPLIFIED BIBLE, Old Testament copyright© 1965, 1987 by the Zondervan Corporation. The Amplified New Testament copyright ©1958, 1987 by the Lochman Foundation. Used by permission.

18. The Zondervan Corporation, The Amplified Bible, Expanded Edition, John 10:10 page 1227. (La Habra, California: The Zondervan Corporation and the Lochman Corporation, 1965) Scripture taken from THE AMPLIFIED BIBLE, Old Testament copyright© 1965, 1987 by the Zondervan Corporation. The Amplified New Testament copyright ©1958, 1987 by the Lochman Foundation. Used by permission.

19. The Zondervan Corporation, The Amplified Bible, Expanded Edition, Exodus 20:1-3 page 91. (La Habra, California: The Zondervan Corporation and the Lochman Corporation, 1965) Scripture taken from THE AMPLIFIED BIBLE, Old Testament copyright© 1965, 1987 by the Zondervan Corporation. The Amplified New Testament copyright ©1958, 1987 by the Lochman Foundation. Used by permission.

20. The Zondervan Corporation, The Amplified Bible, Expanded Edition, Ephesians 1:19 page 1373. (La Habra, California: The Zondervan Corporation and the Lochman Corporation, 1965) Scripture taken from THE AMPLIFIED BIBLE, Old Testament copyright© 1965, 1987 by the Zondervan Corporation. The Amplified New Testament copyright ©1958, 1987 by the Lochman Foundation. Used by permission.

21. The Zondervan Corporation, The Amplified Bible, Expanded Edition, John 21:17 page 1246. (La Habra, California: The Zondervan Corporation and the Lochman Corporation, 1965) Scripture taken from THE AMPLIFIED BIBLE, Old Testament copyright© 1965, 1987 by the Zondervan Corporation. The Amplified New Testament copyright ©1958, 1987 by the Lochman Foundation. Used by permission.

22. The Zondervan Corporation, The Amplified Bible, Expanded Edition, Psalm 139: 6-14 page 693. (La Habra, California: The Zondervan Corporation and the Lochman Corporation, 1965) Scripture taken from THE AMPLIFIED BIBLE, Old Testament copyright© 1965, 1987 by the Zondervan Corporation. The Amplified New Testament copyright ©1958, 1987 by the Lochman Foundation. Used by permission.

23. The Zondervan Corporation, The Amplified Bible, Expanded Edition, Romans 5:17-21 page 1304. (La Habra, California: The Zondervan Corporation and the Lochman Corporation, 1965) Scripture taken from THE AMPLIFIED BIBLE, Old Testament copyright© 1965, 1987 by the Zondervan Corporation. The Amplified New Testament copyright ©1958, 1987 by the Lochman Foundation. Used by permission

24.Tommy Nelson, *The Holy Bible, International Children's Bible* Nashville Tennessee,1999) Scriptures quoted from the International Children's Bible®, copyright © 1986, 1988, 1999 by Thomas Nelson, Inc. Used by permission. Mark 3:29 page 1066.

25. The Zondervan Corporation, The Amplified Bible, Expanded Edition, Psalm 119:143 page 686. (La Habra, California: The Zondervan Corporation and the Lochman Corporation, 1965) Scripture taken from THE AMPLIFIED BIBLE, Old Testament copyright© 1965, 1987 by the Zondervan Corporation. The Amplified New Testament copyright ©1958, 1987 by the Lochman Foundation. Used by permission.

26. The Zondervan Corporation, The Amplified Bible, Expanded Edition, 2 Corinthians 1:3-4 page 1346. (La Habra, California: The Zondervan Corporation and the Lochman Corporation, 1965) Scripture taken from THE AMPLIFIED BIBLE, Old Testament copyright© 1965, 1987 by the Zondervan Corporation. The Amplified New Testament copyright ©1958, 1987 by the Lochman Foundation. Used by permission.

27. The Zondervan Corporation, The Amplified Bible, Expanded Edition, Romans 8:28 page 1309. (La Habra, California: The Zondervan Corporation and the Lochman Corporation, 1965) Scripture taken from THE AMPLIFIED BIBLE, Old Testament copyright© 1965, 1987 by the Zondervan Corporation. The Amplified New Testament copyright ©1958, 1987 by the Lochman Foundation. Used by permission.

28. The Zondervan Corporation, The Amplified Bible, Expanded Edition, 2 Peter 3:7 page 1461. (La Habra, California: The Zondervan Corporation and the Lochman Corporation, 1965) Scripture taken from THE AMPLIFIED BIBLE, Old Testament copyright© 1965, 1987 by the Zondervan

Corporation. The Amplified New Testament copyright ©1958, 1987 by the Lochman Foundation. Used by permission.

29. The Zondervan Corporation, The Amplified Bible, Expanded Edition, 2 Peter 2:13 page 1460. (La Habra, California: The Zondervan Corporation and the Lochman Corporation, 1965) Scripture taken from THE AMPLIFIED BIBLE, Old Testament copyright© 1965, 1987 by the Zondervan Corporation. The Amplified New Testament copyright ©1958, 1987 by the Lochman Foundation. Used by permission.

30. The Zondervan Corporation, The Amplified Bible, Expanded Edition, Matthew 25:41 page 1113. (La Habra, California: The Zondervan Corporation and the Lochman Corporation, 1965) Scripture taken from THE AMPLIFIED BIBLE, Old Testament copyright© 1965, 1987 by the Zondervan Corporation. The Amplified New Testament copyright ©1958, 1987 by the Lochman Foundation. Used by permission.

31. The Zondervan Corporation, The Amplified Bible, Expanded Edition, John 5:28, 29 page 1216. (La Habra, California: The Zondervan Corporation and the Lochman Corporation, 1965) Scripture taken from THE AMPLIFIED BIBLE, Old Testament copyright© 1965, 1987 by the Zondervan Corporation. The Amplified New Testament copyright ©1958, 1987 by the Lochman Foundation. Used by permission.

32. Regency publishing House, The Holy Bible Old and New Testaments in the King James Version, John 5:29 page 1563. Nashville Tennesee, ©1976 Thomas Nelson Inc 33.The Zondervan Corporation, The Amplified Bible, Expanded Edition, Joshua 24:15 page 283. (La Habra, California: The Zondervan Corporation and the Lochman Corporation, 1965) Scripture taken from THE AMPLIFIED BIBLE, Old Testament copyright© 1965, 1987 by the Zondervan Corporation. The Amplified New Testament copyright ©1958, 1987 by the Lochman Foundation. Used by permission.

34. The Zondervan Corporation, The Amplified Bible, Expanded Edition, Matthew 1:21 page 1074. (La Habra, California: The Zondervan Corporation and the Lochman Corporation, 1965) Scripture taken from THE AMPLIFIED BIBLE, Old Testament copyright© 1965,1 987 by the Zondervan Corporation. The Amplified New Testament copyright ©1958, 1987 by the Lochman Foundation. Used by permission.

35. The Zondervan Corporation, The Amplified Bible, Expanded Edition, Matthew 22:37 page 1108. (La Habra, California: The Zondervan Corporation and the Lochman Corporation, 1965) Scripture taken from THE AMPLIFIED BIBLE, Old Testament copyright© 1965, 1987 by the Zondervan Corporation. The Amplified New Testament copyright ©1958, 1987 by the Lochman Foundation. Used by permission.

36. The Zondervan Corporation, The Amplified Bible, Expanded Edition, Ecclesiastes 12:13 page 745. (La Habra, California: The Zondervan Corporation and the Lochman Corporation, 1965) Scripture taken from THE AMPLIFIED BIBLE, Old Testament copyright© 1965, 1987 by the Zondervan Corporation. The Amplified New Testament copyright ©1958, 1987 by the Lochman Foundation. Used by permission.

37. The Zondervan Corporation, The Amplified Bible, Expanded Edition, Romans 1:19-20 page 1298. (La Habra, California: The Zondervan Corporation and the Lochman Corporation, 1965) Scripture taken from THE AMPLIFIED BIBLE, Old Testament copyright©

1965, 1987 by the Zondervan Corporation. The Amplified New Testament copyright ©1958, 1987 by the Lochman Foundation. Used by permission.

38. The Zondervan Corporation, The Amplified Bible, Expanded Edition, John 3:16-21 page 1211. (La Habra, California: The Zondervan Corporation and the Lochman Corporation, 1965) Scripture taken from THE AMPLIFIED BIBLE, Old Testament copyright© 1965, 1987 by the Zondervan Corporation. The Amplified New Testament copyright ©1958, 1987 by the Lochman Foundation. Used by permission.

39. The Zondervan Corporation, The Amplified Bible, Expanded Edition, John 3:36 page 1212. (La Habra, California: The Zondervan Corporation and the Lochman Corporation, 1965) Scripture taken from THE AMPLIFIED BIBLE, Old Testament copyright© 1965, 1987 by the Zondervan Corporation. The Amplified New Testament copyright ©1958, 1987 by the Lochman Foundation. Used by permission.

40. The Zondervan Corporation, The Amplified Bible, Expanded Edition, Romans 10:13 page 1312 (La Habra, California: The Zondervan Corporation and the Lochman Corporation, 1965). Scripture taken from THE AMPLIFIED BIBLE, Old Testament copyright© 1965, 1987 by the Zondervan Corporation. The Amplified New Testament copyright ©1958, 1987 by the Lochman Foundation. Used by permission.

www.ingramcontent.com/pod-product-compliance
Lightning Source LLC
Chambersburg PA
CBHW081104170526
45165CB00008B/2317